COSMOLOGY AND ELEMENTARY PARTICLES

Proceedings of the Second Winter School of Physics

COSMOLOGY AND ELEMENTARY PARTICLES

University of Puerto Rico, Puerto Rico
27 March – 5 April 1991

Editors

Daniel R. Altschuler
Physics Department, University of Puerto Rico and NAIC - Arecibo Observatory

José F. Nieves
Jaime Ponce de Leon
Marcelo R. Ubriaco
Physics Department, University of Puerto Rico

World Scientific
Singapore • New Jersey • London • Hong Kong

Published by

World Scientific Publishing Co Pte Ltd
5 Toh Tuck Link, Singapore 596224
USA office: 27 Warren Street, Suite 401-402, Hackensack, NJ 07601
UK office: 57 Shelton Street, Covent Garden, London WC2H 9HE

British Library Cataloguing-in-Publication Data
A catalogue record for this book is available from the British Library.

COSMOLOGY AND ELEMENTARY PARTICLES
Proceedings of the Second Winter School of Physics

ISBN-13 978-981-02-0808-0
ISBN-10 981-02-0808-1

TABLE OF CONTENTS

Preface

This volume contains the notes of the lectures delivered at the Second Winter School of Physics, which took place at the Río Piedras Campus of the University of Puerto Rico from March 27 to April 5 of 1990. In addition to the University of Puerto Rico, the school was supported by funds provided by the National Astronomy and Ionosphere Center (NAIC, Arecibo Observatory) and by the Instituto de Educación Universal, which we gratefully acknowledge.

During those honorable days for our institution, a distinguished group of lecturers and invited speakers addressed the audience of about forty partcipants from several countries. Following the lead established by the first school, the main subjects of the lectures were Cosmology and Elementary Particles. On this occasion we decided to include a course on the interface of Field Theory and Condensed Matter, a field that has become the subject of such intense interest that we are planning to devote the next school exclusively to it. We thank the speakers for their effort in providing us with the written version of the lectures. Unfortunately, the notes of the lectures delivered by Marciano and Turok are not included in this volume because we had not received them by the time the manuscript had to be sent to the publisher.

It is gratifying to have an opportunity to express once more our sincere appreciation to the various people at the University of Puerto Rico and the other sponsoring institutions, without whose decisive support and encouragement this activity would not have taken place. We wish to express our sincere thanks to the Chancellor of the Río Piedras Campus, Dr. Juan Fernandez; the Legal Counsel of the Chancellor, Lic. Ileana Lacot, ; the Assistant to the Chancellor, Dr. Cesar Cordero; the Dean of the Faculty of Natural Sciences, Dr. Rafael Arce; the Dean of Studies, Dr. Eduardo Rivera Medina; the Chairman of the Department of Physics, Dr. Alfredo Torruella; the Director of NAIC, Dr. Tor Hagfors; and the President of the Instituto de Educación Universal, Mr. Angel Ruíz Rivera. For their help with the inevitable administrative chores that an event like this one brings, we thank the school secretary Marilyn Rodríguez, Carmelo Figueroa and Ileana García, of the administrative staff of the Deparment of Physics.

We are grateful to all those participants from Latin America who decided to attend the school, in spite of the fact that we were not able to reimburse their expenses as we had promised, due to the new, unreasonable visa requirements imposed by the U.S. Immigration Department.

Daniel R. Altschuler
José F. Nieves
Marcelo R. Ubriaco
Jaime Ponce de León
December 1991

COSMOLOGY AND ELEMENTARY PARTICLES

LARGE SCALE STRUCTURE AND PECULIAR MOTIONS IN THE UNIVERSE

RICCARDO GIOVANELLI[1]
National Astronomy and Ionosphere Center[2], Cornell University
Arecibo Observatory, Box 995, Arecibo, PR 00613, USA

CONTENTS

---

[1] Current Address: Dept. of Astronomy, Space Sciences Bldg., Cornell University, Ithaca, NY 14853, USA

[2] The National Astronomy and Ionosphere Center is operated by Cornell University under a management agreement with the National Science Foundation.

1. Introduction

That of redshift surveys has been one of the most succesful "industries" in the history of modern astronomy. After the pioneering work of Slipher, Humason, Mayall and others in the earlier half of the century, the near completion of the redshift data base for the *Shapley–Ames Catalog* yielded the first three-dimensional maps of the Local Universe. In the last fifteen years, advances in detector and spectrometer technology at both optical and radio wavelengths have spurred a tremendous explosion in the galaxy redshift tally. The resulting view of the universe has proved quite complex, and has fostered considerable effort to explain both the topology of the large scale structure and its origin. Following the early work of de Vaucouleurs, who first parametrized our infall and rotation in the Local Supercluster, the study of deviations from smooth Hubble flow, produced by large–scale mass density enhancements, has now painstakingly been extended to redshifts on the order of 10^4 km s^{-1}, and currently aims at the mapping of the overall matter distribution in the local universe.

In this series of lectures, we shall first review the observational achievements of modern redshift surveys, on a key that emphasizes the differences among survey strategies. In the process, we shall qualitatively illustrate the main characteristics of the large scale distribution of luminous mass. We shall next approach in quantitative terms the details of the large scale structure of luminous matter, reviewing what we've learned in terms of clustering, topology, luminosity function and segregation properties of the galaxian population. In the final part, we shall review the Newtonian theory of small perturbations, and survey the modern evidence for peculiar motions on scales up to about $100h^{-1}$ Mpc. Throughout these notes, we adopt $H_o = 100h$ km s^{-1} Mpc^{-1}, unless otherwise specified.

2. Overview of Redshift Surveys

By any standards of human activity, the redshift industry is amongst the most succesful, as it can boast a sustained growth rate (expressed as the yearly increment of known redshifts) in excess of 10% per year over its whole 80–year history, and promises the potential to maintain its growth for the foreseeable future. While the majority of redshifts are obtained via optical spectroscopy, a significant contribution derives from the 21 cm H I line surveys of spirals and gas–rich dwarfs. Surveys conducted in the 21 cm line arrived relatively late as contributors to the redshift industry, but improvements in receiver and spectrometer technology led radio astronomers to a protagonic role. In the early 1970's, only 150 or so galaxies had a measured 21 cm redshift (Roberts 1975) but, since then, radio astronomers have rapidly increased their productivity. Of the 30,000 or so galaxies with a measured redshift at the beginning of this decade, those that have been observed at 21 cm add to over 12,000. Recent reviews of redshift surveys have been given by Huchra (1988), Geller and

Huchra (1988) and Giovanelli and Haynes (1991). Recent reviews on the topic of the large–scale distribution of galaxies have been given by Oort (1983) and Rood (1988).

Until a few years ago, large–scale redshift efforts were concentrated in the northern hemisphere but recent progress made by southern observers, as described by da Costa (1991) elsewhere in these proceedings, has brought the overall coverage closer to equipartition. The steep growth rate in measured redshifts makes a precise census difficult, especially because a significant fraction of the data base is still in unpublished form. A good part of that fraction is widely if informally circulated. In this section we present a brief view of our own 1990 redshift census. While current redshift coverage is inhomogeneous and incomplete to varying degrees from region to region, it nonetheless provides an informative three-dimensional picture of the large–scale structure of the nearby universe.

Figure 2.1 displays a heliocentric radial velocity histogram, in 500 km s^{-1} bins, up to 30,000 km s^{-1}. With the caveats of the inhomogeneity of the data base and of its incompleteness given our lack of access to some unpublished results and their rapid rate of growth, the histogram in Figure 1 resembles the redshift distribution expected of a sample with a limiting magnitude of $m_{pg} \sim 15$, and a mean depth of about $75h^{-1}$ Mpc. To complement Figure 2.1, Figures 2.2(a) through 2.2(f) provide an impression of the extent of the observational effort. Symbols represent individual galaxies with known redshift, superimposed on a grid of equatorial coordinates in an Aitoff projection. In order to provide equal graphic relevance to each galactic hemisphere, the projection is centered at $R.A. = 6^h$. Each panel corresponds to a separate, non–overlapping window of heliocentric radial velocity.

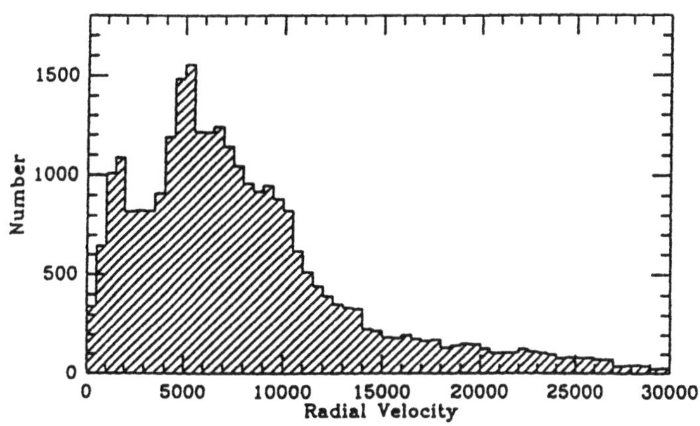

Figure 2.1: Histogram of the distribution of heliocentric radial velocities currently known, up to of 30000 km s^{-1}.

In figure 2.2(*a*), the large asymmetry between northern and southern hemisphere coverage is not as much the result of sample selection or completion bias, as of the preference for nearby galaxies to be located in the northern galactic hemisphere towards the central regions of the Local Supercluster. This asymmetry was well known already a century ago by Helmholtz and Herschel (thanks to A. Sandage for pointing this out), and its nature is now understood in terms of our location in the Local Supercluster, in great measure thanks to the pioneering work of G. de Vaucouleurs (1953,1958). The most conspicuous concentration in Fig. 2.2(*a*) is of course the Virgo cluster. The supergalactic plane concentration of nearby galaxies is easily visible passing through Virgo. The two large southern clusters between 3^h and 4^h are Eridanus and Fornax. Beyond the Local Supercluster, several features are apparent as seen in panel 2.2(*b*). North of the equator in the southern galactic cap, the filamentary structure known as the Pisces–Perseus supercluster dominates. The total extent of this supercluster is unknown, because it "disappears" to the North and East into the zone of avoidance [3]. Nearly symmetric to this supercluster, south of the equator and in the northern galactic cap, the Hydra and Centaurus clusters are conspicuous in a region that is under intense scrutiny in searches for the "Great Attractor". In panel 2.2(*c*), the Pisces-Perseus supercluster begins to fade, while the extensive enhancement associated with the Coma–A1367 clusters becomes dominant near the north galactic pole. The deeper sampling along the Center for Astrophysics constant declination slices becomes particularly noticeable here.

In panel 2.2(*c*) as in the following ones, another detail becomes prominent, i.e. the different depths with which the two hemispheres have been sampled. Southern samples, although rapidly growing in scope and depth, are still somewhat shallower than northern ones, by about half a magnitude. The depth and completeness of available galaxy catalogs also varies, contributing to the inhomogeneity in the sampling across the sky (Lynden–Bell and Lahav, 1988). One of the least well covered regions is that between $-2°$ and $-18°$ in declination: while to the north of it galaxian taxonomy is well represented by the *Uppsala General Catalogue* (UGC: Nilson 1973) and by the *Catalogue of Galaxies and Clusters of Galaxies* (CGCG: Zwicky *et al.* 1963-68), and to the south by the *ESO Uppsala Survey of the ESO(B) Atlas* (ESO–UGC: Lauberts 1982) and its recent complement, the *Surface Photometry Catalogue of the ESO-Uppsala Galaxies* (Lauberts and Valentijn 1989), the intermediate region is only served by the *Morphological Catalogue of Galaxies* (MCG: Vorontsov–Velyaminov *et al.* 1962-68), which is less complete and less photometrically and astrometrically accurate than the others.

In panel 2.2(*d*), the Hercules supercluster is seen to extend across the region between $+10°$ and $+40°$ near 16^h. The concentration around A2151, the canonical Hercules cluster, connects via a low density bridge to the northern enhancement around A2197/A2199. In the remaining panels the greatest concentration occurs in

[3] However, a gallant effort by the *ZOA Collaboration*^{TM} (Pantoja *et al.*), currently underway, will soon allow our perspective to pierce the cosmic fog of the galactic anticenter region.

Fig. 2a.

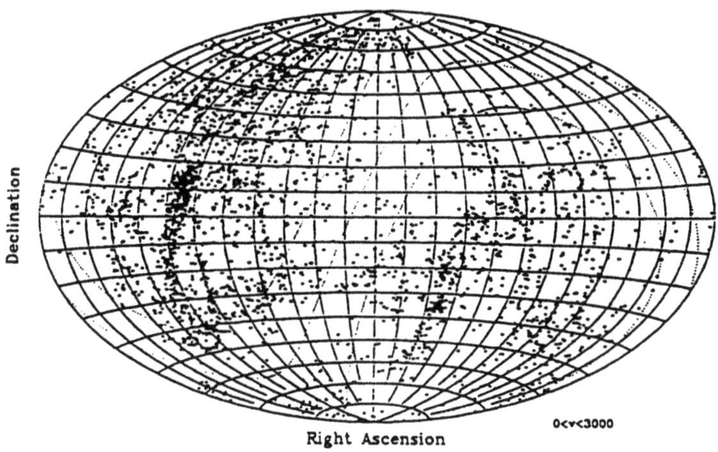
Declination
Right Ascension
0<v<3000

Fig. 2b.

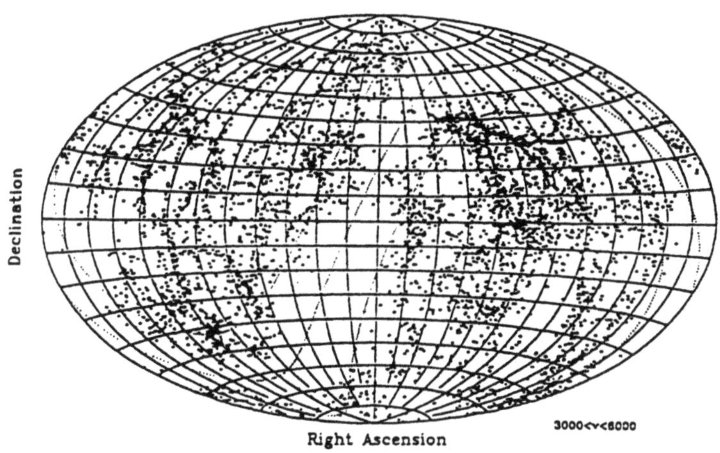
Declination
Right Ascension
3000<v<6000

Fig. 2c.

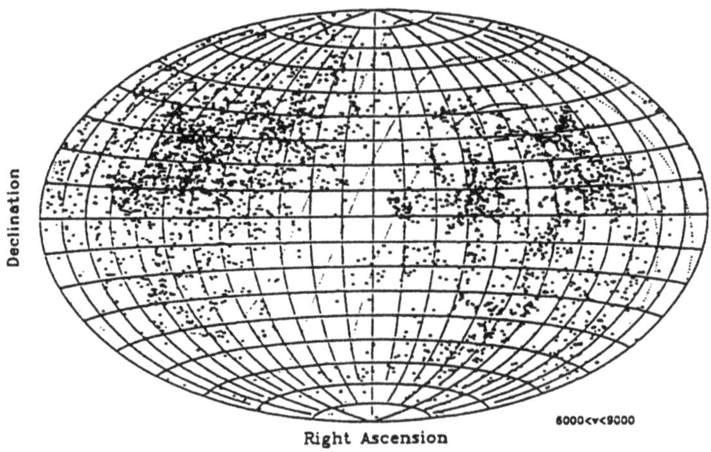
Declination
Right Ascension
6000<v<9000

Fig. 2d.

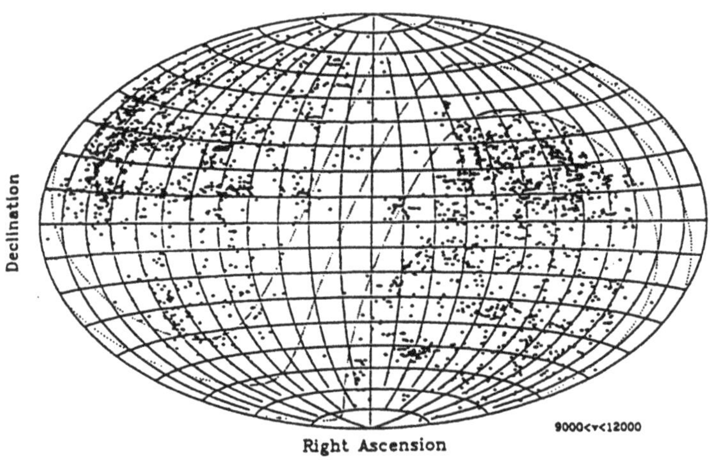
Declination
Right Ascension
9000<v<12000

Fig. 2e.

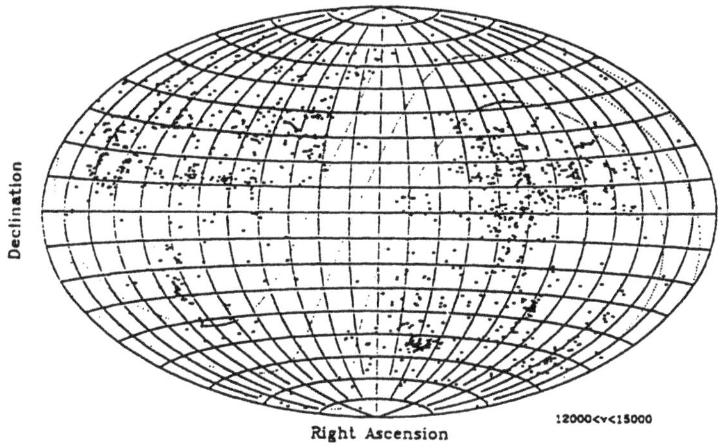

Fig. 2f.

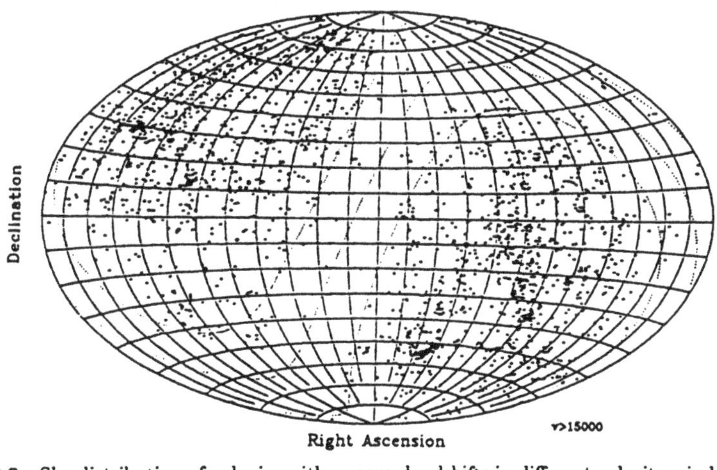

Figure 2.2: Sky distribution of galaxies with measured redshifts in different velocity windows. The displayed grids are equal area Aitoff projections of equatorial coordinates centered on R.A. = 6^h. Individual objects are shown. Solid grid lines are plotted each hour in Right Ascension and each 10° in Declination. Dotted lines trace the galactic latitudes b = 0°, -20° and +20°. The redshift windows corresponding to each panel are: (a) $V_\odot < 3000$ km s^{-1}; (b) 3000 km s^{-1} $< V_\odot <$ 6000 km s^{-1}; (c) 6000 km s^{-1} $< V_\odot <$ 9000 km s^{-1}; (d) 9000 km s^{-1} $< V_\odot <$ 12000 km s^{-1}; (e) 12000 km s^{-1} $< V_\odot <$ 15000 km s^{-1}; (f) $V_\odot >$ 15000 km s^{-1}.

the Horologium region, 35° from the South Pole.

While omissions undoubtedly affect this overview in some details, its general characteristics are representative of the state of surveys within $z < 0.05$. Beyond that value, at this time only small area surveys are significant, a circumstance that is rapidly changing as a result of ongoing or already planned efforts.

3. Redshift Survey Strategies

3.1 General Considerations

An ideal redshift survey would sample a fair volume of the universe with a well-understood selection function. The selection function represents a measure of the probability that a galaxy at distance r is included in the sample. If the luminosity function of galaxies $\phi(L)$ is universal — see section 4 —, then the selection function of a complete, flux-limited sample can be defined as a pure function of distance, $s(r) = \int_{L_m}^{\infty} \phi(L)dL / \int_0^{\infty} \phi(L)dL)$, where $L_m(r)$ is the minimum luminosity detected by the survey at the distance r. A "fair" sample will be one that stretches over a volume such that the inferred mean properties are representative of the universe as a whole. In practice, we still don't know whether the largest well-sampled volumes are a good approximation to fairness. A quick look at figures 2(a) through 2(c) supplies reasons for caution, as *(a)* very large scale features are discernible there, even after projection of fairly deep samples on the plane of the sky; *(b)* similarly, fig. 3 shows how deep probes widely separated in the sky can occasionally yield remarkably similar views of the universe even in the replication of peculiar features.

On the other hand, however, one can find reassuring indications of homogeneity on the largest scales. The Cosmic Microwave Background radiation field exhibits surprising smoothness and isotropy; its well-defined dipole moment is ascribed to the motion of the Local Group as we discuss in the final section. This indication of homogeneity was anticipated by the early experiments of Hubble (1934), who found that galaxy counts up to magnitude m, varied just as expected in a homogeneous universe, where the effects of curvature could be neglected, i.e. as

$$n(m) \propto 10^{0.6m}. \tag{3.1}$$

Hubble carried out the experiment in many directions and reached fairly deep flux limits ($m_{pg} \sim 19.6$, and even deeper in later work), finding that the counts at fainter m were in fair agreement with those extrapolated from brighter values using eqn. (3.1).

In practice, redshift surveys are plagued by an inadequate understanding of the degree of fairness of the chosen sample, and by problems such as catalog incompleteness, limited accuracy of catalogued parameters, instrumental biases and loosely defined survey criteria, which adversely affect the definition of the selection function.

In discussing survey strategies, a few numbers are handy. For example, the mean sky surface density of galaxies to $m_{pg} \simeq 15.5$ is very nearly one per square degree, or ~ 25 to magnitude 18, and so on. The typical depth of galaxy catalogs that have been most commonly used over the last one or two decades, and that list positions and parameters of individual galaxies over at least one hemisphere, varies between $m_{pg} \sim 14.5$ and $m_{pg} \sim 15.7$. Modern surveys of selected areas, obtained digitizing Schmidt plates, are considerably deeper, reaching magnitudes of about $m_J \sim 20.5$.

The most widely used indication of the clustering properties is the two–point correlation function of the galaxian distribution, $\xi(r)$. It is defined by the joint probability of finding a galaxy in each of two volume elements dV_1 and dV_2, separated by a distance r:

$$dP = \bar{n}^2 \; dV_1 dV_2 \; [1 + \xi(r)] \tag{3.2}$$

where $\bar{n}$ is the mean space density of galaxies in the sample under study. In a Poisson distribution, homogeneous on all scales, $\xi(r) \equiv 0$. Thus ξ gives an indication of the excess probability, over random, of finding a companion galaxy at distance r. Peebles (1980) treats extensively the correlation formalism, and has reviewed recent related developments in the last Rio Piedras school (cf. Altschuler and Nieves 1989, p. 177). The two–point correlation function can be described by a power law

$$\xi(r) = \left(\frac{r}{r_o}\right)^{-\gamma} \tag{3.3}$$

with a slope $\gamma = 1.77 \pm 0.05$ and a scale length $r_o = 5.5 \pm 1.5 \; h^{-1}$ Mpc. Many indications exist that $\xi(r)$ changes slope at separations larger than r_o. The existence of structure at those separations is important because it can be treated using linear theory and it is thought to hold a close relation to primordial perturbations.

To obtain estimates of the "fair" properties of the galaxian distribution, a survey needs to overcome both the noise that arises from the size of the sample, i.e. small–number statistics, and that which arises from small–scale fluctuations in the distribution itself, as described by $\xi(r)$. Suppose that the goal of a survey is that of obtaining the behavior of $\xi(r)$ at large values of r. In that case, Kaiser (1986) has shown that the survey efficiency is optimized by sampling sparsely, e.g. by obtaining the redshift of one of every 20 objects, drawn randomly from a very large and complete, flux–limited sample. Why that is the case can be intuitively understood as follows. $1 + \xi(r)$ is estimated by counting pairs at each characteristic separation $(r, r + \Delta r)$, and dividing by the number that would be counted in a Poissonian distribution of the same mean density $\bar{n}$. At separations $r > r_i$, where r_i is large and therefore ξ is small, the error on ξ is given by

$$\epsilon_\xi = N_p^{-1/2} \left[1 + 4\pi\bar{n} \int_0^{r_i} r^2 \xi(r) \, dr\right] \tag{3.4}$$

where N_p is the number of independent pairs used for the estimate. Now, in a Poisson distribution, $\xi \equiv 0$ and $\epsilon_\xi = N_p^{-1/2}$. In a clustered distribution, the error is increased by the volume integral of ξ. The numerator of eqn. (3.4) represents the mean number of members of each "cluster" of size $r < r_i$ in the sample; the amplitude of $\xi(r > r_i)$ can be understood as obtainable from the number of independent pairs of clusters, and therefore, given two such clusters A and B, all pairs of galaxies drawn one from A, the other from B, yield redundant information. This "oversampling" does not significantly contribute to improving the accuracy of the determination of $\xi(r)$. Greater accuracy in the estimate would be achieved by including a larger number of clusters in the sample, rather than oversampling each cluster. On the other hand, if a main goal of the survey is the definition of the topology of the large–scale structure, then complete sampling of a deep magnitude–limited, wide solid angle catalog is desirable. These considerations illustrate how, depending on the scientific goals of the survey, different strategic trade–offs can be possible; with them in mind, we next review some recent important achievements.

3.2 Pencil–Beam Surveys

In trying to balance the requirements of depth, completeness and limited telescope time, pencil–beam surveys are the most economical. The selection of the parameters associated with a pencil–beam survey, such as the beam cross section, is often tied to instrumental constraints, e.g. the size of a multi–object spectrograph's aperture plate. In order to overcome the effect of small–scale fluctuations in the galaxian distribution, one would ideally choose the cross section of a pencil–beam survey (at its nominal redshift depth) to be larger than r_o, the scale–length in eqn. (3.3). The sampling density can then be set by matching the associated sampling error with that arising from the clustering properties and the chosen geometry of the sampled region.

A survey spread over many separate fields can afford to sample deeply in the redshift direction while also covering a large solid angle. Thus a combination of pencil beam surveys can economically yield information on very large–scale features in $\xi(r)$. On the other hand, topological details that could be investigated with a wide angle survey tend to be lost. In addition, the approach may complicate the definition of the survey's selection function; for example, a space dependence of the selection function is introduced by coupling the number of redshifts obtainable per field to the multiplexing factor of a multiobject spectrograph, when the survey field size is comparable to that of the spectrograph. The special niche for discrete field surveys of the pencil–beam variety is in the trail–blazing determination of the rough, large–scale properties of the galaxian distribution, ideally followed by more detailed studies over wider solid angles.

In a series of joint papers, Kirshner, Oemler and Schechter — later joined by Schechtman (hence KOSS) —, sampled almost completely to a fairly deep magnitude limit several widely spaced fields, each about one square degree in size. At the mean depth of their survey, one pencil-beam field is somewhat narrower than r_o, so the structure seen in each tends to be dominated by the small-scale fluctuations in the galaxian distribution. Nonetheless, their survey showed considerable deviations from homogeneity on very large scales. In 1981, on the basis of 133 redshifts in three fields $\sim 35°$ apart, KOSS reported the discovery of a large void in Boötes, centered at $cz \sim 15,000$ km s^{-1} and extending over 6,000 km s^{-1} in the radial direction. In 1987, a more elaborate analysis confirmed the existence and size of the void (if not as extreme a density contrast as initially reported); they sampled 283 small fields — each 15' on a side —, spread between 13^h30^m and 16^h30^m, $+30°$ to $+70°$, to $m_{lim} \sim 17$ (see fig. 3.1). They are currently extending their discrete sampling approach in a survey that covers the two galactic caps to a redshift of 0.2, to eventually include a total of 10,000 to 20,000 redshifts. The strategy combines the advantages of a multifiber spectrograph with the large field of view of the Las Campanas 2.5 meter DuPont telescope. They are surveying 200 fields, each $1.5° \times 3°$ in size, covering about 25% of the sky between 10^h15^m and 15^h15^m, $0°$ to $-18°$ for the northern galactic hemisphere, and between 21^h00^m and 04^h30^m, $-30°$ to $-48°$ for the southern one. They are obtaining their own photometric and astrometric catalog by means of a CCD drift survey. A 60–fiber spectrograph can yield 40 to 45 redshifts of galaxies with magnitudes $16.0 < R < 17.3$, per 2^h exposure, and about 4 fields per night can be observed. By March 1991, they had obtained about 3700 redshifts.

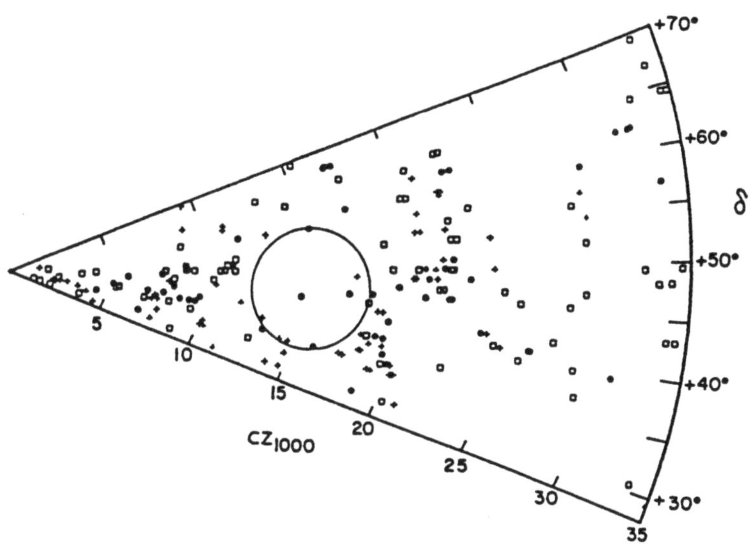

Figure 3.1: Distribution of the galaxies in the KOSS 1987 sample. Different symbols refer to different R.A. ranges. The large circle represents the largest empty sphere that will fit within the void.

The collaboration of groups at Durham University, the Anglo–Australian Observatory (AAT) and the South African Astronomical Observatory has produced several important observational studies, based on accurate photometry derived from the UK Schmidt Telescope (UKST) survey fields, such as those of Metcalfe *et al.* (1989) and Peterson *et al.* (1986). The Durham–led effort has been particularly effective in the study of high redshift objects, as we discuss below.

The study of high redshift galaxies ($z > 0.1$) aims at a number of goals, mainly: (a) the approximation of an elusive fair sample, whereby reliable "mean" properties of the universe can be obtained; (b) the understanding of the evolution of galaxies and of the field luminosity function; and (c) a knowledge of the evolution of clusters and their galaxian content. Recent reviews that emphasize particularly aspect (b), can be found in Ellis (1987) and Koo (1988). Because of the low fluxes, redshift surveys employ different strategies for distant galaxies than for nearby ones. At high z, spectroscopic observations can be time–consuming (at $z \sim 0.5$, a 200 km s^{-1} quality redshift requires the better part of a night's observing on a 4–m class telescope); the surface density of targets, however, is high, and the use of multi–object spectrographs is well–suited for the task. Alternatively, large numbers of galaxies can be surveyed photometrically, and the broad-band spectral information can be used as an indicator of distance, a "photometric" redshift. While economic, the technique of photometric redshifts is inaccurate and uncertain for large–scale structure studies; for further details we address the reader to the excellent review by Koo (1986).

Using a multifiber spectrograph at the AAT, Broadhurst *et al.* (1988) surveyed 200 galaxies of $20.0 < b_j < 21.5$, in five fields, of 200 to 400 arcmin^2 each, at a mean redshift of 0.22. In this very interesting work, the authors tackle the issue of the evolution of the luminosity function at $z < 1$), as we shall see in more detail in Section 5. Also at the AAT, Colless *et al.* (1990) used a multislit spectrograph to probe even deeper ($21 < b_j < 22.5$) obtaining 149 redshifts in three fields. With a multi–slit arrangement at the Kitt Peak 4–m telescope, Koo and Kron (1988) surveyed three fields near the North Galactic Pole; their survey currently (end 1990) includes 401 redshifts. These two complementary efforts have been combined to produce a remarkable view out to $z \sim 0.5$ in the direction of the two galactic poles (Broadhurst *et al.* 1990). The most striking result is the appearance of a strong feature in the clustering at scales of $128h^{-1}$ Mpc (Figs. 3.2 and 3.3). Ongoing efforts will extend the scope of the Durham–AAT and Kitt Peak probes to encompass as many as 30 or 40 pencil beams in both hemispheres, distributed over two patches (each several degrees wide, with one in each hemisphere), for a total of some 1000 redshifts. This combination approach of both pencil beam and sparse sampling emphasizes sensitivity to large transverse structures, since at $z \sim 0.15$, the pencil beam separation is $\sim 10 - 60$ Mpc.

3.3 2-d Surveys

As a compromise between the depth of pencil–beam surveys and the relative shallowness of filled, three–dimensional ones, two–dimensional "slice" surveys offer

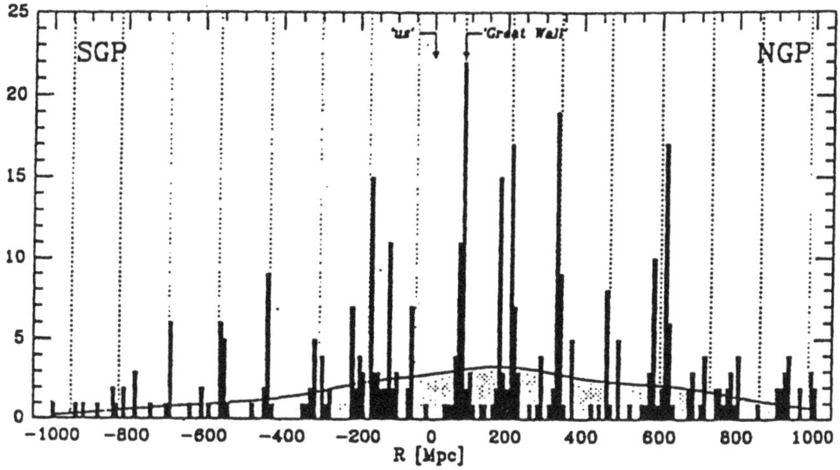

Figure 3.2: Redshift distribution of the Broadhurst et. al. (1990) pencil-beam survey towards the two galactic poles. $\Omega = 1$ and $h = 1$ were used to convert redshift to comoving distance R (x–axis). The smooth line is an estimate of the survey's selection function.

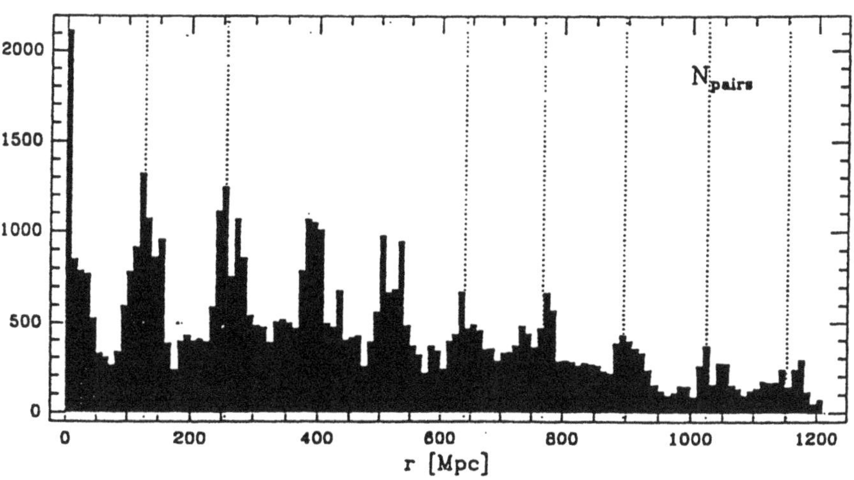

Figure 3.3: One–dimensional pair count correlation for the redshift data displayed in Fig. 3.2. The dashed lines indicate multiples of 128 h^{-1} Mpc (after Szalay et al. 1991).

an efficient means of testing the geometry of the galaxian distribution. Once one chooses the depth of the survey by adopting some sort of flux or angular diameter limit (generally imposed by the available catalogs), the angular thickness of the slice can be chosen to match the correlation scale–length at the median distance of the sample. For example, a sample drawn from the CGCG at a magnitude limit of 15.5, the 6° width of a CGCG strip, conveys a convenient approximation of such optimal thickness. At the Harvard–Smithsonian Center for Astrophysics, the most ambitious effort of this nature is underway by Huchra, Geller and collaborators. While their ultimate goal is the availability of redshifts for all CGCG galaxies with $m \leq 15.5$ and $|b| \geq 40°$ (a sample of about 15,000 galaxies), the completion by slices maximizes scientific returns in terms of the statistical properties of the galaxian distribution. The first of the CfA slices is the 6° by 117° one presented by Huchra *et al.* (1990) and de Lapparent *et al.* (1986) as an extension of the original CfA survey. It covers $8^h \leq R.A. \leq 17^h$ and $26.5° \leq Dec. < 32.5°$. This slice is a magnitude–limited survey of 1100 galaxies, complete to $m_{cgcg} = 15.5$, which crosses some well–studied features such as the Coma cluster. Two other slices are also complete, those between $32.5° \leq Dec. < 38.5°$ and $38.5° \leq Dec. < 44.5°$ (see also de Lapparent *et al.* 1988) and several others are nearing completion. Similar inroads are being made in the completion of their goal in the 20^h to 04^h region. The Center for Astrophysics redshift effort has contributed to the impressive view displayed in Fig. 3.4, a 360° slice which reveals a large scale structure that has been dubbed "the Great Wall" (Geller and Huchra 1989).

Continuing work at Dartmouth College in collaboration with the CfA group aims at a deeper view within the region of the first CfA slice, a 1° × 100° strip to $m_B = 17.5$ between $29° \leq Dec. \leq 30°$. The first two reports on this effort, which will eventually include 2500 galaxies, have been given by Thorstensen *et al.* (1989) and Wegner *et al.* (1990a). The deepest completed effort of this type consists of an even narrower (10') strip across the Coma cluster, extending between $10^h 42^m$ and $15^h 28^m$ and to a magnitude of 18, which has been surveyed by Karachentev and Kopylov (1990) at the 6-m telescope of the Special Astrophysical Observatory of the USSR.

3.4 3-d Surveys

Wide angle surveys designed to sample a specific volume are the best suited for topological studies of the three–dimensional structure. Such surveys can have practical restrictions imposed by galactic extinction, sky coverage of the telescope used, and limits of the target catalog, usually in apparent magnitude, flux or diameter, and may also be restricted by morphology or surface brightness constraints on detectability. Each survey must then be evaluated in light of the biases introduced by the selection process. Below, we use the Pisces–Perseus Supercluster (PPS) survey as a paradigm for the category of 3–d surveys. Elsewhere in these proceedings, L. Da Costa discusses the results of another important observational effort, that of the SSRS in the southern hemisphere. Before we pass on to the description of the

PPS, we should mention the very important efforts of Fisher and Tully (1981), who first used the 21 cm line for an extensive survey of 1787 galaxies in the Local Supercluster, the first CfA survey of Davis *et al.* (1982), Huchra *et al.* (1983), which completed the redshift effort to a limit of $m_{cgcg} = 14.5$ in the northern hemisphere and included 2401 galaxies, and the IRAS survey (Strauss *et al.* 1990), which covers homogeneously 87.6% of the sky and includes 2649 galaxies.

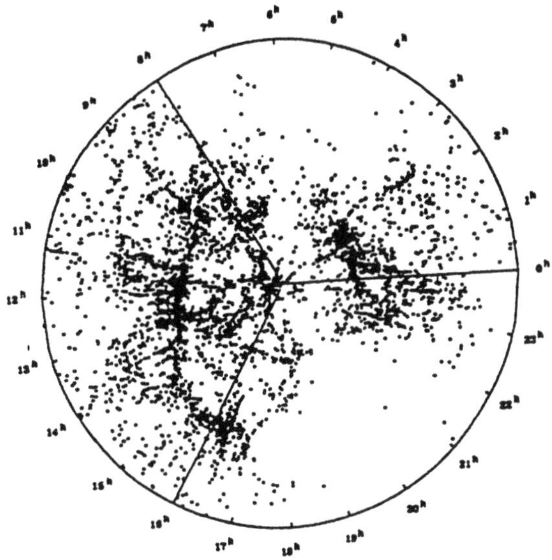

Figure 3.4: A 360° redshift slice, after Geller and Huchra (1989). The High density region between 5000 and 1000 km s^{-1} has been dubbed "the Great Wall".

3.5 The Pisces–Perseus Supercluster Survey

The advent of low–noise receivers, coupled with the great collecting area of the Arecibo reflector and advances in spectrometer technology, made possible the pursuit of ambitious surveys with that instrument even within the non–dedicated framework of a national center. The PPS survey was principally carried out with the Arecibo telescope, using the neutral hydrogen 21 cm line to measure redshifts; that transition is detectable easily in emission from normal spiral galaxies to distances on the order of $150h^{-1}$ Mpc, and radial velocities are measurable with great precision (usually better than 10 km s^{-1}). In addition to radial velocity, the 21 cm profile yields from its integrated flux a relatively accurate estimate of the total HI mass within the galaxy, and from its width, a straightforward measure of the maximum rotational velocities allowed within the gaseous disk, and hence an estimate of the total mass of the galaxy. The well-defined relation between a galaxy's luminosity and its maximum rotational velocity (Tully and fisher 1977) has been used extensively to obtain redshift-independent distances to galaxies, to obtain estimates of H_o and to study deviations from Hubble flow, as we discuss in Section 6.

The maximum velocity cz at which a galaxy of H I mass M_{HI} and velocity width ΔV can be clearly detected — when observed with a single dish of diameter d, a receiver of system temperature T_{sys} and with an integration time t —, is

$$cz \sim (10^4\ kms^{-1})\Big(\frac{h^2 M_{HI}}{4.7\times 10^9 M_\odot}\Big)^{1/2}\Big(\frac{\Delta V}{300\ kms^{-1}}\Big)^{-1/2} \Big(\frac{d}{100m}\Big)\Big(\frac{T_{sys}}{20K}\Big)^{-1/2}\Big(\frac{t}{5\ min}\Big)^{1/4} \tag{3.5}$$

An interference-free five minute on-source observation at Arecibo can detect a normal Sb galaxy of H I mass $4.7\times 10^9 h^{-2} M_\odot$ at $cz \sim 16{,}000$ km s^{-1}. Upon completion of the Gregorian Upgrade [4], the same observation will be possible in between 1/8 and 1/60 of this time, depending on the source's zenith distance.

Not all galaxies contain significant amounts of neutral gas. Most notably, elliptical galaxies, at least the majority of them, do not contain easily detectable amounts of H I. Thus redshift surveys conducted via the 21 cm line concentrate on spiral and irregular objects; galaxies of type S0 and earlier are underrepresented in H I selected samples. Furthermore, spiral galaxies in the cores of rich clusters are severely H I deficient (Haynes *et al.* 1984), presumably as a result of recent environmental alteration. On the other hand, spirals constitute the vast majority of galaxies in the field, and late type, low surface brightness galaxies are H I-rich and even more easily visible to the 21 cm eye than to the optical one. The 21 cm line surveys are thus biased towards late-type, low-surface brightness objects that preferentially occupy the field. In contrast, the high surface brightness objects preferred in magnitude-limited optical redshift surveys show a higher degree of clustering. Radio and optical surveys are thus, to a large degree, complementary.

The general region of the PPS is enclosed by $b \leq 10°$ and $\delta \geq 0°$ (see fig. 3.5). Within those boundaries there are approximately 6000 cataloged galaxies either apparently brighter than $m_{pg} = 15.7$ or with angular diameters larger than 1'.

[4] The Gregorian Upgrade involves the implementation of a new subreflector assembly for the 305-m reflector — that will replace the narrow-band, lossy line feeds that currently correct for spherical aberration of the primary — and the construction of a "ground" screen, 1 km long and 20-m high, that will shield feeds from electromagnetic contamination arising from the 300 K ground surface. These improvements will increase the absolute gain of the antenna, drastically reduce vignetting, allow much lower system temperatures, greatly increase instantaneous observing bandwidths and add interference rejection capacity. The Upgrade got under way in early 1991 and it is scheduled for completion in late 1993.

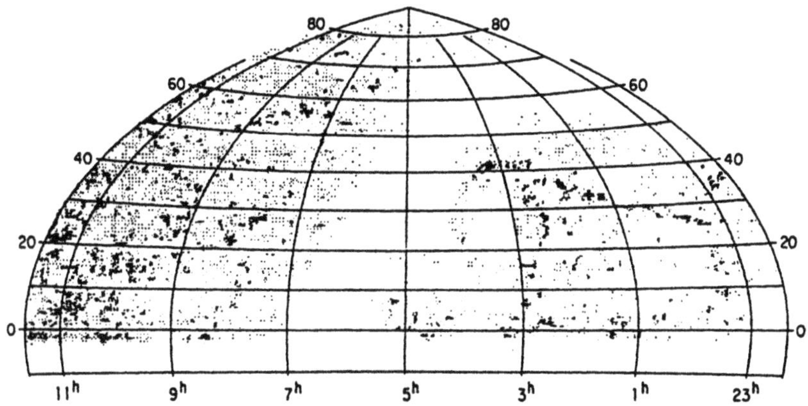

Figure 3.5: Galaxy counts projected on Northern sky, after the CGCG. The extinction associated with galactic plane dust outlines the zone of avoidance. The PPS survey region is bounded by the zone of avoidance to the North and the Equator to the South. The main ridge of the PPS is discernible even in this projected view (after Giovanelli and Haynes 1982, *A.J.* 87, 1355).

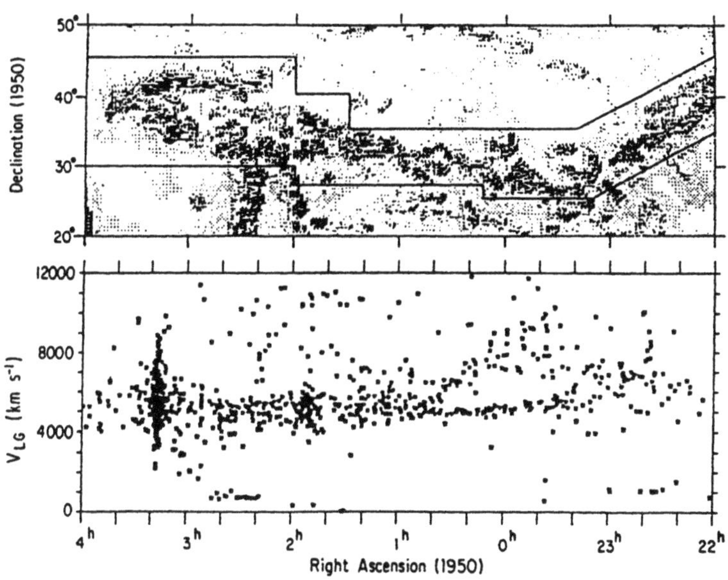

Figure 3.6: The main ridge in the PPS, seen projected on the plane of the sky (integral counts of all CGCG galaxies over outlined region; upper panel), and in redshift space (lower panel). After Giovanelli, Haynes and Chincarini 1986.

Following pioneering work by Gregory *et al.* (1981), over the last decade, Giovanelli, Haynes and their collaborators (see Giovanelli and Haynes 1991 and refs. therein) have worked towards completion in the PPS region, contributing new redshifts for approximately 3500 galaxies, observed primarily with the Arecibo and with the Green Bank 300–foot (*r.i.p.*) telescopes. Pre–existing or ongoing optical observations complement the radio survey effort. In particular, Wegner *et al.* (1991) aim at solving the most severe incompleteness biases of the PPS survey, by concentrating on obtaining redshifts of early–type (i.e. HI poor) galaxies, as well as on those located in the regions north of $\delta = +35°$ and south of $\delta = 3°$, which are most severely affected by zenith distance limitations of the 305–m telescope. Neighboring areas have been recently surveyed by Merighi *et al.* (1986) and the extension of the PPS towards southern declinations has been mapped by Maurogordato *et al.* (1991). Part of this region is also included in the CfA survey effort. The redshift sample assembled by Giovanelli and Haynes in the PPS includes 5500 galaxies. Overall, it is better than 85% complete to $m_{pg} = 15.7$ and better than 90% complete to a major angular diameter of 1'. In the regions of poorer completion ($45° > \delta > +35°$ and $3° > \delta > 0°$), completion rate to $m_{pg} = 15.7$ is better than 65%. A recent summary on the status of this survey is given by Giovanelli (1990) and a survey of results is given in Haynes and Giovanelli (1988). The selection function to be applied for survey incompleteness is then a complicated function of magnitude (or size), sky location and redshift. However, the incompleteness corrections are in all cases quite mild, and the statistical inferences are of highly accurate reliability. More worrysome are the uncertainties that obtain from patchy galactic extinction and inaccuracy of cataloged parameters. As in most cases, a surveyed region is not completed by the exclusive efforts of a single team of observers. The combined, painstaking contributions of many lead to the composition of the three–dimensional picture, a detail sometimes neglected by the hasty writer.

The PPS survey has revealed an extensive network of filamentary structures, dominated by the main body of the PPS, which is displayed in Fig. 3.6. The number density contrast between PPS and foreground low density regions is quite high (> 20), and the linear extent of the main filament exceeds the linear cross section of the surveyed region, at the mean distance of the PPS (about $50h^{-1}$ Mpc). Also notice that the high density region of the PPS is made up of not only aligned cluster cores, but also — and mainly — of a linear, monotonic density ridge.

3.6 Sparse and Targeted Surveys

While the power–law behavior of the two–point correlation function of the galaxian distribution is fairly well established, its signature becomes weak at large scales ($r > 20h^{-1}$ Mpc). The determination of $\xi(r)$ then requires surveys that need be both deep and wide–angle. As mentioned in section 3.1, Kaiser (1986) has convincingly argued that a sparse, rather than a complete survey approach is observationally more adequate, quantifying the optimization. The full advantage of a sparse sampling technique is obtained when a very deep, wide–angle and homogeneous catalog is available, such as those generated by the APM (Maddox *et al.* 1990) and COSMOS (Heydon–Dumbleton *et al.* 1989) machines from UKST survey plates. The coverage of 4300 square degrees in the south galactic cap by the APM survey has produced a catalog of 2×10^6 galaxies brighter than $b_J \sim 20.5$. A sparse survey aimed at obtaining redshifts of one for every 20 APM galaxies in selected areas is currently under way at the AAT.

Numerous surveys include objects that are selected according to criteria that do not conform to flux, size or spatial completeness, for example those aimed at binaries and groups, dwarf, emission–line galaxies, the better definition of supercluster and void outlines, isolated galaxies, the highly obscured regions near the galactic plane, etc. We refer to Giovanelli and Haynes (1991) for a brief review of such efforts.

4 Qualitative Overview of Results on Large–Scale Structure

A common experience to those who analyze the properties of redshift samples has been that of verifying that the largest traceable inhomogeneities in the galaxian distribution "escape" through the boundaries of the sampled region, raising the question of what is a "fair" sample of the universe (see, for example figs. 3.4–3.6). To some degree, the measure of "fairness" for a sample depends on the statistical tools that are to be applied to determine its properties. While the scalelength of the galaxy–galaxy correlation function $\xi(r)$ over small scales is about the same in any sample (hovering near $r_o = 5h^{-1}$ Mpc), the sizes of the largest structures in 2–d and 3–d surveys, be they voids, or connected high density regions, are nonetheless comparable to those of the sampled volumes, typically $20h^{-1}$ to $60h^{-1}$ Mpc. How much power is there at the large scales or, up to what scale is there any measurable power? A useful analogy to illustrate the ambiguities associated with these measurements might be the determination of the size of oceans on Earth: although a single body of water can be continuously traced around the planet's surface, we characterize the oceanic structure by a scale–length related to the mean separation of continents. Thus, the definition of "structure" and the measurement of its size are strongly linked to topology. Topological descriptions of the galaxian distribution have evolved from the visual, highly subjective appreciation of morphologies in poorly sampled data, to more objective numerical approaches applied to progressively richer samples. Panoramic surveys have played an important role in forming our views of the cosmic fabric, as in establishing the prejudices that permeate the numerical approaches that strive for its objective description.

The concept of "supercluster" gained acceptance during the late 1970's. The early studies of the Coma (Chincarini and Rood 1975; Gregory and Thompson 1978), Hercules (Tarenghi *et al.* 1980), and Perseus regions (Gregory *et al.* 1981; Einasto *et al.* 1981) identified features that extend well beyond the boundaries of clusters (see the review by Oort 1983). Kirshner *et al.* (1981) and Davis *et al.* (1982) brought attention to the existence of large, underdense volumes with typical sizes of tens of Mpc that, the first CfA slice suggested (de Lapparent *et al.* 1986), might be outlined by a bubble–like galaxian distribution. The Arecibo Pisces–Perseus supercluster survey on the other hand revealed the existence of very extensive, linear, filament–like structures (Giovanelli *et al.* 1986). The length of such filaments, $> 80h^{-1}$ Mpc, and their topology, propose serious obstacles to gravitational clustering scenarios. Topology–discriminating algorithms have been developed (e.g. Gott *et al.* 1987, Ryden *et al.* 1989) and applied to a variety of redshift catalogs (Gott *et al.* 1989); they indicate that, when the galaxian distribution is smoothed to scales as large as the correlation length, r_o, the topology appears sponge–like in all samples, an appearance that is consistent with the standard model in which today's structure has grown from small, random noise fluctuations in the early universe. However, when the galaxian distribution is smoothed to lengths smaller than r_o, the character of the topology shifts towards a "meatball" (filaments, in this context, are indistinguishable from meatballs) rather than a "bubble" character (i.e. one where volumes are dominated by voids surrounding the enhancements in the galaxian distribution, rather than one where voids are surrounded by closed surfaces of enhanced galaxian density. See section 5.6 for details). Reservations on the adequacy of this type of analysis have been expressed by Geller and Huchra (1988).

The ability of a survey to trace a structure from one end of the volume sampled to the other does not exclude that the volume has approximated a fair sample. In a cellular network, for example, connected structures that stretch across the sampled region would be seen, no matter how large the sampled volume is; yet a fair sample will be approximated once the volume spans a few cells. Has such a point been reached with nearby, wide–area redshift surveys, or have we not run yet the gamut of large-scale inhomogeneities? The open-ended aspect of the Pisces-Perseus supercluster as surveyed over nearly two radians (fig. 3.6), and the galaxian distribution in "the Great Wall" (Geller and Huchra 1989), shown in fig. 3.4, might appear to suggest that the answer to that question is negative. Even more impressively, the finding of Broadhurst *et al.* (1990) of a strong clustering feature with a periodicity of $128h^{-1}$ Mpc indicates that structure may exist on scales much larger than could have been found by wide-area surveys. Karachentsev (1984) pointed out a similar feature in the clustering spectrum, with a periodicity of about $70h^{-1}$ Mpc, although his result, based on only 92 redshifts, was less convincing than that of Broadhurst *et al.* .

While the existence of structure with very large scales is well accepted, that of a characteristic periodicity in such structure remains statistically not very firm. Kaiser and Peacock (1990) warn that in pencil-beam surveys, apparent clustering on large scales might result from aliasing of 3-D clustering on small scales, a form of clustering noise. It is useful to maintain in perspective that, over volumes of size comparable with the largest scales discussed here, several aspects of the galaxian distribution suggest that mean properties have been grasped: e.g. the identity of slopes of galaxy counts $\log N(m)$ in all directions, known already from the work of Hubble, pose a strong case in favor of fairness over sizes on the order of $100h^{-1}$ Mpc. Nonetheless, the perspective that the large-scale structure of the galaxian distribution might be dominated by a network of structures with characteristic scale on the order of $100h^{-1}$

Mpc is most tantalizing, and it remains to be established whether the amplitude of density perturbations with the increasing size of inhomogeneities (found and suggested) can be accomodated with the constant slope of galaxy counts. The issue of tight periodicity, as exhibited in figs. 3.2 and 3.3, poses worrisome concerns for theory and demands urgent corroborating observations. Much debate will be focused on the issue of the adequacy of previously favored theoretical and modeling schemes, in the description of the largest scale clustering features. Weinberg and Gunn (1990) have examined the structures expected to be seen when varying magnitude limits are imposed on the numerical results of biased cold dark matter simulations. They propose that wide angle surveys to limiting magnitudes fainter than 16.5 should reveal an even greater wealth of structure. Providing a measure of sobriety, they also underscore the resilience of current theoretical schemes to the threats of, at first sight, hostile observational results: "...the existence of $50--150h^{-1}$ Mpc structures is not in itself an argument against gravitational instability, Gaussian fluctuations, and cold dark matter". But strict periodicity would most likely provide a powerful such argument.

5. On Topology and Statistical Tools

5.1 The Luminosity Function of Galaxies

The topic has been recently reviewed by Binggeli, Sandage and Tammann (1988). Let the number of galaxies within the volume element dV, centered at position (x, y, z), and with luminosity within $(L, L+dL)$ be a separable function

$$\nu(x, y, z, L)dV\ dL = \phi(L)\ D(x, y, z)\ dL\ dV \tag{5.1}$$

with $\int_0^\infty \phi(L)dL = 1$. Under this commonly made assumption, the *luminosity function* (LF) $\phi(L)dL$ represents the fraction of galaxies with luminosity within $(L, L+dL)$, while the *density function* $D(x, y, z)$ is the number density of all galaxies at (x, y, z). The LF is generally parametrized in the form first given by Schechter (1976)

$$\phi(L)dL \sim \left(\frac{L}{L^*}\right)^\alpha e^{-L/L^*} dL \tag{5.2}$$

The LF is also often expressed in terms of the absolute magnitude

$$\phi(M)dM \sim 10^{-0.4(\alpha+1)M} e^{-10^{0.4(M^*-M)}} dM \tag{5.3}$$

At faint magnitudes, $\phi(M)$ is exponential with slope $-0.4(\alpha+1)$, while on the bright end, $\phi(M)$ drops quickly as a double exponential, after turnover at the "knee" near $M = M^*$.

Under the assumptions of homogeneity and that $v = H_o r$, i.e. that radial velocity corresponds to distance (perfect Hubble flow), the number of galaxies with magnitudes in the range (m_1, m_2) and with velocities between v and $v + dv$ can be written as

$$N(v)dv = \omega v^2 H_o^{-2} \int_{m_1}^{m_2} \int_{-\infty}^{\infty} \phi(M) D(r)\, dm\, dM \tag{5.4}$$

where $r = 10^{0.2(m-M-5)}$. This expression is used to estimate the expected velocity distribution of a given sample of galaxies, provided that curvature effects are neglected (z of a few tenths or less).

If $\phi(M)$ is known, one can also estimate the number of galaxies per square degree in the direction r, with absolute magnitude within (M_1, M_2), that would be counted within the apparent magnitude interval $(m, m + dm)$:

$$A(m)dm = \frac{4\pi}{41253} \int_{M_1}^{M_2} r^2 \phi(M) D(r) dr dM \tag{5.5}$$

with $m - M = 5 \log r - 5$.

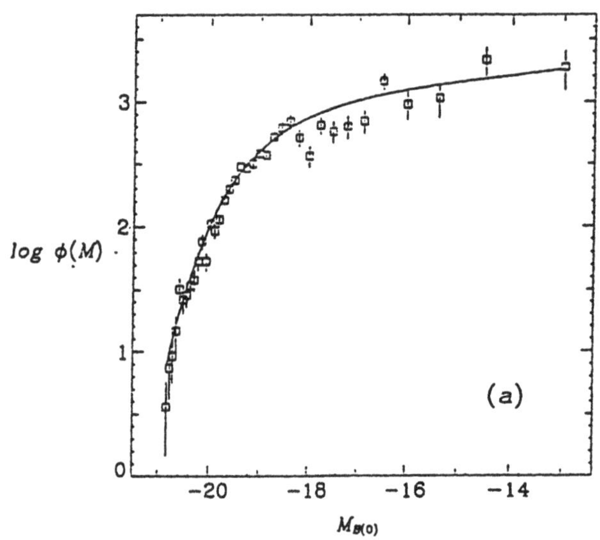

Figure 5.1: The Luminosity Function obtained by de Lapparent *et al.* (1989) from one of the CfA slices. The fit corresponds to a Schechter function with $M_B^* = -19.2$ and $\alpha = -1.1$, for $h = 1$.

Problems associated with the determination of the LF are extensively discussed in the review of Binggeli *et al.* (1988). Here, we shall briefly comment on two questions, associated with the intrinsic characteristics of the galaxian distribution. (1) When the LF is obtained from relatively shallow surveys ($m \sim 15$), complete to their flux limit and over a wide solid angle, our location on the periphery of the Local Supercluster biases the results in the sense that the space density of faint galaxies — which can only be seen at relatively small distances from us — tends to be underestimated; the problem is progressively alleviated as the magnitude (flux) limit of the survey used to obtain the LF becomes progressively fainter. (2) With deep, narrow pencil-beam surveys, on the other hand, a different problem arises, associated with clustering noise. Derived redshift distributions are quite irregular, as large-scale structures that exceed the cross section of the pencil beam appear as large enhancements in an $N(v)\Delta v$ plot, alternating with adjacent regions of very low counts (e.g. see the Broadhurst *et al.* figure in section 3). The unbiased determination of the parameters of the LF is another example where the sparse sampling in deep, wide angle surveys advocated by Kaiser (1986), presents obvious advantages. With techniques that attempt a correction for the mentioned (and other) biases, two important determinations of the LF have been carried out following the review of Binggeli *et al.* (1988), : those by Efstathiou *et al.* (1988) and de Lapparent *et al.* (1989). The two results are fairly consistent with each other, and yield $M_B^* \simeq -19.2 \pm 0.1$, $\alpha \simeq -1.1 \pm 0.1$, for $h = 1$. A plot of the LF with representative error estimates is shown in Fig. 5.1.

5.2 Galaxy Counts

Galaxy counts $N(m)dm$, i.e. the number of galaxies with apparent magnitude within $(m, m + dm)$ and within some solid angle ω, constitute a tool used to discriminate among cosmological models, for evolutionary studies and to estimate the luminosity density of the universe. The expected values can be estimated from eqn. (5.5), as $N(m)dm = \omega \int A(m)dm$. At high redshift, spatial curvature needs to be taken into account; that is achieved by replacing the product $r^2 dr$ within the integral in eqn. (5.5) by the expression

$$\frac{y^2 dy}{[1 - (h_\circ y)^2 (2q_\circ - 1)]^{1/2}}$$

where

$$h_\circ y = \frac{1 - q_\circ + q_\circ z + (q_\circ - 1)(1 + 2q_\circ z)^{1/2}}{q_\circ^2 (1 + z)}$$

Other corrections are required to account for the shift of the galaxy spectral energy distribution (K–correction) as well as for the evolution that may have occurred with the galaxian population. As we mentioned in section 3, in the 1930's, Hubble used $N(m)$ counts to test the homogeneity of the universe over large scales. These counts are, of course proportional to the mean value of the density function, $\bar{D}$; this quantity is often referred to as ϕ^*, as it is a frequent practice not to separate ϕ and D, as in eqn. (5.1), in which cases the expression "luminosity function" is reserved to the product we have referred to as ν, and ϕ^* yields a mean density scale factor. The inferred values for ϕ^* are quite uncertain, varying by factors of several between different determinations. Averaging from several different determinations, Efstathiou *et al.* (1988) adopt

$$\phi^* = (1.56 \pm 0.34) \times 10^{-2} h^3 \mathrm{Mpc}^{-3} \tag{5.6}$$

The *mean luminosity density of the universe*, which is obtained by calculating the first moment of eqn. (5.1) and substituting the proper values for α, L^* and ϕ^*, is

$$\bar{L}^B = \phi^* \Gamma(\alpha + 2) L^* = (1.93^{+0.8}_{-0.6}) \times 10^8 h \ \ L_\odot \tag{5.7}$$

where Γ is the incomplete gamma function defined by $\Gamma(n) = \int_0^\infty e^{-x} x^{n-1} dx$ and the recursion relation $\Gamma(n+1) = n\Gamma(n)$, and an absolute magnitude of $M_B = 5.48$ has been assumed for the Sun. Recalling that the mass density required to close the universe is $\rho_{crit} = 3H_0^2/8\pi G$, the critical mass-to-light density of the universe is then

$$(M/L)_{crit} = (1500^{+700}_{-400}) h \ \ (M_\odot/L_\odot) \tag{5.8}$$

which is much larger than the values that the Virial theorem or other mass estimators obtain for the mass-to-light ratios of groups and clusters of galaxies — $(50 - 400)h^{-1}$ $(M_\odot/L_\odot)$ —, of individual galaxies — $(5 - 20)h^{-1}$ $(M_\odot/L_\odot)$ —, or of the solar neighborhood in our Galaxy — $\sim 2.7h^{-1}$ $(M_\odot/L_\odot)$.

5.3 Evolution of the Luminosity Function

Next, we tackle the problem of the evolution of the galaxian population between $z = 1$ and now. Figure 5.2 displays differential galaxy counts — $dN(m)$ in our definition — over the magnitude range covered by several medium-deep photometric surveys, as summarized by Ellis (1987). The solid line corresponds to the expectation obtained from eqn. (5.5) and the aforementioned corrections for curvature and shift of the energy spectral distribution of the sampled galaxies with redshift, but *no evolution* of the galaxian population. The overabundance of faint galaxies, with respect to the number expected from a no-evolution extrapolation of the counts at brighter magnitudes, is more than half an order of magnitude at $B_J \simeq 24$. Tyson (1988) and co-workers have pushed to even fainter magnitudes, and found that the gap between the no-evolution curve and the data keeps widening, as shown in Fig. 5.3. The count excess could be attributed to luminosity evolution, i.e. galaxies get brighter with look-back time, so that at every further step in z, the LF shifts bright-side. However, Kron (1982) cautioned that, without redshift information,

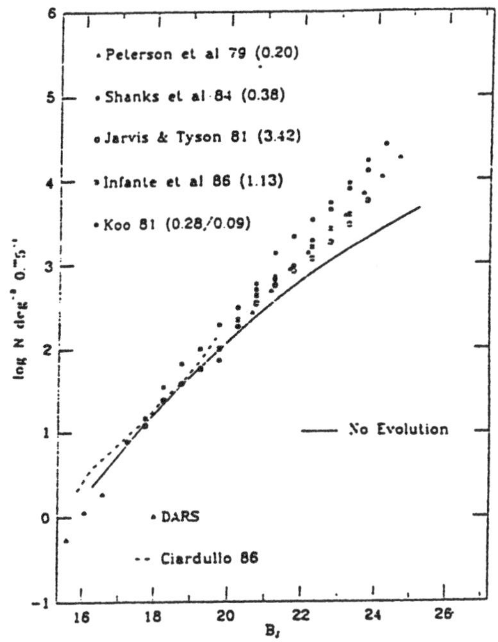

Figure 5.2: Galaxy counts after Ellis (1987).

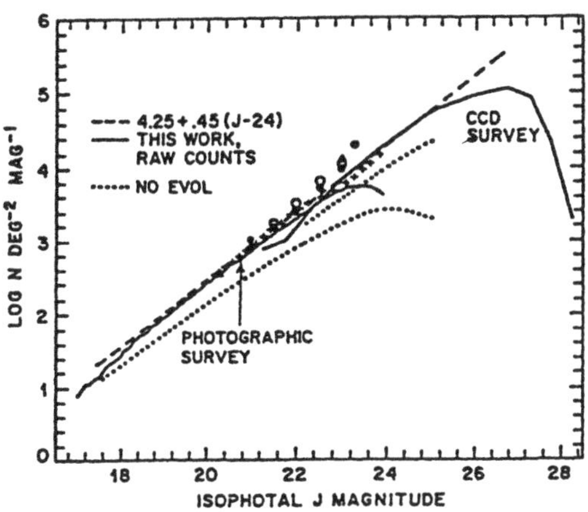

Figure 5.3: Galaxy counts after Tyson (1988).

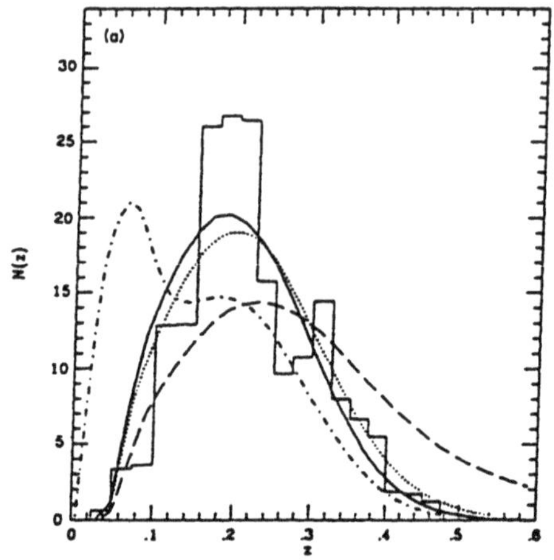

Figure 5.4: Redshift distribution of the Broadhurst *et al.* (1988) survey (histogram). Curves are for model predictions: no evolution (solid), luminosity evolution adjusted to match the counts (see figs. 5.2, 5.3) slope (dash), extension of the field LF adjusted to match the counts slope (dot–dash); luminosity dependent evolution as shown in fig. 5.5 (dot).

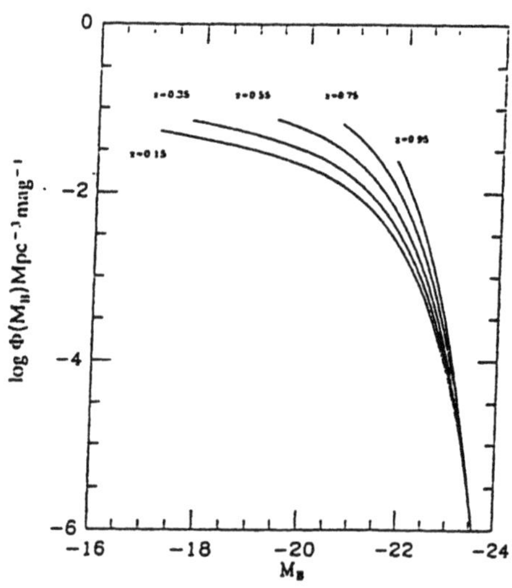

Figure 5.5: Broadhurst *et al.* (1988) LF evolution model, which matches both counts (fig. 5.2) and redshift distribution (fig. 5.3).

it was impossible to exclude that the effect could instead be due to a population of nearby, blue, low luminosity field galaxies that may have escaped local surveys. Broadhurst *et al.* (1988) used the DARS survey (see section 2) to test that hypothesis. Figure 5.4 shows the histogram of the DARS counts, plotted as a function of redshift. Superimposed on it is the expected distribution allowing no evolution of the LF (solid curve), using eqn. (5.4) with curvature and K–corrections. Surprisingly, the data are well fit by the expectation. On the other hand, if the LF is allowed to change with z in the simplest way that fits the excess in the differential counts shown in Fig. 5.2, the resulting expected redshift distribution (dashed line in fig. 5.4) is widely off the data histogram; a similar discordance is obtained if the LF is forced to fit the slope of the galaxy counts (dash–dot line).

The results of Broadhurst *et al.* first establish that the excess counts at $b_J > 21$ are not due to a population of previously undetected, nearby low luminosity galaxies. The excess objects are distant, and they need be explained by some change in the LF, but not simply one of wholesale shift: the evolution must be *luminosity dependent.* They propose an empirical model for the evolution of the LF that would fit both the counts and the redshift distribution. The model is illustrated in Fig. 5.5: there is little or no evolution of the bright end of the LF, while the slope of the faint end keeps rising with increasing redshift. The redshift distribution predicted by this model is plotted as a dotted line in Fig. 5.4. Broadhurst *et al.* suggest that the excess in the faint magnitude galaxy counts, which is modeled via a steepening slope of the LF with z, is produced by low luminosity galaxies undergoing bursts of star formation.

5.4 Is there a Universal LF at $z = 0$?

The assumption of separability of the LF into two parts, one that depends exclusively on the spatial coordinates and one that depends only on the luminosity bin — as expressed by eqn. (5.1) — has been supported by the observed similarity in shape of the LF's of cluster and field galaxies. More recently, the characteristics of the large scale structure in the galaxian distribution (LSS) have provided reinforcing evidence. If the LF were different in high and low density regions, for example, or if significant luminosity bias were present in the LSS of the galaxian distribution, the topology of the LSS would appear different depending on the galaxies used to trace it. It is generally agreed that large differences in the characteristics of the LSS with galaxy luminosity or surface brightness do not exist, with the corollary that dwarf galaxies do not fill the LSS voids (Thuan *et al.* 1987; Eder *et al.* 1989; Geller and Huchra 1988).

A number of lines of thought does, on the other hand, suggest that separation as expressed by eqn. (5.1) should perhaps not be expected. Scenarios for the formation of structure in the universe yield descriptions of the distribution of mass; we can readily observe the distribution of *visible* matter, basically through galaxy counts; it is by no means obvious that the observed $\delta n/n$ map faithfully the generally unobserved $\delta\rho/\rho$. After Kaiser's (1984) suggestion, it is in fact customary to assume that "biasing" in the rise of structure may exist, which in its simplest form can be described by the linear expression

$$\frac{\delta\rho}{\rho} = b^{-1}\frac{\delta n}{n} \tag{5.9}$$

The parameter b is referred to as the *bias* parameter. A simple statement of bias would be that more massive — thus brighter galaxies — favor high density regions. Thus, schemes of galaxy formation that postulate some form of biasing would also in most cases yield a luminosity function with a shape that is density dependent. Approaching the problem on another tack, if (a) segregation exists of galaxies of different morphological types as a function of the local density $D(x, y, z)$, and (b) the luminosity function differs from type to type, then one should expect to observe differences in the shape of the LF in regions of different density, i.e. a breach of the postulate of separability expressed by eqn. (5.1) should occur. Evidence for morphological segregation exists certainly in clusters of galaxies (Dressler 1980 and refs. therein), and has also been reported over scales well in excess of that of clusters (e.g. Giovanelli, Haynes and Chincarini 1986). The Las Campanas survey of the Virgo cluster (see Binggeli, Sandage and Tammann 1988 and refs. therein) has shown incontrovertibly that the LF's of individual types differ substantially from each other. Thus the shape of the LF should have a density dependence of the form

$$\phi(L,\rho)\, dL = \sum_t \phi_t(L) f_t(\rho) \tag{5.10}$$

where the subscript t refers to morphological type and $f_t(\rho)$ is the population fraction at density ρ appropriate for type t.

By synthesizing LF's with widely different population fractions f_t, e.g. using the individual type LF's given by Sandage, Binggeli and Tammann (1986, cf. BST88), one can easily satisfy oneself that the overall shape of $\phi(L)$ is fairly stable: an L–conspiracy of sorts. It is this conspiratorial resiliency to varying morphological type make–up that's responsible for the lack of dramatic, observed differences between, say, cluster and field LF's and, ultimately for the statement of separability illustrated by eqn. (5.1). The conspiracy may however not be perfect. For example, the case can easily be made for the fact that while, as mentioned above, dwarf irregular galaxies trace the same general structure as the bright population, they do so much less sharply, i.e. the *density contrast* of the LSS is a function of luminosity class (e.g. Salzer, Murray and Gavazzi 1990). If indeed that were the case, and given the wide luminosity range in L bridged by the individual ϕ_t, the pessimist would apply Murphy's law to expect an explicit dependence of ϕ on ρ not only because of eqn. (5.10), but also because the individual ϕ_t would depend on ρ. Using the PPS data base, Iovino *et al.* (1991) have tentatively concluded that that may indeed be the case. The problem is still open to debate and needs corroboration by independent data bases.

5.5 On the Two-Point Correlation Function

As mentioned in Section 3.1, the two-point correlation function, $\xi(r)$, is usually expressed in terms of a power law, $\xi(r) = (r/r_o)^{-\gamma}$, with $\gamma \sim 1.8$ and $r_o \sim 5h^{-1}$ Mpc. Were that expression to be valid for an infinite set, we would say it is characterized by a fractal dimension $D \simeq 3 - \gamma \simeq 1.2$. The power law behavior of the two-point correlation function is well established at small separations r; at larger separations, a clear divergence from the small-scale behavior is observed (Shanks *et al.* 1983; Martinez and Jones 1990) and the possibility that some kind of scale invariance be present also on scales $r \gg r_o$ has been a matter of controversy (e.g. Szalay and Schramm 1985). The clustering properties at separations well in excess of the clustering length r_o are of particular interest, since the perturbations they describe may still be treated by linear theory and they are thought to have retained "memory" of the recombination shape of the spectrum of fluctuations. This is so because all Fourier modes in the linear regime grow at the same rate, and thus preserve unchanged the pattern imprinted at the epoch of galaxy formation. Let $1 + \xi_\rho(r)$ be the autocorrelation function of the density field ρ; its Fourier transform, is the power spectrum, i.e. $|\delta_k|^2$. A distribution with autocorrelation function $1 + \xi_\rho(r) \propto r^{D-3}$ will then be characterized by a power spectrum given by

$$|\delta_k|^2 \propto \int_0^\infty r^{D-3} \frac{\sin kr}{kr} r^2 \, dr \quad \propto k^{-D} \tag{5.11}$$

If the density field at recombination had a power-law spectrum $|\delta_k|^2 \propto k^n$ and random phases, the corresponding matter distribution in real space would show fractal properties. Thus, by exploring the fractal properties (if any) on large scales we would expect to encounter some memory of the mechanism that produced the primordial perturbations.

In the analysis of the Giovanelli and Haynes redshift survey sample of the Pisces–Perseus supercluster sample, Guzzo *et al.* (1991) have identified a break in the power law behavior of $1 + \xi(r)$, near $r_b \simeq 3.5h^{-1}$ Mpc; at $r < r_b$ the standard slope of correlation dimension $D = 3 + \gamma \simeq 1.2$ is observed, while for $r_b < r < 30h^{-1}$ Mpc, the power law is much shallower, with $D \simeq 2.2$, as shown in fig. 5.x. This break, already suggested in previous studies (see Guzzo *et al.* for refs.), may mark the transition between large-scale fluctuations in the linear regime, and the collapsed regions where gravitational amplification has forced a departure from the primordial spectrum, at the small scales. Dekel and Aarseth (1984) suggested that a bimodal power law behavior with a large-scale correlation dimension near 2 may be evidence for flat pancake-like structures at large scales; however, a gaussian density field with a power-law spectrum $\propto k^{-2}$ with galaxies tracing the mass would also yield $D = 2$, without galaxies being constrained on 2-d surfaces; Gott *et al.* (1989) have in fact presented convincing evidence that the galaxy distribution is sponge-like.

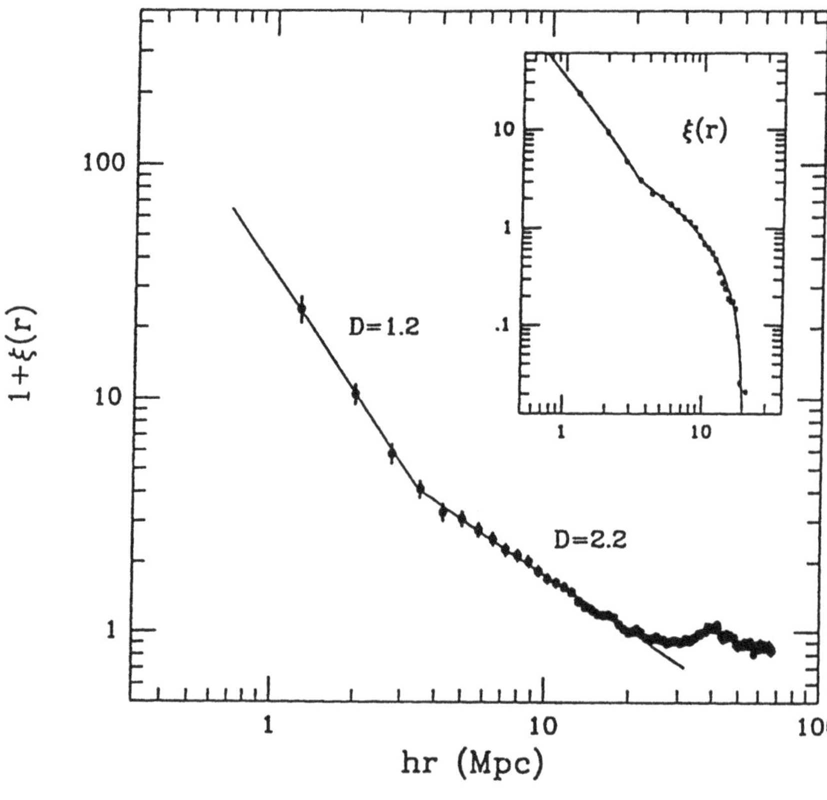

Figure 5.6: $1+\xi(r)$ obtained from a volume-limited subsample of the Pisces-Perseus catalog, with $M < -19.0$ and $cz < 7900$ km s^{-1}, containing 1092 galaxies. The solid lines are best fits of two power laws in the ranges 1-3.5h^{-1} Mpc and 4-20h^{-1} Mpc. The inset shows $\xi(r)$ for the sample, while the solid line represents the curve obtained by subtracting 1 from the fit to $1+\xi(r)$ (after Guzzo *et al.* 1991).

A distribution approximating the properties of a 'fractal dust' might be a better description of the observed large-scale distribution. It is interesting to note, for example, that a primordial $D \sim 2$ is suggested by phase transition physics, a possible pathway towards understanding the origin of the fluctuation spectrum (Schulman and Seiden 1986).

Another way of investigating the existence of structure at very large scales, using the correlation function formalism, is that of considering the distribution of clusters of galaxies. Neta Bahcall (1988) has recently reviewed the subject. The two-point correlation function of a complete sample of nearby Abell clusters (reaching a volume of radius $\simeq 300h^{-1}$ Mpc) was found to exhibit a power law behavior similar to that of galaxies, consistent with the same exponent of the galaxy $\xi(r)$, but with a much larger (by about a factor of 20) amplitude (Bahcall and Soneira 1983):

$$\xi_{cc}(r) = \left(\frac{r}{26}\right)^{-1.8} = 360\, r^{-1.8}, \tag{5.12}$$

and $\xi_{cc}(r)$ also appears to remain positive out to $50h^{-1}$ Mpc or more. Such a strong correlation is in conflict with "popular" scenarios that assume that structure arose from scale-invariant Gaussian density fluctuations, such as $\Omega = 1$ cold dark matter models. The latter in fact predict that ξ_{cc} should become negative at scales larger than about $20h^{-2}$ Mpc, because at large scales ξ_{cc} should mimick the behavior of the galaxy-galaxy ξ. On the other hand, a large ξ_{cc} could be consistent with other schemes, such as the explosion scenario. Sutherland (1988) suggested that contamination of the cluster catalog due to projection effects (clusters overlapping in sky projection) would spuriously enhance the measured amplitude of ξ_{cc}, by as much as a factor 3. Olivier *et al.* (1990 and refs. therein) have studied in detail the effect of projection contamination. They analyzed the *angular* two-point correlation function, $w_{cc}(\theta)$, both for the Abell (1958) cluster catalog (North) and the Abell-Corwin-Olowin (1989) catalog (South); they used the angular correlation function, rather than the spatial one, because redshifts for all the clusters in the two samples are not available, and used scaling relations to infer the behavior of $\xi_{cc}(r)$. If the spatial correlation function is a power law as eqn. (3.3), Limber equation predicts that the amplitude of $w_{cc}(\theta)$ should scale with the mean sample depth as $d^{-\gamma}$, provided that no contamination effects are present. They concluded that contamination in cluster samples is particularly severe at the smallest scales , $\theta < 1^\circ$ (factor of 2). At scales in the range $1^\circ < \theta < 3^\circ$ ($10 - -30h^{-1}$ Mpc), however, the effect is quite mild (producing an overestimate of amplitudes on the order of 30%). At even larger scales, the effect of contamination is negligible, but the analysis of southern and northern cluster catalogs yield different results: while in the North ξ_{cc} stays significantly positive to rather large scales, in the South it is indistinguishable from zero at $r > 30h^{-1}$ Mpc. Sutherland and Efstathiou (1991) confirm that contamination is responsible for earlier claims of large amplitude and positive correlation on large scales for ξ_{cc}; they base their conclusions on the analysis of other cluster catalogs, in addition to that of Abell. They find that spurious line–of–sight effects severely affect correlation measurements in all catalogs; when corrections are applied for these effects, the value of $r_{o,cc}$ hovers between $9h^{-1}$ and $14h^{-1}$, and that no significant evidence exists for positive correlation on large scales, say comparable with $100h^{-1}$. These results suggest a measure of caution in weighing the evidence for large scale superclustering on the grounds of correlation analysis.

5.6 Some Remarks on Topology

While correlation functions have yielded fundamental information on the clustering properties of the LSS, the full characterization of the galaxian distribution — theoretically provided by the set of N–correlation functions to a large N — has not been reached with that approach, as there are practical limitations to its applicability. Other diagnostic tools, more sensitive to topological details, have been devised, in particular the "void probability" statistic (e.g. Fry *et al.* 1989) and the genus–threshold density method, pioneered by Gott and co–workers (Gott, Melott and Dickinson 1986, GMD). In this section we shall introduce the latter method and its findings.

When analyzing the topology of a distribution of galaxies, we deal with a discrete distribution, where galaxies are associated with points in 3–d space. Since we are interested in the LSS, we derive a density distribution, smoothed over a length somewhat larger than the mean galaxy–galaxy separation, and study the topological properties of the isodensity surfaces derived from the smoothed distribution. GMD have devised an analysis scheme that invokes the relation between the *genus* and the *threshold density* of the smoothed distribution, as follows.

The genus G of a surface can be defined as the number of closed curves that may be drawn on the surface without cutting it into 2 separate pieces, or as the number of "holes" in the surface. For example a sphere has a genus of 0; a doughnut has a genus of 1; the genus of two separate spheres is -1. The Gauss–Bonnet theorem states that the genus of a 2–d surface is given by

$$G = 1 - \frac{1}{4\pi}\int \kappa dS \tag{5.13}$$

where the integral is over the surface and κ is the Gaussian curvature of the surface at the element dS; κ is defined as

$$\kappa = \frac{1}{a_1 a_2} \tag{5.14}$$

where a_1 and a_2 are the two principal radii of curvature at that location [5]. Now let

[5] Another intuitive way of characterizing the topology of a surface is by counting the "number of handles", H: e.g. a closed classic water jar can be visualized as a sphere with two handles, a doughnut as a sphere with a single handle, etc. It is also customary to characterize the topology of a surface by means of its *Euler characteristic*, defined as $\chi = \int \kappa dS/2\pi$, with $\chi = (G-1)/2 = 2 - 2H$.

$\bar{n}$ be the mean density of a density of counts array $n(x, y, z)$, and σ_n its standard deviation. Measure the density level in units

$$\nu = \frac{n - \bar{n}}{\sigma_n}, \tag{5.15}$$

where ν is refer to as the *threshold density*. The measurement of the genus G of isodensity surfaces at different values of ν yields GMD's genus–threshold density relation.

For a standard big bang inflationary model, where the power spectrum of density fluctuations $P(k)$ is a power law (Harrison–Zeldovich spectrum), the arising structure should be expected to yield a $G(\nu)$ relation of the kind

$$G(\nu) = A(1 - \nu^2)e^{-\nu^2/2} \tag{5.16}$$

where

$$A = \frac{1}{(2\pi)^2}\left[\frac{\int k^2 P(k) d^3k}{3\int P(k) d^3k}\right]^{3/2} \tag{5.17}$$

Notice that while the amplitude of the genus depends on $P(k)$, the shape of the curve is independent from it, and it is symmetric around the maximum at $\nu = 0$. That shape is represented by the smooth curve in figure 5.7. It is interesting to anticipate what would be expected for $G(\nu)$ for a non–Gaussian field. We can consider two extreme cases: (1) a density regime of several isolated spheres, separated by a mean distance comparable to the sphere radius, and (2) the "negative" of such a distribution, that can be described as a number of bubbles in an otherwise homogeneously filled space. The two examples are sometimes familiarly referred to as a "meatball" case and a "swiss cheese" topology. In the first case, it is intuitive that at large positive values of ν, the isodensity contours will consist of many detached spheres; thus $G(\nu)$ will be constant (depending on the number of spheres) and negative; as ν decreases towards values somewhat smaller zero, the volume subtended by the isodensity contours fills most space, with holes connecting the spheres: $G(\nu)$ becomes positive and large; as ν further decreases, isodensity surfaces will surround again detached volumes, surrounding pockets in the center of spaces between spheres, and $G(\nu)$ becomes large and negative again. In this meatball topology, then the $G(\nu)$ peak is shifted towards negative values of ν. Similarly, one can verify that in case (2), $G(\nu)$ is symmetrically shifted in the other direction. Thus, the genus–threshold density relation can be a good discriminator between different topologies that have been tentatively identified by visual means, e.g. the "bubble" topology that was strongly suggested by the first slice of the new CfA survey (de Lapparent *et al.* 1986), and, say filamentary structures as suggested by the PPS (hotdogs and meatballs are topologically identical). The application of the $G(\nu)$ technique to observational samples can also be compared with the results obtained from its application to numerical

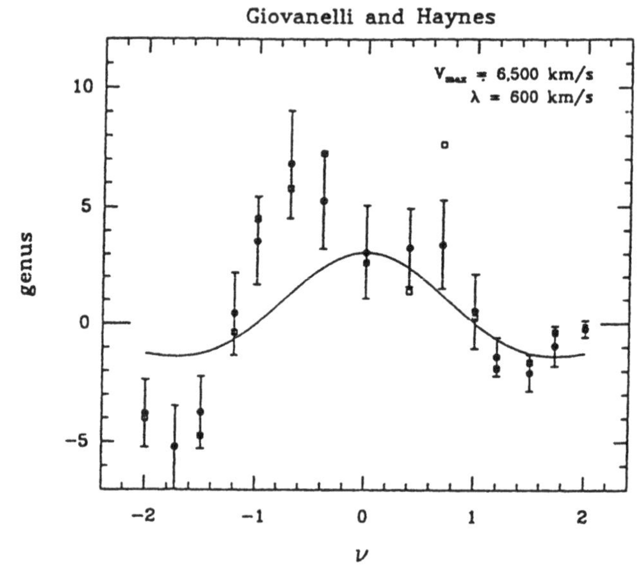

Figure 5.7: $G(\nu)$ for the PPS sample, bound to a v_{max} = 6500 km s^{-1} and with a Gaussian smoothing length of 650 km s^{-1}, after Gott *et al.* (1989).

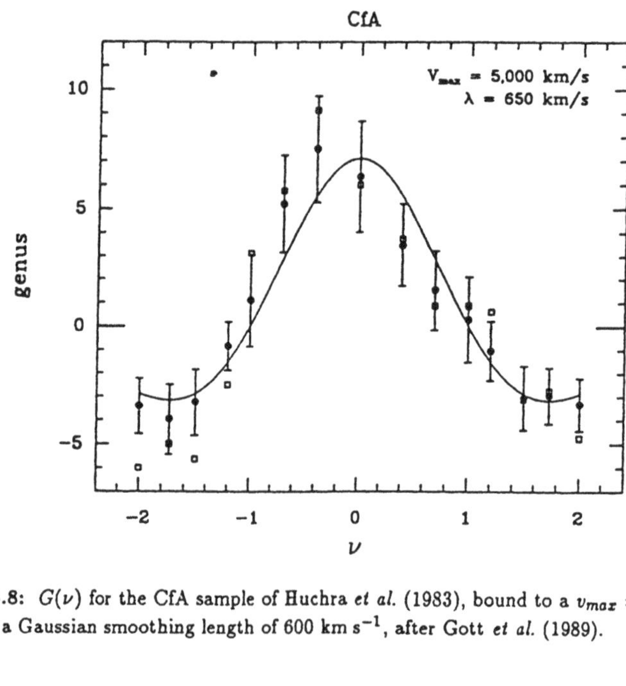

Figure 5.8: $G(\nu)$ for the CfA sample of Huchra *et al.* (1983), bound to a v_{max} = 5000 km s^{-1} and with a Gaussian smoothing length of 600 km s^{-1}, after Gott *et al.* (1989).

simulations based on different theoretical realizations. Gott *et al.* (1989) have applied the technique to a variety of samples and simulations. Figures 5.7 and 5.8 show the results of this analysis applied to two samples, both smoothed with a Gaussian function of length as indicated in the captions. The two samples indicate a clear preference for "meatball"-type topologies. In each case, the smooth curve identifies the "sponge-like" topology described by eqn. (5.16). The comparisons with numerical models favor most strongly cold dark matter hierarchical clustering models.

6. Peculiar Motions

In this final section, we are concerned with the deviations, in the motions of galaxies in the local universe, from a smooth Hubble flow. These motions, more concisely dubbed as "peculiar", are generally thought to arise from gravitational forces originating from large amplitude mass perturbations. Their study does provide dynamical information, therefore, which can lead to the overall mass content of the universe, and which cannot be obtained by a simple census of the luminous matter alone. Historically, the study of large-scale peculiar motions emerges together with the rise of the concept that the Milky Way and the Local Group are part of a much larger complex of galaxies with the Virgo cluster near its center. The early efforts were then directed towards trying to measure the perturbation induced in the motion of the Local Group by the mass concentration in the Local Supercluster ("Virgo infall"). In the late 1970's, the discovery of a dipole component in the Cosmic Microwave Background (CMB) radiation field indicated that the Local Group is moving at a velocity and with a direction that were quite different from those suggested by Virgo infall. Towards the mid 1980's, it appeared clear that the Local Supercluster as a whole is in motion with respect to the CMB rest frame, and several groups pointed at supercluster structures in the direction of Hydra–Centaurus as the cause for such motion. Only within the last three years, thanks to energetic work that led to the assembly of deeper samples of galaxies with redshift-independent distance measurements, has it become apparent that Hydra–Centaurus itself, a supercluster enhancement at a redshift of about 3000 km s^{-1}, may itself partake of the same large-scale flow that involves the Local Group, caused by an even more conspicuous source of gravitational attraction, farther than Hydra–Centaurus.

In this section, we first develop the formal tools for the analysis of kinematic information, starting with an introduction to the newtonian theory of small perturbations. We follow a treatment developed in full by Weinberg (1972) and Peebles (1980). Maintaining a loose historical sequence, we then tackle the problem of infall towards a centralized mass perturbation, such as the Local Supercluster; in the process, we firm up our knowledge of the relative motion of the Milky Way with respect to its immediate neighborhood, i.e. the Local Group and its surroundings. After a discussion of the models that describe motions within the Local Supercluster in linear theory, we step through the complications that are met when perturbations depart so much from the mean that nonlinear treatment becomes necessary. The limits of a study of peculiar motions that invokes perturbations produced by a single attractor, such as the Local Supercluster, become clear when its results are compared with measurements of the CMB dipole in the analysis. Next, we review models that incorporate the role of a distant "Great Attractor" and, finally, we append a brief description of the methods that are used for the all-important task of inferring the characteristics of the density field.

6.1 Newtonian Theory of Small Perturbations

6.1.1 Basic Equations

In an expanding universe, the development of irregularities in the matter distribution can be approximated by classical mechanics, provided that the size of the region considered is small in comparison with the Hubble length cH^{-1} and large compared with the Schwarzschild radius of collapsed objects — say $10^{14} \ll r \ll 10^{28}$ cm — and provided that velocities are non-relativistic. Thus it is said that the treatment is possible in the Newtonian approximation. If the energy density is dominated by non-relativistic particles — so that their mass density largely exceeds the pressure ($\rho \gg p/c^2$) —, the fluid of density ρ, pressure p, velocity $\vec{v}$, gravitational acceleration $\vec{a}$ (potential ϕ), is governed by the equation of continuity

$$\frac{\partial \rho}{\partial t} + \nabla \cdot (\rho \mathbf{v}) = 0, \tag{6.1}$$

the Euler equation

$$\frac{\partial \mathbf{v}}{\partial t} + (\mathbf{v} \cdot \nabla)\mathbf{v} = -\frac{1}{\rho}\nabla p + \mathbf{a}, \tag{6.2}$$

and the gravitational field equation

$$\nabla \cdot \mathbf{a} = -\nabla^2 \phi = -4\pi G \rho, \tag{6.3}$$

where the gravitational field is conservative, i.e.

$$\nabla \times \mathbf{a} = 0 \tag{6.4}$$

and it therefore can be expressed as the gradient of a potential ($\nabla \phi = -\mathbf{a}$, as in eqn. 6.3). We have also assumed that the cosmological constant $\Lambda = 0$ (if it weren't, it would appear as an additive term on the right of eqn. 6.3)).

A simple, spatially uniform solution of equations (6.1 – 6.4) is that given by

$$\rho = \rho_o [R_o / R(t)]^3 \tag{6.5}$$

$$\mathbf{v} = \mathbf{r}\ [\dot{R}(t)/R(t)] \tag{6.6}$$

$$-\mathbf{a} = \nabla\phi = (4/3)\pi G\rho\mathbf{r}, \tag{6.7}$$

where $R(t)$ is the Robertson–Walker scale factor which satisfies the Friedmann equation

$$\dot{R}^2 + k = (8/3)\pi G\rho R^2 \tag{6.8}$$

[As a reminder, the current value of the ratio $\dot{R}/R$ is the Hubble constant $H_\circ = \dot{R}(t_\circ)/R(t_\circ)$, so that eqn. (6.6) for the *mean* or *background* solution is Hubble's law; also from eqn. (6.8) one can derive the critical mass density $\rho_c(t_\circ)$ which, at this time, would just close ($k = 0$) the universe, i.e. $\rho_c = 3H_\circ^2/8\pi G$].

6.1.2 Perturbation Equations

We now add a *small* perturbation to the background solution as described above:

$$\rho = \bar{\rho}(1 + \delta)$$
$$\mathbf{v} = \bar{\mathbf{v}} + \mathbf{u}$$
$$\phi = \bar{\phi} + \delta\phi$$
$$p = \bar{p} + \epsilon,$$

with $\delta \ll 1$, $\mathbf{u} \ll \mathbf{v}$, $\delta\phi \ll \phi$ and $\epsilon \ll p$. Replacing these in eqns. (6.1 – 6.4) and keeping only the linear terms in the perturbations, we get

$$\frac{\partial\delta}{\partial t} + \frac{\dot{R}}{R}\mathbf{r}\cdot\nabla\delta + \nabla\cdot\mathbf{u} = 0 \tag{6.9}$$

$$\frac{\partial\mathbf{u}}{\partial t} + \frac{\dot{R}}{R}(\mathbf{r}\cdot\nabla)\mathbf{u} + \frac{\dot{R}}{R}\mathbf{u} = -\frac{\nabla\epsilon}{\bar{\rho}} - \nabla\delta\phi \tag{6.10}$$

$$\nabla^2(\delta\phi) = 4\pi G\bar{\rho}\delta \tag{6.11}$$

The Eulerian coordinates $(\mathbf{r}, t)$ can be converted to comoving coordinates $(\mathbf{x}, t)$, with $\mathbf{x} \equiv \mathbf{r}/R(t)$ [for two points that move exactly with the Hubble flow, $\mathbf{x}(t)$ is a constant]. Transforming differentials from Eulerian to comoving coordinate systems, $(\partial/\partial t) \rightarrow \partial/\partial t - (\dot{R}/R^2)\mathbf{r}\cdot\nabla_c$, and $\nabla \rightarrow [1/R(t)]\nabla_c$, where the subscript $_c$ indicates that the gradient should be taken with respect to comoving coordinates. Eqns. (6.9–6.11) then become

$$\frac{\partial\delta}{\partial t} + \frac{1}{R}\nabla_c\cdot\mathbf{u} = 0 \tag{6.12}$$

$$\frac{\partial \mathbf{u}}{\partial t} + \frac{\dot{R}}{R}\mathbf{u} = -\frac{1}{R}\frac{\nabla_c \epsilon}{\bar{\rho}} - \frac{1}{R}\nabla_c(\delta\phi) \tag{6.13}$$

$$\frac{1}{R^2}\ \nabla_c^2 \delta\phi = 4\pi G\bar{\rho}\delta \tag{6.14}$$

where $\delta, \mathbf{u}, \epsilon$ and $\delta\phi$ are now functions of $(\mathbf{x}, t)$ and $R \equiv R(t)$. Taking the divergence of eqn. (6.13) and eliminating $\nabla_c \cdot \mathbf{u}$ via eqn. (6.12) and $\nabla_c^2(\delta\phi)$ via eqn. (6.14), we get

$$\ddot{\delta} + 2\frac{\dot{R}}{R}\dot{\delta} - 4\pi G\bar{\rho}\delta - \frac{1}{R^2}\frac{\nabla_c^2 \epsilon}{\bar{\rho}} = 0 \tag{6.15}$$

This is the standard equation of linear perturbation theory; it describes the evolution of small pertubations in an expanding universe. For adiabatic perturbations, for example,

$$\epsilon = \frac{\partial p}{\partial \rho}\bar{\rho}\delta = c_s^2 \bar{\rho}\delta,$$

where c_s is the adiabatic sound speed. Then,

$$\ddot{\delta} + 2\frac{\dot{R}}{R}\dot{\delta} - 4\pi G\bar{\rho}\delta - \frac{c_s^2}{R^2}\nabla_c^2 \delta = 0 \tag{6.16}$$

As in an expanding universe the energy density of radiation drops as R^{-4}, while that of matter drops only as R^{-3}, eqn. (6.16) becomes applicable after the onset of the matter dominated era, i.e. when p/c^2 drops well below ρ. The solution of eqn. (6.16) can be expressed as the sum of plane waves

$$\delta(\mathbf{x}) = \sum_{\mathbf{k}}\ \delta_{\mathbf{k}}\ e^{i\mathbf{k}\cdot\mathbf{x}} \tag{6.17}$$

which, when replaced in eqn. (6.16), yields

$$\ddot{\delta}_{\mathbf{k}} + 2\frac{\dot{R}}{R}\dot{\delta}_{\mathbf{k}} - (4\pi G\bar{\rho} - \frac{c_s^2 k^2}{R^2})\delta_{\mathbf{k}} = 0 \tag{6.18}$$

For $R(t) \equiv const$ and $k' = k/R$, eqn. (6.18) reduces to Jeans' equation, where for wavenumbers k' below the critical value of $k_J = (4\pi G\bar{\rho}/c_s^2)^{1/2}$, perturbations can grow or decay. Jeans assumed a static medium, thus perturbations with $k'^2 \ll k_J^2$ grow or decay exponentially, with an e-folding rate $\sim c_s k_J$. In an expanding universe, however, a growing perturbation has to fight Hubble expansion, which occurs at a rate $\dot{R}/R$, which is also comparable with $c_s k_J$. Lifshitz (1946: cf Weinberg 1972) first showed that perturbations with $k \ll k_J$ grow not exponentially, but like a power law of t; for $k^2 < 4\pi G\bar{\rho}R^2/c_s^2$, eqn. (6.18) allows a solution that is a linear combination of two modes, one that grows like $\delta_+ \propto t^{2/3}$ and one that decays like $\delta_- \propto t^{-1}$.

6.1.3 The Peculiar Velocity Perturbation

Now we obtain the behavior of the pertubation velocity **u**. Note that the solution to eqn. (6.14) for the perturbing potential $\delta\phi$ is

$$\delta\phi(\mathbf{x}) = -GR^2 \int d^3x' \, \frac{\bar{\rho}\delta(\mathbf{x}')}{|\mathbf{x}' - \mathbf{x}|} \tag{6.19}$$

where the integral is over the volume element d^3x' in the comoving coordinates **x**. The peculiar acceleration **g** of the gravity field can be written as

$$\mathbf{g}(\mathbf{x}) = -\frac{\nabla(\delta\phi)}{R} = GR\bar{\rho}\int d^3x' \, \delta(\mathbf{x}')\frac{\mathbf{x}' - \mathbf{x}}{|\mathbf{x}' - \mathbf{x}|^3} \tag{6.20}$$

Note that a solution of eqn. (6.12) is

$$\mathbf{u} = R\,\frac{\partial}{\partial t}\left(\frac{\mathbf{g}}{4\pi G\bar{\rho}R}\right) \tag{6.21}$$

as can be proved by substituting **u** as above in eqn. (6.12), with the replacements allowed by eqn. (6.14). Inspection of eqn. (6.14) also shows that the time dependence of $\delta\phi$, and therefore of $\mathbf{g}R$, is the same as that of $\bar{\rho}\delta R^2$; therefore

$$\frac{1}{(\mathbf{g}/\bar{\rho}R)}\frac{\partial}{\partial t}(\mathbf{g}/\bar{\rho}R) = \frac{1}{\delta}\frac{d\delta}{dt} \tag{6.22}$$

Substituting now eqn. (6.22) in eqn. (6.21), we get

$$\mathbf{u} = \frac{\mathbf{g}}{4\pi G\bar{\rho}}\,\frac{1}{\delta}\frac{d\delta}{dt} \tag{6.23}$$

Using the definitions $H(t) = \dot{R}(t)/R(t)$ and $\Omega = (8/3)\pi G\bar{\rho}/H^2(t)$, then eqn. (6.23) can be rewritten as

$$\mathbf{u} = \frac{\mathbf{g}H(t)f}{4\pi G\bar{\rho}} = \frac{2}{3}\frac{f\mathbf{g}}{\Omega H(t)} \tag{6.24}$$

where $f = (R/\delta)(d\delta/dt)$. The function f has a complicated expression that can be well approximated by $f \simeq \Omega^{0.6}$, so

$$\mathbf{u} = \frac{2}{3}\Omega^{-0.4}H^{-1}(t)\mathbf{g} \tag{6.25}$$

6.2 Supercluster Infall

After the formal interlude of section 6.1, we now approach the more "practical" concern of investigating the evidence — and, in passing, further elaborating some of the tools developed in section 6.1 — for the existence of peculiar motions in the local universe. It has been well accepted for over two decades that the Local Group is part of the same large scale agglomeration of galaxies known as the Local Supercluster (LS), of which the Virgo cluster constitutes the densest region. It is therefore natural to expect that that region, constituting the most conspicuous source of extragalactic light (see fig. 2.2a), will have an influence on the local dynamics. Before we tackle the treatment of supercluster infall, however, we'll take a quick look at the very local neighborhood, and put the annoyance of local motions behind us.

6.2.1 Motion of the Galaxy in The Local Group

It is well known that M31, the other main member of the Local Group (LG), has a large negative velocity with respect to the Milky Way ($V_\odot \simeq -299$ km s^{-1}); this velocity remains negative even after removal of the component of the Sun's galactic rotation. M31 and the Milky Way move with respect to each other at about 100 km s^{-1}, an observation that has been used to infer that the LG has recently passed the time of maximum separation, and gravity has now succeeded in locally reversing the Hubble expansion flow. With M31 at $\sim$ 750 kpc, the crossing time in the LG is comparable with the Hubble time. It is customary to remove the motion of the Milky Way within the LG from heliocentric velocities of distant galaxies by means of a standard (de Vaucouleurs *et al.* 1976)

$$V_{gal} = V_{gal}^{\odot} + 300 \sin\, l \cos\, b \tag{6.26}$$

where $V_{gal}^{\odot}$ is the heliocentric velocity of the galaxy and (l, b) its galactic coordinates. There have been several estimates of the motion of the Sun with respect to the Local Group, recently summarized by Lynden–Bell and Lahav (1988), including those of Sandage (1986) and Lynden–Bell and Lahav themselves. Given in terms of the components respectively toward the galactic center, in the direction of $l = 90°$ and in that of the north galactic pole, the velocity vectors of the Sun with respect to the LG as determined respectively by de Vaucouleurs, Sandage and Lynden–Bell and Lahav are (0,300,0), (-37,291,-28) and (-30,297,-27). Errors on the velocities are comparable with the differences between the modern determinations and eqn. (6.26), so use of the latter is justified and preferred for the sake of standardization.

6.2.2 The Local Anomaly

Sandage (1986) investigated the deceleration produced by a mass concentration within the LG, and concluded that such an effect is not discernible for galaxies at distances larger than 2 Mpc, even for an upper limit to the mass of the LG of 4×10^{12} $M_{\odot}$.

Similarly, no bulk motion of the LG with respect to nearby galaxies (less than a few Mpc) has been positively detected, i.e. whichever peculiar motion is associated with the LG, it is shared *in toto* by all galaxies within a radius of 700–800 km s^{-1} from the LG. This is well illustrated in the review of Faber and Burstein (1988). This region has come to be referred to as the *Local Anomaly*. It might at first appear peculiar that what is in part a lack of peculiar motions be found to be an anomaly. As we shall see presently, however, the evidence is quite strong for infall of galaxies towards the center of the Local Supercluster; thus, a seemingly quiescent region extending over a distance comparable to that to the Virgo cluster and coherently deviating from supercluster infall does qualify for notice. This region appears associated with the so–called Coma–Sculptor cloud of galaxies, a relatively flattened system of well–determined membership.

6.2.3 De Vaucouleurs' Model of LS Infall.

De Vaucouleurs (1958) pioneered the work on deviations from Hubble flow induced by the Local Supercluster. He reasoned that the mass concentration associated with the LS should be expected to locally slow down the Hubble expansion, and because the galaxy distribution in the LS appears flattened (see fig. 2.2a), he proposed to look for global rotation of the system. As a function of distance r from the center of the Virgo cluster — assumed to be at the center of the LS mass concentration —, the two components of the flow, as suggested by de Vaucouleurs, would be

$$V_r = H_o r \; [1 - \exp(-r/r_o)] \tag{6.27a}$$

$$\omega_r = \omega_o \exp[-(r/r_o)^2 + 1] \tag{6.27b}$$

He also estimated values $V_r(r_o) = V_V = H_o r_o / e \simeq 550$ km s^{-1} and $r_o \omega(r_o) \simeq 500$ km s^{-1}, where r_o is the distance of the LG from the center of the LS. From this section on, all velocities are assumed to have been corrected by eqn. (6.26), unless otherwise specified.

6.2.4 LS Peculiar Motion in Linear Theory

Peebles (1976; cf Davis and Peebles 1983) developed a more sophisticated treatment of the flow analysis applying the linear theory described in Section 6.1, in a manner summarized as follows. Let's assume that the mass excess associated with the LS within the radius r, δM, is distributed with spherical symmetry. At a distance r from the center of the distribution it will induce a gravitational acceleration $\mathbf{g} = G\delta M \mathbf{r}/r^3$; substitution of this in eqn. (6.25) yields

$$V_r = \frac{2G\delta M}{3H_o r^2 \Omega^{0.4}} \tag{6.28}$$

If one assumes that *mass is clustered like galaxies*, then we can express V_r in terms of the galaxy count contrast, which is related to δM as

$$\delta M = \frac{4}{3}\pi\bar{\rho}r^3\frac{\delta N}{\bar{N}} \tag{6.29}$$

where $\bar{N}$ is the mean number of galaxies that one should expect to find within a sphere of radius r in a homogeneous universe, $\bar{\rho}$ the mean density and $\delta = \delta M/\bar{\rho} = \delta N/\bar{N}$. The fractional excess in the galaxy counts, $\delta N/\bar{N}$, is in principle a measurable quantity, and V_r can then be expressed as

$$V_r = \frac{1}{3}H_o r\ \Omega^{0.6}\ \frac{\delta N}{\bar{N}} = \frac{1}{3}H_o r\ \Omega^{0.6}\ \delta \tag{6.30}$$

Several estimates (not necessarily correct, as density variations in the galaxian population are estimated from redshifts as distance indicators, which are clearly distorted by peculiar motions) suggest that $\delta N/\bar{N}$ within r_o is $\simeq 2$; for a mean redshift to Virgo $H_o r_o \simeq 1000$ km s^{-1}, we get $670\Omega^{0.6}$ km s^{-1} for the peculiar velocity of the LG induced by the mass of the LS. Notice also that eqn. (6.30) allows an estimate of Ω, independently on any assumption on the value of H_o (because we measure recession velocities, which are approximated by $H_o r$, rather than distances r).

The radial dependence of δN has been determined for various Abell clusters, for which a power law has been found to be an adequate representation: $\delta N \propto r^{-\gamma}$, with $\gamma \sim 2$. Assuming a similar empirical law for the LS, one finds that $V_r \propto r^{1-\gamma}$, or an inverse distance law for $\gamma = 2$. In such a model, the radial velocity of any galaxy results from a perturbation superimposed on the Hubble flow, which results from the vector sum of the galaxy's peculiar velocity and the peculiar velocity of the LG, projected along the line of sight to the given galaxy, i.e. (Schechter 1980)

$$V_{lmod} = V_{Virgo}x - V_{r_o}(\cos\theta - x)\,[1 - (x^2 - 2x\cos\theta + 1)^{-\gamma/2}] \tag{6.31}$$

where V_{Virgo} is the observed velocity of the Virgo cluster, corrected for the peculiar velocity of the Sun and that of the LG, V_{r_o} is the peculiar velocity field amplitude at the LG position, as given by eqn. (6.30), x is the distance to the galaxy in units of r_o, the distance of the LG to the Virgo cluster, and θ is the angular separation between the Virgo cluster and the galaxy under consideration. Notice that at small values of θ, eqn. (6.31) is a cubic, and three different values of the observed velocity are possible.

6.2.5 Nonlinear Treatment of LS Peculiar Motion

The linear approximation that yields eqn. (6.31) assumes that the density enhancement within radius r is small in comparison with $\bar{\rho}$. In fact, if the estimates of $\delta N/\bar{N} \sim 2$ at r_o, and $\delta N \propto r^{-2}$, the density contrast grows steeply in the inner regions of the supercluster. Therefore, while eqn. (6.31) would yield a fairly accurate representation of the LS simplified redshift field outside r_o, the inner regions of the LS will be inadequately described by it. Following an earlier treatment by Silk (1977), Schechter (1980) approaches the nonlinear case as follows.

Each shell of radius r is treated separately, with its own mean interior density $\bar{\rho}(r) \propto 3G\int_0^r r'^{2-\gamma}\,dr'/(4\pi r^3)$ and its own Hubble parameter $H(r) = d\ ln\ r/dt$ and density parameter $\Omega(r) = 8\pi G\bar{\rho}(r)/rH^2(r)$. The equations of motion for $r(t)$ are exactly the same governing the evolution of the scale factor $R(t)$ in a Friedmann cosmology (see eqn. 6.8):

$$\dot{r}^2 - 2E = 2GM/r \tag{6.32}$$

where M is the effective gravitating mass within the comoving radius associated with r, and E can be interpreted as the associated total energy per unit mass (spatial curvature $k = -2E$ to obtain eqn. 6.8). The solution of eqn. (6.32) can be expressed in parametric form

$$r = A(1-\cos\eta), \qquad t = B(\eta - \sin\eta) \qquad for\ \ E<0\ \ (\Omega>1) \tag{6.33a}$$

$$r = A(\cosh\eta - 1), \qquad t = B(\sinh\eta - \eta) \qquad for\ \ E>0\ \ (\Omega<1) \tag{6.33b}$$

where η is the "arc parameter" of the Friedmann solution, with

$$A = \frac{4}{3}\pi G\bar{\rho}r^3|2E|^2, \qquad B = A|2E| \tag{6.34}$$

Substituting eqns. (6.33) and (6.34) in eqn. (6.32), we find

$$\begin{aligned} 1+\cosh\eta &= 2\Omega^{-1} && (\Omega>1)\\ \cosh\Omega + 1 &= 2\Omega^{-1} && (\Omega<1)\\ (BH)^2 &= \Omega^{-1}/(4|\Omega^{-1}-1| \end{aligned} \tag{6.35}$$

We now define the parameters h and Δ, characteristic of the *velocity* and *mass* of the shell, evaluated at $t = t_o$ for $\Omega \le 1$:

$$h(t_o) = \frac{\dot{r}}{H_o r} = \frac{\sinh\eta(\sinh\eta - \eta)}{(\cosh\eta - 1)^2}\frac{1}{H_o t_o}$$

$$1+\Delta(t_\circ)=\frac{M}{\frac{4\pi}{3}\left(\frac{3H_\circ^2\Omega}{4\pi G}\right)r^2(t_\circ)}=\frac{2}{\Omega(H_\circ t_\circ)^2}\frac{(\sinh\eta-\eta)^2}{(\cosh\eta-1)^3} \tag{6.36}$$

The equivalent expressions for $\Omega > 1$ can be obtained by substituting η with $i\eta$ in eqn. (6.36):

$$h(t_\circ)=\frac{\sin\eta(\eta-\sin\eta)}{(1-\cos\eta)^2}\frac{1}{H_\circ t_\circ}$$

$$1+\Delta(t_\circ)=\frac{(\eta-\sin\eta)^2}{(1-\cos\eta)^3}\frac{2}{\Omega(H_\circ t_\circ)^2} \tag{6.37}$$

Now, if η_u is the present universal value of the arc parameter and η_g is the present value of that quantity for the shell associated with a particular galaxy, the ratios of the "local" Hubble parameter and average interior density are given by $h(\eta_g)/h(\eta_u)$ and $[1+\Delta(\eta_g)]/[1+\Delta(\eta_u)]$, respectively. Correspondingly, we indicate with an $_l$ subscript the values of h and $1+\Delta$ at the location of the Local Group. The composition of velocities for the nonlinear case, analogous to that in eqn. (6.31) yields:

$$\begin{aligned}V_{nlmod} &= V_{Virgo}\Big[\frac{h_u}{h_l}x-\frac{(h_u-h_l)}{h_l}\cos\theta-\frac{(h_u-h_g)}{h_l}(x-\cos\theta)\Big]\\ &= V_{Virgo}\Big[\cos\theta+\frac{h_g}{h_l}(x-\cos\theta)\Big],\end{aligned} \tag{6.38}$$

which can be used to fully map the 3–dimensional reedshift field around the LS. The ratios h_u/h_g and h_l/h_g can be computed if we know the appropriate values of η. Those can be derived from *(i)* the assumption that the density in the LS perturbation varies as $r^{-\gamma}$, i.e. $[1+\Delta(\eta_g)]/[1+\Delta(\eta_u)]=r^{-\gamma}$, *(ii)* our estimate that $[1+\Delta(\eta_l)]/[1+\Delta(\eta_u)]\sim 3$ and eqns. (6.36) and (6.37). Figure 6.1 (after Davis and Peebles 1983) shows the difference between the linear and the nonlinear models. The ordinate in fig. 6.1 is the ratio of the peculiar velocity to the Hubble velocity (that would be measured in the absence of a density perturbation); solid lines illustrate runs of that ratio for different values of $(M-\bar{M})/\bar{M}$, the fractional mass excess within radius r ($\bar{M}$ is the expectation for mass within r in a homogeneous universe) in the *linear* approximation, while the dashed lines illustrate the analogous ratio in the *nonlinear* case. Regõs and Geller (1989) have computed the exact solution, equivalent to eqn. (6.30), in the nonlinear case, and examined approximate solutions given in the literature. They pointed out that the dependence of V_r on Ω is nearly the same for any value of δ (for mass condensations with δ of up to 1000, the exponent of Ω remains between 0.55 and 0.61). The dependence on δ is more marked: an approximate solution given by Yahil (1985)

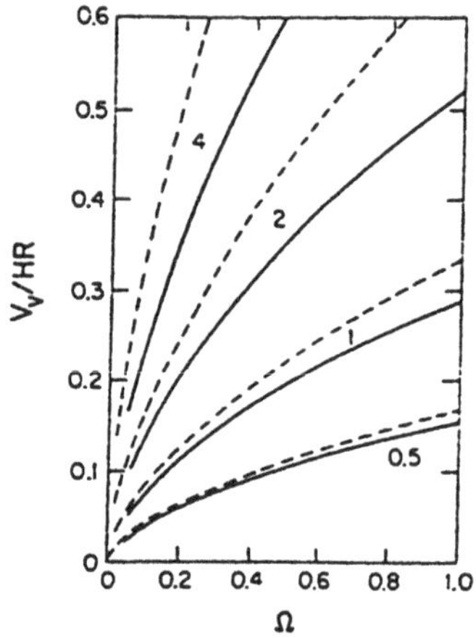

Figure 6.1: Ratio of the peculiar to Hubble velocity at the distance HR from the center of a spherical mass pertubation, plotted versus the cosmological density parameter, for different values of $\delta \sim (M - \bar{M})/\bar{M}$, the fractional mass excess within the radius R. Solid lines correspond to calculations that use the linear approximation, dashed lines refer to the nonlinear case (after Davis and Peebles 1983).

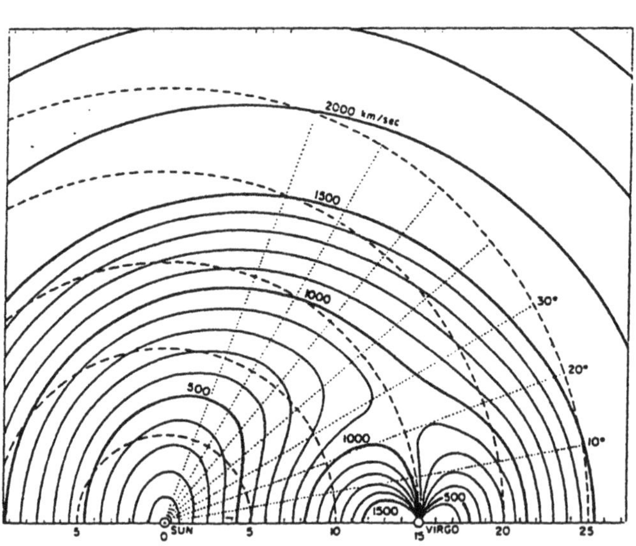

Figure 6.2: Redshift field around the Local Supercluster, computed using the nonlinear spherical model of eqn. (4.38), $\gamma = 2$, $\Delta_l = 2$ and an infall velocity of the Local Group towards Virgo of 400 km s^{-1} (after Tonry and Davis 1981).

$$V_r = \frac{1}{3} H_o r \, \Omega^{0.6} \, \frac{\delta}{(1+\delta)^{0.25}} \tag{6.39}$$

mimicks quite well the exact solution for values of δ as high as 30. For higher values of δ, eqn. (6.39) progressively overestimates the magnitude of the peculiar velocity: for $\delta \sim 300$, eqn. (6.39) is about 40 % higher than the exact solution.

Figure 6.2 (after Tonry and Davis 1981) shows the redshift field around the LS, computed using the nonlinear spherical model of eqn. (6.38), adjusted for $\gamma = 2$, $\Delta_l = 2$ and a peculiar velocity of the LG towards Virgo of 400 km s^{-1}. Notice that at large distances from Virgo, the velocity field in (6.38) tends to look like a simple dipole, reflecting a pure Hubble flow only perturbed by the LG infall component toward Virgo. For small values of r, the flow is more complicated: within $\theta \sim 20°$, the redshift field is tripled–valued.

There are various limits to the adequacy of the treatment described here. First, remember that we made the assumption of a spherically symmetric density enhancement; the distribution of galaxies in the LS is obviously *not* spherical. Nonetheless, even for a rather non spherically symmetric mass distribution, the gravitational field tends to be roughly radial: in the extreme case of a *flat* homogeneous spheroid, the acceleration at the rim is ~ 2.4 that at the periphery of a sphere with the same radius and mass, and the acceleration at the same distance as the edge, but perpendicular to the plane, is only ~ 0.64 of that produced by the spherical mass distribution. Therefore, the deviations produced by the asphericity of the LS are probably relatively mild, as discussed in detail by Yahil *et al.* (1980).

A more serious source of velocity field distortion is the instability of the flow against shear. The latter becomes severe when the displacements produced by the peculiar motions are large, in comparison with the size of the perturbed region, i.e. when $\delta r/r \sim (\rho/\bar{\rho})^{1/3} - 1 > 1$. At the LG position, estimates of $\rho/\bar{\rho} \sim 3$ imply $\delta r_o/r_o \sim 0.4$. For $\gamma \sim 2$, at $r \sim r_o \sim 0.5$ the perturbation displacement $\delta r/r \sim 1$, and orbits should be expected to start departing significantly from radial directions.

We have made the assumption that *light traces mass*; if that were not the case, then we'd use the incorrect value of δ. If light is more clumped than mass, the application of eqns. (6.30) or (6.39) using galaxy counts to gauge the perturbation could lead, for example, to an underestimate of Ω.

At large values of r, the importance of other large–scale inhomogeneities grows with respect to the perturbation produced by the LS. The gravitational disturbance produced even by relatively distant clumps of matter has been re–evaluated and deemed important in recent years. We will consider the growing evidence for the presence of multiple *attractors* in the local universe, at the end of these lectures. In the remainder of this section, we shall examine what is known on the effects induced by the LS on the redshift field.

6.2.6 Results on Peculiar Motions Purely in the LS

As can be seen from eqns. (6.36) or (6.37), the ratios h_g/h_l are equivalent to those between the Hubble parameters corresponding to the shells that contain, respectively, the galaxy $_g$ and the LG. In other words, if we knew the *ratio* of distances r_g/r_0, from simple geometry and straightforward measurement of redshifts, we could infer the ratio h_g/h_l. As we shall see, a variety of methods exist to estimate redshift–independent distances of galaxies; more correctly, for relatively distant galaxies these methods can yield accurate *ratios* of their distances to those of nearby calibrators. If one assumes that the LS is the source of all significant peculiar motions (which, as we discussed briefly above and shall see later in greater detail, is incorrect) then one should expect to observe a discrepancy between the ratio of redshifts to the Virgo cluster and to a remote field of galaxies at large r, and the corresponding ratio of their respective distances. The difference can be attributed to the LG peculiar velocity in the LS. Aaronson *et al.* (1980) carried out such an experiment, which yielded the large value of $V_{Virgo} = 480 \pm 75$ km s^{-1}.

Later, Aaronson *et al.* (1982) refined their treatment of the problem, fitting the whole field obtained from redshift and redshift–independent distance measurements of 306 spiral galaxies within 3000 km s^{-1} to a a model. They improved on eqn. (6.38) by incorporating non–radial peculiar motions, allowing a de Vaucouleurs' type of rotational motion V_{rot} as described by eqn. (6.27b). They also allowed an additional velocity component (i.e one of *random* motion) w superimposed to the infall and rotational components of the LG, so that eqn. (6.38) could be rewritten as

$$V_{mod} = V'_{Virgo}\,[\cos\theta + \frac{h_g}{h_l}\,(x - \cos\theta)] - \mathbf{w}\cdot\frac{\mathbf{r}}{r} + V_{rot} \qquad (6.40)$$

where

$$V'_{Virgo} = V_{Virgo} - \mathbf{w}\cdot\frac{\mathbf{r}_0}{r_0}$$

If they allowed no rotation, the best fit obtained an infall of the LG towards the center of the LS of 250 ± 64 km s^{-1}; the component of the random motion towards Virgo was however rather large in this model —, on the order of 80 km s^{-1}— so that the total velocity of the LG towards the LS was 331 km s^{-1}. On the other hand, if rotation of the system was allowed, the amplitude of the infall pattern at the LG was obtained as 281 km s^{-1}, for a total velocity towards Virgo of 303 km s^{-1}. The rotational velocity at the LG resulted a rather large 180 km s^{-1}. A high value of the LG infall was also excluded by Kraan–Korteweg (1985) and Tammann and Sandage (1985), who favored a value on the order of 200–210 km s^{-1}. More recent determinations of LS infall had to contend with the realization of the importance of other sources of gravitational disturbance.

Before we proceed to a more complex view of the local gravitational field, however, it is instructive to apply some of these preliminary results to obtain estimates of a couple of basic cosmological quantities: Ω and H_o. The mean observed velocity to the Virgo cluster, after correction for the motion of the Galaxy in the LG via eqn. (6.26), is 1073 km s^{-1} (Huchra 1985). In absence of infall, the redshift of the cluster would be between 1280 and 1350 km s^{-1}, according to the estimates quoted above. Modern estimates of the distance to the cluster range between 14 and 22 Mpc (see Richter 1985 for an interesting historical note), which yield values for H_o ranging between 55 and 95 km s^{-1} Mpc^{-1}. Application of eqn. (6.39), with $\delta = 2$, yields values of Ω that hover between 0.1 and 0.2.

6.3 The CMB Dipole

The Cosmic Microwave Background radiation exhibits a clear dipole anisotropy. With the most recent measurements, obtained with the COBE satellite, the best fitted dipole has an amplitude of 3.3 ± 0.2 mK in the direction $(l, b) = (265° \pm 2°, 48° \pm 2°$ (Smoot *et al.* 1991). If the anisotropy is due to the Doppler shift of the Earth's motion relative to the CMB radiation field with $\beta = v/c$, the observed temperature at an angle θ with respect to the direction of the motion can be written as

$$\begin{aligned} T(\theta) &= T_o \frac{\sqrt{1-\beta^2}}{1-\beta\cos\theta} \\ &= T_o[1 + \beta\cos\theta + \frac{\beta^2}{2}\cos 2\theta + O(\beta^3)] \end{aligned} \tag{6.41}$$

where T_o is the temperature of the CMB at rest, 2.735 K, the first order term in β is the dipole, while no quadrupole component has been identified above the $\Delta T/T \simeq 3 \times 10^{-4}$ level. Assigning the dipole detection to a Doppler shift as in eqn. (6.41), the heliocentric velocity with respect to the CMB inferred is one of 365 ± 18 km s^{-1}. After subtraction of the motion of the Sun with respect to the LSR (20 km s^{-1} towards $(l, b) = (57°, 23°)$) and a galactic rotation of the LSR of 220 km s^{-1}, the motion of the Galaxy with respect to the CMB is 547 ± 17 km s^{-1} towards $(l, b) = (266° \pm 2°, 29° \pm 2°)$. If, on the other hand, the solar motion with respect to the LG is subtracted, one obtains a motion of the LG with respect to the CMB of 622 ± 20 km s^{-1} in the direction $(l_{cmb}, b_{cmb}) = (277° \pm 2°, 30° \pm 2°)$. This direction is about 45° away from the Virgo cluster.

6.4 Multiple Attractors

6.4.1 The "Great Attractor" and Bi–infall Models

A simple Virgo infall model does not agree with either the magnitude or the direction of the motion implied by the CMB dipole anisotropy. A pure infall motion of the type discussed in 6.2.5 would have an unexplained residual component of 400 to 500 km s^{-1} along (l_{cmb}, b_{cmb}). Even a solution with LS rotation does not alleviate the problem. This disagreement, known for several years, was a forewarning that perturbing agents besides the LS are at work. However, large uncertainties in the extragalactic distance measurement methods, and contrasting signals obtained from idependent bodies of data (e.g. Rubin *et al.* 1976), left matters undecided through the early 1980's. As the complexity of the large–scale structure of the galaxian distribution started to emerge, Shaya (1984) and Tammann and Sandage (1985) pointed out that the perturbations produced by nearby superclusters could play an important role in the local velocity field. With much foresight, they also suggested that the Hydra–Centaurus supercluster at a redshift of about 3000 km s^{-1} or supercluster structures hidden in the "zone of avoidance" in the general direction of (l_{cmb}, b_{cmb}) could play a role as important as the LS in determining the gravitational field at the LG. Fig. 6.3, after Tammann and Sandage (1985), illustrates the geometry just described.

Lilje, Yahil and Jones (1986) used the Aaronson *et al.* (1982) data set of about 300 galaxies to detect the tidal field induced on the LS by supercluster structures external to it. They detected significant shear, in the sense of a motion of expansion of the LS pointing towards $(l_q, b_q) \simeq (308°, 13°)$ (and its antipode), and a motion of contraction along the axis pointing towards $(l, b) \simeq (65°, 55°)$ (and its antipode). Quadrupole signature of this kind can be the result of inflow toward an external, localized source of gravity. The expansion axis points in the general direction of the Hydra–Centaurus supercluster. The data set used had little depth, in the sense that a small fraction of objects in it was at distances comparable with that of that supercluster, and therefore the regions where the infall signature might be expected to be the strongest are not accessible with this sample. These studies were however sufficient to argue quite convincingly that large–scale density inhomogeneities, predominantly in the region of the Hydra–Centaurus supercluster, are responsible for the external contribution to the gravitational field in the LS. Aaronson *et al.* (1986) expanded their observations to include 11 clusters at distances well beyond that of Hydra–Centaurus. With respect to this new sample, which with clusters at redshifts of between 4000 and 11000 km s^{-1} represents a closer approximation to a rest frame than previous ones, the large component of the LG peculiar motion which is indicated by the CMB dipole, besides Virgo infall, was detected. A precise directional analysis was however not possible, because all the distant clusters of Aaronson *et al.* fall within the Arecibo telescope declination range, which unfortunately comes no closer than about one radian to the Hydra–Centaurus region.

Dressler *et al.* (1987), utilizing a sample of about 400 elliptical galaxies that covered the whole sky and extended in redshift to about 6000 km s^{-1}, showed that Hydra–Centaurus could not be the gravitational attractor mainly responsible for the motion that causes the CMB dipole, *because the supercluster itself participates in the flow, with a motion of about 1000 km* s^{-1} *with respect to the CMB.* This is illustrated in figure 6.4, which illustrates the peculiar motions of the E–galaxy sample in the CMB rest frame. Using the same sample, Lynden–Bell *et al.* (1988) carried the analysis one step further. They excercised their data with essentially two models : (a) one with a Virgo infall velocity at the LG of 250 km s^{-1} and *bulk motion of the*

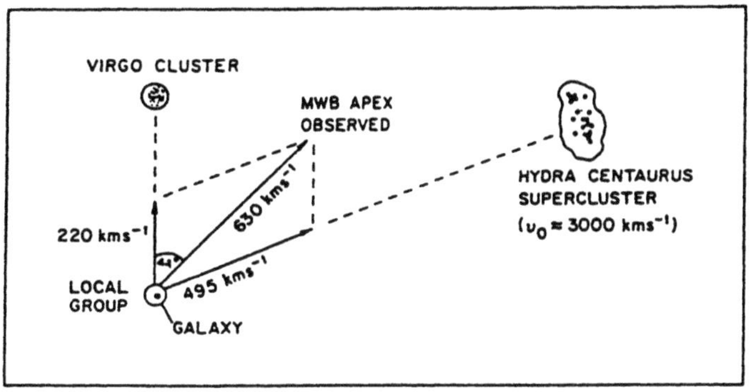

Figure 6.3: Suggested decomposition of the motion of the Local Group with respect to the CMB, after Tammann and Sandage (1985).

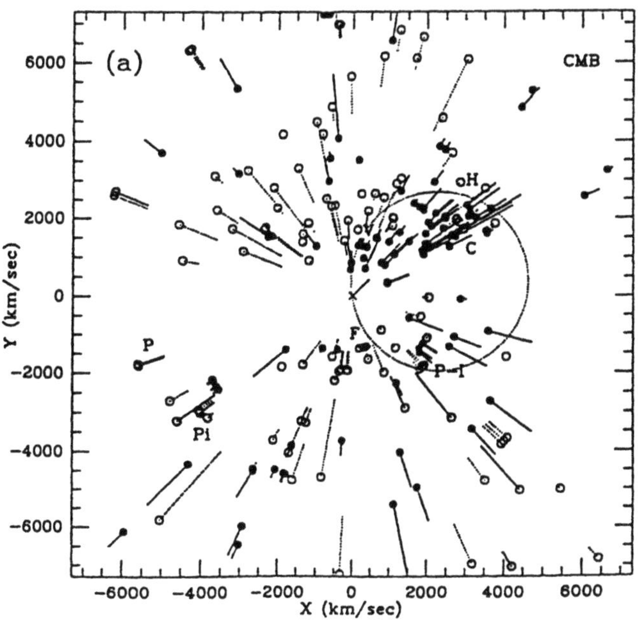

Figure 6.4: Peculiar radial velocity field for E-galaxies within 45° of the supergalactic plane in the rest frame of the CMB. Vector lengths are indicative of the magnitude of the peculiar motion; solid lines indicate recession, dashed ones indicate approach. Labels indicate respectively Virgo (V), Fornax (F), Perseus (P), Pisces (Pi), Pavo-Indus (P-I), Centaurus (C) and Hydra (H) (after Lynden-Bell *et al.* 1988).

entire sample towards direction (l_b, b_b), with a bulk velocity with respect to the CMB V_b; (b) the second model includes also a mass concentration external to the LS, at a distance R_m, towards which the infall field at the LG has a velocity V_m. The second model fitted best with a smaller Virgo infall velocity, as we shall see.

The most favored bulk flow model (a) of Lynden–Bell *et al.* yielded $V_b = 521 \pm 89$ km s^{-1} towards $(l_b, b_b) = (307° \pm 11°, 9° \pm 8°)$. They also detected the quadrupole signature reported by Lilje, Yahil and Jones, with the expansion axis roughly in the same direction $(l_q, b_q) = (336°, 18°)$. The near coincidence of the direction of the bulk flow and of the quadrupole main axis was reinforced. Their mass concentration model required the two additional degrees of freedom R_m and V_m, and an assumption on the shape of the velocity field induced by the mass concentration; in accordance with the arguments presented in section 6.2.4 and after some corroborating experimentation, they chose $\gamma = 2$, which yields a velocity field where the peculiar velocity falls with the inverse of the distance from the center of the mass concentration. Significant improvements were obtained in the quality of the fit with respect to models of type (a); the best results largely reduced the size of the bulk flow component to values barely distinguishable from the errors and also required a significantly reduced Virgo infall component. In other words, the distant mass concentration appeared to account for most of the peculiar motions in the sample. It was natural then to go the extra step of simplifying the model by removing the *ad hoc* bulk flow component and fitting only Virgo infall and distant mass concentration. Their results of such an experiment yielded, for a Virgo infall component at the LG of only 100 km s^{-1},$V_m = 570$ km s^{-1} at a distance of $R_m = 4350$ km s^{-1}. This large mass concentration, surprisingly, would fall farther than the Hydra–Centaurus supercluster which, as we saw, lies near 3000 km s^{-1}. The name of "Great Attractor" has since been firmly tagged to it.

Before we investigate further the nature of the Great Attractor (GA), it is interesting to ask whether the existence of a GA truly reduces the Virgo infall peculiar motion to a merely marginal phenomenon, as found by Lynden–Bell *et al.* . Faber and Burstein (1988) reiterate that result, suggesting that the magnitude of the Virgo infall at the LG might be as small as 85 km s^{-1}. On the other hand, re–analyzing the Aaronson *et al.* (1982) sample within the framework of a bi–infall model (Virgo plus GA), Han and Mould (1990) find optimal solutions with Virgo infall velocities on the order of 200 km s^{-1}, and GA infall solutions of the order of 400 km s^{-1}. Figure 6.5 shows contours of constant observed velocity in the supergalactic plane, as computed from the best fit bi–infall model of Han and Mould (their fig. 1c). The question of the true relative magnitude of Virgo and GA infall should be considered still open.

Han and Mould analysis of infall into two centers is a natural extension of Schechter's (1980) model, as expressed in eqn. (6.31), and because of its usefulness in immediate applications, we address the reader to its description in their Appendix.

6.4.2 Do we see the Great Attractor?

The field of peculiar motions places the Great Attractor (GA) near $(l_{ga}, b_{ga}) =$

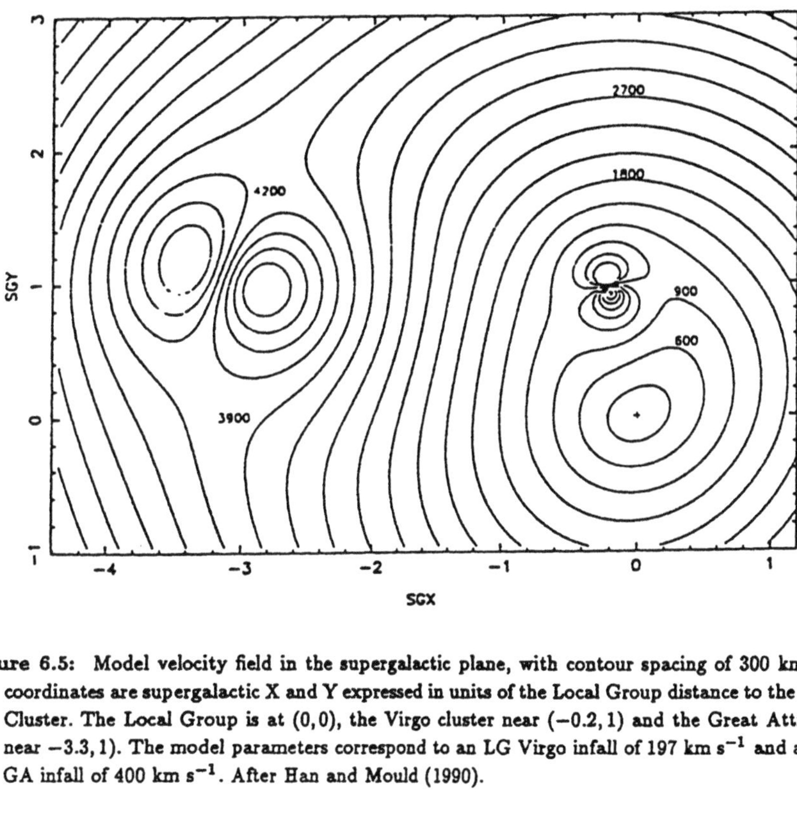

Figure 6.5: Model velocity field in the supergalactic plane, with contour spacing of 300 km s^{-1}; coordinates are supergalactic X and Y expressed in units of the Local Group distance to the Virgo Cluster. The Local Group is at (0,0), the Virgo cluster near (−0.2,1) and the Great Attractor near −3.3,1). The model parameters correspond to an LG Virgo infall of 197 km s^{-1} and an LG GA infall of 400 km s^{-1}. After Han and Mould (1990).

(309°, 18°), where the surface density enhancement of galaxies associated with the supergalactic plane becomes obscured by the galactic disk. However, keep in mind that the "traditional" picture of the LS is a relatively modest structure extending perhaps 20 Mpc, centered at the Virgo cluster. Here we are searching for a structure located at a distance which is about twice that size. There is an obvious enhancement in the number counts of galaxies in the region, over a relatively large solid angle but, surprisingly, without the clear presence of a dominant cluster; note that the conspicuous Centaurus and Hydra are in the foreground (see figure 2.2 and figure 6.6).

Dressler (1988) conducted a redshift survey in a wide region centered roughly on (l_{ga}, b_{ga}), i.e. $-35° < b < +45°$, $290° < l < 350°$. Its results, when compared with those of a wider survey with similar selection criteria (SSRS: see da Costa 1991, these proceedings), underscore quite a significant overdensity of galaxies in the redshift range between 2000 and 5000 km s^{-1} (fig 6.7) in the GA region. Next we might ask: is the overdensity of the enhancement of the right amplitude to cause the large peculiar motions? The overdensity of the GA mass concentration at the LG can be obtained from eqn. (6.30), after eliminating the dependence on Ω by scaling with the LS. If we allow a Virgo infall on the order of 200 km s^{-1} at the LG, which is $H_o r_o \simeq 1350$ km s^{-1} away from the cluster, then a GA infall of 570 km s^{-1} at a distance of 4350 km s^{-1} obtains an overdensity $\delta_{ga} \sim 0.8\delta_v$; for an LS overdensity at the LG of about 2, we get $\delta_{ga} \sim 1.6$. From galaxy counts and fig. 6.7, Dressler (1988) infers a galaxy count overdensity of about 1.5, which is in good agreement with the expectation from dynamics. The same reasonment gets into some trouble when applied to estimating the overdensity at the distance from the GA at which the Centaurus cluster is located. Centaurus appears to be moving with respect to the GA with a large peculiar velocity of about 1000 km s^{-1}; application of eqn. (6.30) implies that at Centaurus — located at a Hubble distance from the GA of about 2000 km s^{-1}—, the overdensity should be about $1.5\Omega^{-0.4}$, or about 4–6 for $\Omega \sim 0.1 - 0.2$. On the other hand, counts of galaxies suggest a lower value for δ, of perhaps 3. Agreement could be obtained by assuming a larger value of Ω, but it then would be relatively unpalatable that the indicative values of Ω *decrease* with size, implying antibiasing. Another possibility is that a simple GA + LS model does not properly explain the velocity field; nearby voids might, for example nudge galaxies away, while other great attractors, such as the Perseus–Pisces supercluster which is nearly opposite the GA in the sky, might introduce further complexity in the velocity field. This type of "soft ground" provides an indication of the the high degree of caution necessary in entering this field.

Scaramella *et al.* (1989) have noted a large concentration of clusters near 14,000 km s^{-1}, in the general region near (l_{ga}, b_{ga}), and postulate that infall into that overdense region may play an important role in the determination of the local velocity field. A simple way to decide whether a GA exists near 4300 km s^{-1} that is responsible for most of the local peculiar velocity field or whether a farther, more formidable great attractor plays an important role, is that of measuring the velocity field at

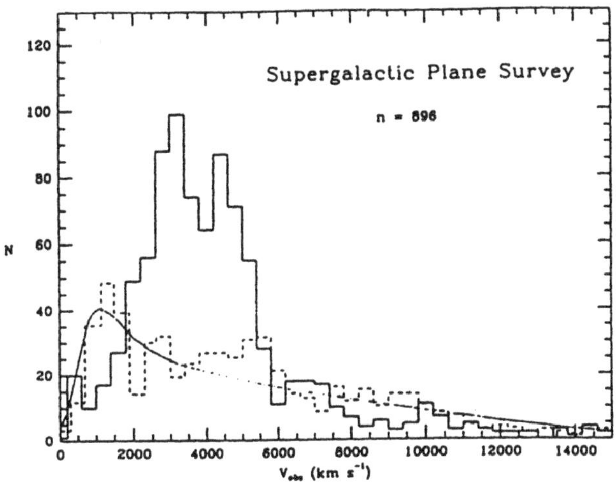

Figure 6.6: Distribution in galactic coordinates of 413 galaxies with measured distance, within 60° of $l = 309°, b = 319°$; the inset shows an expanded view of the central part of the Centaurus region. After Burstein *et al.* (1990).

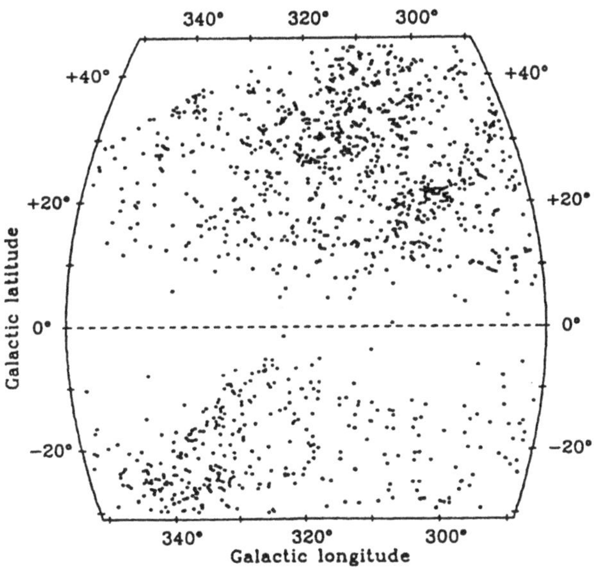

Figure 6.7: Velocity distribution of 896 galaxies in Dressler's (1988) supergalactic plane survey (SPS) (solid histogram), superimposed on that of all galaxies in the SSRS, scaled to the size of the SPS region (dashed histogram) and its smoothed rendition. If the SSRS represents a close approximation to a fair sample, the region between 2000 and 5000 km s^{-1} in the SPS exhibits a large overdensity. After Dressler (1988).

intermediate regions, i.e. between 4500 and 14,000 km s^{-1}. If a distant supercluster were perturbing the local velocity field, all local studies would only detect *bulk* flow, while if the perturber is within the region sampled by observations of peculiar motions, they might be able to reveal *backflow* into the perturber. The preliminary indications, based on new data reported by Dressler and Faber (1990), suggest that backflow into the GA at around 4000 km s^{-1} is observed in galaxies that lie in its general direction, but behind it, between 4500 and 7000 km s^{-1}. While the farther structure reported by Scaramella *et al.* could be a conspicuous perturber of the velocity field at distances beyond 6000 km s^{-1}, the Dressler and Faber result, if confirmed, would exclude that the 14,000 km s^{-1} supercluster may account for the main characteristics of the local flow.

6.4.3 A Note on Model Fitting

The procedure most commonly used in fitting velocity fields models to the observed redshif–distance sample applies a maximum likelihood method. A velocity field model is characterized by a set of free parameters, i.e. number of attractors, density profile of each attractor, distance of the LG to the center of each attractor, infall velocity of the LG towards each attractor, random motion component; in addition, the parameters in the velocity width – magnitude or internal velocity dispersion – magnitude relations (slope and intercept in Fisher–Tully or Faber–Jackson relations) might be added to the list of free parameters. Each galaxy in the sample is associated with a set of observables, e.g. radial velocity V, magnitude m and line width or internal velocity dispersion W. The sample represents a certain distribution in the three–dimensional space $(V, m, \log W)$.

First one seeks the *a priori* probability that, while observing in a given direction $\hat{\mathbf{r}}$, a galaxy i be found within $(V, V + dV)$, $(m, m + dm)$, $(\log W, \log W + d \log W)$: $p_i(V, m, \log W|\hat{\mathbf{r}})$. This conditional probability can be estimated from the characteristics of the velocity field model, the distance determination relation, the density field model and the luminosity function of galaxies (usually assumed to be universal, i.e. position independent). The giant probability of finding all the sample galaxies in their observed distribution is then

$$P = \prod_i p_i(V, m, \log W|\hat{\mathbf{r}}) \tag{6.42}$$

The likelihood function is then defined as

$$\mathcal{L} = \ln P = \sum_i \ln p_i(V, m, \log W|\hat{\mathbf{r}}) \tag{6.43}$$

The principle of maximum likelihood states that the free parameters of the model should be chosen so that $\mathcal{L}$ is maximized.

It should be stressed that the optimal model thus found does reflect reality only to the extent that one is justified in selecting a given model, which reflects a set of relatively simple, intuitive choices. Ideally, one would like to be able to extract from the data set the true density field, freed from constraining assumptions regarding the geometry of the density fluctuations. The first steps in that direction have been taken, and we discuss them briefly in the next section.

6.5 Recovery of the Density Field

Several approaches have been followed in the reconstruction of the density field. Using the condition (6.25), that peculiar velocity is proportional to gravity perturbation, one can use the three-dimensional distribution of galaxies as given to the first order by a redshift survey, then compute from this the peculiar gravity field $\mathbf{g}(\mathbf{r})$, and match $\mathbf{u} \cdot \hat{\mathbf{r}}$ as given by eqn. (6.25) to the observed peculiar velocity field. An alternative method takes advantage of the fact that the velocity field is expected to be irrotational, i.e. the velocity field is the gradient of a scalar potential Φ. Eqn. (6.25) then tells us that the peculiar velocity potential Φ is proportional to the perturbation gravitational potential $\delta\phi$; in principle $\Phi(\mathbf{r})$ can be obtained by integrating the radial component of the peculiar velocity along the line of sight $\hat{\mathbf{r}}$ from zero to distance r, then eqn. (6.11) will allow the recovery of the density field. This approach has been adopted by Bertschinger and Dekel (1989), and applied to the analysis of the combination of spiral (Aaronson *et al.* 1982) and elliptical galaxy (that used by Lynden-Bell etal 1988) samples by Bertschinger *et al.* (1990). Figure 6.8 shows the reconstructed mass density field in the supergalactic plane, as obtained by Bertschinger *et al.* (1990). A major problem in the application of their method is encountered with the sparseness of the data, forcing them to resort to rather extreme smoothing of the velocity data. This results in a severely smoothed resulting density field, but it also garantees that the treatment can well be handled within linear perturbation theory. The GA is identified in fig. 6.8 by the large enhancement at positive supergalactic Y and negative X, near the edge of the reconstructed region; the LS is the very minor lump near the center of the region. A major surprise is the rather unimpressive prominence of the Pisces–Perseus supercluster, opposite the GA in the reconstructed region; this region appears dynamically negligible, while it is extremely conspicuous in the galaxy distribution maps (see fig. 2.2). Lack of good sampling of the peculiar velocity field in that region is to a large extent thought to be the cause for the effect.

Rather than applying severe smoothing, Stebbins and Kaiser (1989) follow a similar approach to that of Bertschinger and Dekel (1989), but they fit a model to the reconstructed density field, given by a Fourier expansion of the density contrast:

$$\delta(\mathbf{r}) = \sum_{k} (a_k \cos \mathbf{k}\cdot\mathbf{r} + b_k \sin \mathbf{k}\cdot\mathbf{r}), \qquad (6.44)$$

whereby the goal of the fitting procedure is to best adjust the parameters a_k, b_k. Fig. 6.9 shows a contour plot of the density field obtained by their method; again, this represents a slice parallel to the supergalactic plane. The GA is the feature to the far

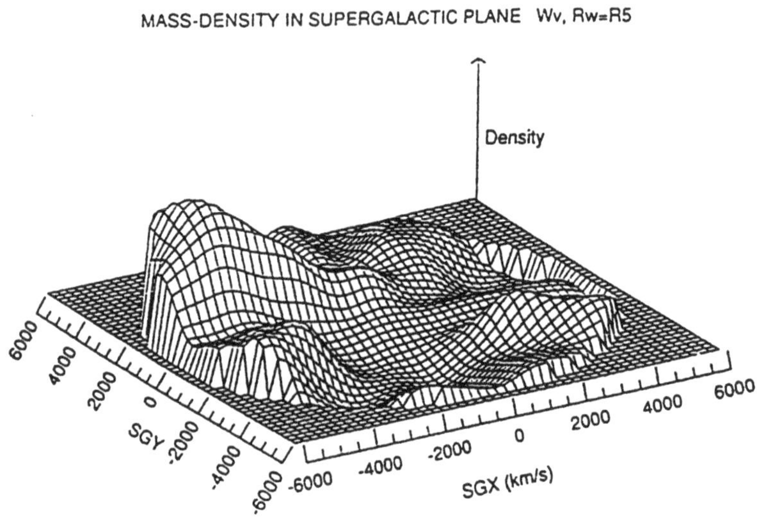

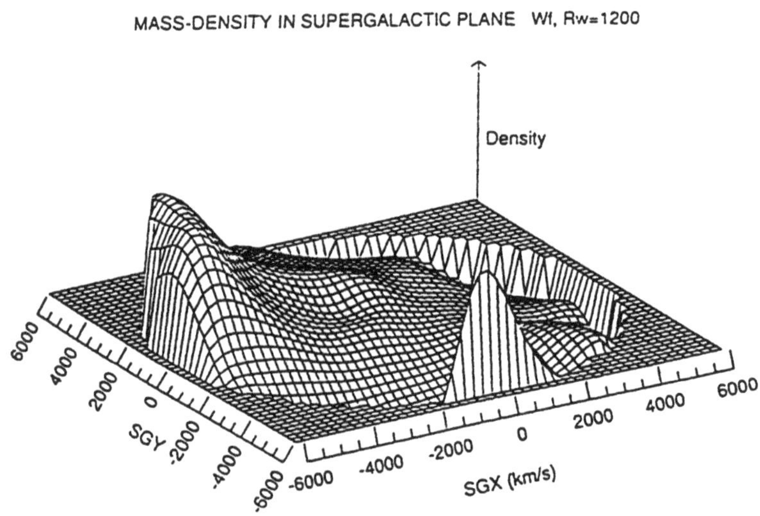

Figure 6.8: Reconstructed density field in the supergalactic plane. The LG is in the center, the GA is the large enhancement at positive Y, negative X, and the Virgo cluster is the minor extension of that enhamcement closer to the center of the reconstructed region. After Bertschinger *et al.* (1990).

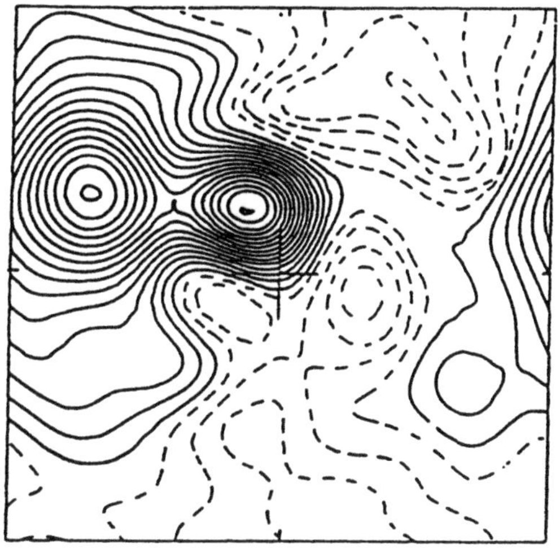

Figure 6.9: Reconstructed density field in the supergalactic plane, after Stebbins and Kaiser (1991). The contour spacing is in units of $\delta = 0.1$. Solid contours identify overdense regions, dashed ones do so for underdense regions. The LG is at the center; the largest feature to the left is the GA, while the Virgo cluster is the sharp peak above and slightly to the left of center; the other positive feature at the right margin is the Pisces–Perseus supercluster.

left, and a strong distinguishing characteristic from the Bertschinger *et al.* results is the conspicuous prominence of the LS. Dashed contours refer to underdense regions, and the Pisces–Perseus supercluster is the large, poorly sampled region to the far right.

An interesting fall–out of these methods that attempt a reconstruction of the density field is that, within the uncertainties associated with the fitting process, they allow an estimate of the cosmological density parameter Ω, independently of the density contrast δ as done with the application of eqn. (6.39). Remember how the procedures that apply such expression need an estimate of δ, usually derived from galaxy counts. Interestingly, recent applications of these density recovery methods seem to be consistent with higher values of Ω than previous estimates based on the data. As Schramm discusses in this Lecture Series, there are strong heuristic and theoretical reasons to expect that $\Omega = 1$. Gratifyingly to many, observations may be exhibiting the uncharacteristically conformist trend of inching in towards the expectations of theory.

The author thanks the organizers of the Rio Piedras Winter School for their kind invitation to participate, and wishes them a resounding success in the organization of the third episode of this highly deserving initiative.

REFERENCES

Aaronson, M., Huchra, J., Mould, J., Schechter, P.L. and Tully, R.B. 1982, *Ap. J.*, **258**, 64.

Aaronson, M., Bothun, G.D., Mould, J.R., Huchra, J., Schommer, R. and Cornell, M. 1986, *Ap. J.*, **3002**, 536.

Abell, G.O. 1958, *Ap. J. Suppl.*, **3**,211.

Abell, G.O., Corwin, H.C. and Olowin, R. 1989, *Ap. J. Suppl.*, **70**, 1.

Altschuler,D.R. and Nieves, J.F. 1989, editors of *Cosmology and Elementary Particles*, World Scientific, Singapore.

Bahcall, N.A. 1988, *Ann. Rev. Astr. Ap.*, **26**, 631.

Bahcall, N. and Soneira, R.M. 1983, *Ap. J.*, **270**, 20.

Bertschinger, E. and Dekel, A. 1989, *Ap. J. (Letters)*, **336**, L5.

Bertschinger, E., Dekel, A. Faber, S.M., Dressler, A. and Burstein, D. 1990, *Ap. J.*, **364**,370.

Binggeli, B., Sandage, A. and Tammann, G.A. 1988, *Ann. Rev. Astr. Ap.*, **26**, 509.

Broadhurst, T.J., Ellis, R.S. and Shanks, T. 1988, *M.N.R.A.S.*, **235**, 827.

Broadhurst, T.J., Ellis, R.S., Koo, D.C. and Szalay, A.S. 1990, *Nature*, **343**, 726.

Burstein, D., Faber, S.M. and Dressler, A. 1990, *Ap. J.*, **354**, 18.

Chincarini, G.L. and Rood, H.J. 1975, *Nature*, **257**, 294.

Colless, M., Ellis, R.S., Taylor, K. and Hook, R.N. 1990, /MNRAS 244, 408.

da Costa, L.N. 1991, these proceedings.

Davis, M., Huchra, J., Latham, D.W. and Tonry, J. 1982, *Ap. J.*, **253**,423.

Davis, M. and Peebles, P.J.E. 1983, *Ann. Rev. Astr. Ap.*, **21**, 109.

Dekel, A. and Aarseth, S.J. 1984, *Ap. J.*, **283**, 1.

de Lapparent, V., Geller, M.J. and Huchra, J.P. 1986, *Ap. J. (Letters)*, **302**, L1.

de Lapparent, V., Geller, M.J. and Huchra, J.P. 1988, *Ap. J.*, **342**, 44.

de Vaucouleurs, G. 1953, *Astron. J.*, **58**, 30.

de Vaucouleurs, G. 1958, *Astron. J.*, **63**, 252.

de Vaucouleurs, G., de Vaucouleurs, A. and Corwin, H.G. 1976, *Second Reference Catalog of Bright Galaxies*, Univ. of Texas Press, Austin.

Dressler, A. 1980, *Ap. J.*, **236**, 351.

Dressler, A. 1988, *Ap. J.*, **329**, 519.

Dressler, A. and Faber, S.M. 1990, *Ap. J. (Letters)*, **354**, L45.

Dressler, A., Faber, S.M., Burstein, D., Davies, R.L., Lynden-Bell, D., Terlevich, R.J. and Wegner, G. 1987, *Ap. J. (Letters)*, **313**, L37.

Eder, J.A., Schombert, J.M., Dekel, A. and Oemler, Jr., A. 1989, *Ap. J.*, **340**, 29.

Efstathiou, G., Ellis, R.S. and Peterson, B.A. 1988, *M.N.R.A.S.*, **232**,431.

Einasto, J., Joeveer, M. and Saar, E. 1981, *M.N.R.A.S.*, **193**, 353.

Ellis, R.S. 1987, in *Observational Cosmology*, ed. by A. Hewitt, G. Burbidge and L.Z. Fang, Reidel, Boston, p. 367.

Faber, S.M. amd Burstein, D. 1988, in *Large-Scale Motions in the Universe*, ed by V.C. Rubin and G.C. Coyne, Princeton Univ. Press, Princeton, p. 115.

Fisher, J.R. and Tully, R.B. 1981, *Ap. J. Suppl.*, **47**, 119.
Fry, J.N., Giovanelli, R., Haynes, M.P.,Melott, A.L. and Scherrer, R.J. 1989, *Ap. J.*, **340**, 11.
Geller, M.J. and Huchra, J.P. 1988, in *Large-Scale Motions in the Universe*, ed by V.C. Rubin and G.C. Coyne, Princeton Univ. Press, Princeton, p. 3.
Geller, M.J. and Huchra, J.P. 1989, *Science*, **246**, 897.
Giovanelli, R. 1990, in *Large-Scale Structures and Peculiar Motions in the Universe*, ed. by L.N. da Costa and D.W. Latham, PASP, San Francisco.
Giovanelli, R., Haynes, M.P. and Chincarini, G.L. 1986, *Ap. J.*, **300**, 77.
Giovanelli, R. and Haynes, M.P. 1991, *Ann. Rev. Astr. Ap.*, **29**, in press.
Gott, J.R. III, Melott, A.L. and Dickinson M. 1986, *Ap. J.*, **306**, 341.
Gott, J.R. III, *et al.* 1989, *Ap. J.*, **340**, 625.
Gregory, S.A. and Thompson, L.A. 1978, *Ap. J.*, **222**, 784.
Gregory, S.A., Thompson, L.A. and Tifft, W.G. 1981, *Ap. J.*, **243**, 411.
Guzzo, L., Iovino, A., Chincarini, G.L., Giovanelli, R. and Haynes, M.P. 1991, *Ap. J.*, in press.
Han, M. and Mould, J. 1990, *Ap. J.*, **360**, 448.
Haynes, M.P. and Giovanelli, R. 1986, *Ap. J. (Letters)*, **306**, L55.
Haynes, M.P. and Giovanelli, R. 1988, in *Large-Scale Motions in the Universe*, ed by V.C. Rubin and G.C. Coyne, Princeton Univ. Press, Princeton, p. 31.
Haynes, M.P., Giovanelli, R. and Chincarini, G.L. 1984, *Ann. Rev. Astr. Ap.*, **22**, 445.
Heineken, A., Adams, S. and Medalla, B. 1991, in *Organizing Workshops with Gusto*, **1**, 1.
Heydon-Dumbleton, N.H., Collins, C. and MacGillivray, H.T. 1989, *M.N.R.A.S.*, **238**, 379.
Hubble, E. 1934, *Ap. J.*, **79**, 8.
Huchra, J.P. 1988, in *Large Scale Structure and its Relation to Clusters of Galaxies*, ed. by J. Dickey, PASP, S. Francisco.
Huchra, J., Davis, M., Latham, D. and Tonry, J. 1983, *Ap. J. Suppl.*, **52**,89.
Huchra, J.P., Geller, M.J., de Lapparent, V. and Corwin, H.G. 1990, *Ap. J. Suppl.*, **72**, 433.
Iovino, A., Giovanelli, R., Haynes, M.P., Chincarini, G.L. and Guzzo, L. 1991, preprint.
Kaiser, N. 1984, *Ap. J.*, **284**, L89.
Kaiser, N. 1986, *M.N.R.A.S.*, **219**, 785.
Kaiser, N. and Peacock, J.A. 1990, Preprint.
Karachentsev, I.D. 1984. *Sov. Astron. Lett.* 10:72.
Karachentsev, I.D. and Kopylov, A.I. 1990, *M.N.R.A.S.*, **243**, 390.
Kirshner, R.P., Oemler, Jr., A., Schechter, P.L. and Schectman, S.A. 1981, *Ap. J. (Letters)*, **248**, L57.
Kirshner, R.P., Oemler, Jr., A., Schechter, P.L. and Schectman, S.A. 1987, *Ap. J.*, **314**, 493.
Koo, D.C. 1986, *Astron. J.*, **90**, 418.
Koo, D.C. 1988, in *Large-Scale Motions in the Universe*, ed by V.C. Rubin and G.C. Coyne, Princeton Univ. Press, Princeton, p. 514.
Koo, D.C. and Kron, R.G. 1988, in *Understanding Galaxies at High Redshift*, eds. R.G. Kron and A. Renzini, Kluwer, Dordrecht.
Kraan-Korteweg, R. 1985, in *The ESO Workshop on the Virgo Cluster*, ed. by O.-G. Richter and B. Binggeli, p. 397.
Kron, R.G. 1982, *Vistas in Astron.*, **26**, 37.
Lauberts, A. 1982, *The ESO/Uppsala Survey of the ESO(B) Atlas*, European Southern Observatory,

Garching.
Lauberts, A. and Valentijn, E.E. 1989, *Surface Photometry Catalogue of the ESO–Uppsala Galaxies*, European Southern Observatory, Garching.
Lilje, P.B., Yahil, A. and Jones, B.T. 1986, *Ap. J.*, **307**, 91.
Lynden–Bell, D. and Lahav, O. 1988, in *Large–Scale Motions in the Universe*, ed by V.C. Rubin and G.C. Coyne, Princeton Univ. Press, Princeton, p. 199.
Lynden–Bell, D., Faber, S.M., Burstein, D., Davies, R.L., Dressler, A., Terlevich, R.J. and Wegner, G. 1988, *Ap. J.*, **326**, 19.
Maddox, S.J., Sutherland, W.J., Efstathiou, G. and Loveday, J. 1990, *M.N.R.A.S.*, **243**,692.
Martinez, V.J. and Jones, B.J.T. 1990, *M.N.R.A.S.*, **242**, 517.
Maurogordato, S., Proust, D. and Balkowski, C. 1991, preprint.
Merighi, R. Focardi,P., Marano, B. and Vettolani, G. 1986, Å160, 398.
Metcalfe, N., Fong, R., Shanks, T. and Kilkenny, D. 1989, *M.N.R.A.S.*, **236**, 207.
Nilson, P. 1973, *Uppsala General Catalogue of Galaxies, Acta Univ. Ups.* Ser V:A, Vol 1, Uppsala.
Olivier, S., Blumenthal, G.R., Dekel, A., Primack, J.R. and Stanhill, D. 1990, *Ap. J.*, **356**, 1.
Oort, J.H. 1983, *Ann. Rev. Astr. Ap.*, **21**,373.
Peebles, P.J.E. 1980, *The Large–Scale Structure of the Universe*, Princeton Univ. Press, Princeton.
Peebles, P.J.E. 1989, *cf.* Altschuler and Nieves 1989.
Peterson, B.A., Ellis, R.S., Efstathiou, G., Shanks, T., Bean, A.J., Fong, R. and Zen–Long, Z. 1986, *M.N.R.A.S.*, **221**, 233.
Regös, E. and Geller, M.J. 1989, *Astron. J.*, **98**, 755.
Roberts, M.S. 1975, in *Galaxies and the Universe*, vol IX of *Stars and Stellar Systems*, ed. by A. Sandage, M. Sandage and J. Kristian, Univ. of Chicago Press, Chicago.
Rood, H.J. 1988, *Ann. Rev. Astr. Ap.*, **26**, 245.
Rubin, V.C., Ford, W.K., Thonnard, N., Roberts, M.S. and Graham, J.A. 1976, *Astron. J.*, **81**, 687.
Ryden, B.S., Melott, A.L., Craig, D.A., Gott, J.R., Weinberg, D.H., Scherrer, R.S., Bhavsar, S.P. and Miller, J.M. 1989, *Ap. J.*, **340**, 647.
Salzer, J.J., Murray Hanson, M. and Gavazzi, G. 1990, *Ap. J.*, **353**, 39.
Sandage, A. 1986, *Ap. J.*, **307**, 1.
Scaramella, R. Baiesi–Pillastrini, G., Chincarini, G., Vettolani, G. and Zamorani, G. 1989, *Nature*, **338**, 562.
Schechter, P.L. 1976, *Ap. J.*, **203**, 297.
Schechter, P.L. 1980, *Astron. J.*, **85**, 801.
Schulman, L.S. and Seiden, P.E. 1986, *Ap. J.*, **311**, 1.
Shanks, T., Bean, A.J., Efstathiou, G., Ellis, R.S., Fong, R. and Peterson, B.A. 1983, *Ap. J.*, **274**, 529.
Shaya, E.J. 1984, *Ap. J.*, **280**, 470.
Silk, J. 1977, Å59, 53.
Smoot, G.F. *et al.* (28 authors) 1991, *Ap. J. (Lett.)*, in press.
Stebbins, A. and Kaiser, N. 1991, cf paper by Kaiser in in *Large Scale Structures and Peculiar Motions in the Universe*, ed. by L.N. da Costa and D. Latham, in press.
Strauss, M.A., Davis, M., Yahil, A. and Huchra, J.P. 1990, *Ap. J.*, **00**,000.
Sutherland, W. 1988, *M.N.R.A.S.*, **234**, 159.

Sutherland, W. and Efstathiou, G. 1991, *M.N.R.A.S.*, **248**, 159.

Szalay, A.S., Ellis, R.S., Koo, D.C. and Broadhurst, T.J. 1991, preprint.

Szalay, A.S. and Schramm, D.N. 1985, *Nature*, **314**, 718.

Tammann, G.A. and Sandage, A. 1985, *Ap. J.*, **294**, 81.

Thorstensen, J.R., Wegner, G.A., Hamwey, R., Boley, F., Geller, M.J., Huchra, J.P., Kurtz, M.J. and McMahan, R.K. 1989, *Astron. J.*, **98**, 1143.

Thuan, T.X., Gott, J.R. and Schneider, S.E. 1987, *Ap. J. (Letters)*, **315**, L93.

Tonry, J.M. and Davis, M. 1981, *Ap. J.*, **246**, 680.

Tully, R.B. and Fisher, J.R. 1977, Å54,661.

Tyson, J.A. 1988, in *Towards Understanding Galaxies at High Redshift*, ed. by R.G. Kron and A. Renzini, Kluwer, Dordrecht, p. 187.

Vorontsov–Velyaminov, V.A. Krasnogarskaja, A.A. and Arhipova, V.P. 1962–68, *Morphological Catalog of Galaxies*, Moscow State Univ., Moscow, four vols.

Weinberg, D.H. and Gunn, J.E. 1990, *Ap. J. (Letters)*, **352**, L25.

Wegner, G., Thorstensen, J.R., Kurtz, M.J., Geller, M.J. and Huchra, J.P. 1990a, *Astron. J.*, **100**, 1405.

Wegner, G., Haynes, M.P. and Giovanelli, R. 1991, preprint.

Weinberg, S. 1972, *Gravitation and Cosmology: Principles and Applications of the General Theory of Relativity*, Wiley, New York.

Yahil, A. 1985, in *The ESO Workshop on the Virgo Cluster*, ed. by O.–G. Richter and B. Binggeli, p. 359.

Yahil, A., Sandage, A. and Tammann, G.A. 1980, *Ap. J.*, **242**, 448.

Zwicky, F., Herzog, E., Karpowicz, M., Kowal, C.T. and Wild, P. 1963–68, *Catalogue of Galaxies and of Clusters of Galaxies*, Calif. Inst. of Techn., Pasadena, six vols.

LUIZ DA COSTA

MAPPING THE GALAXY DISTRIBUTION IN THE SOUTHERN HEMISPHERE

L. NICOLACI DA COSTA
Harvard-Smithsonian Center for Astrophysics
60 Garden Street, Cambridge, MA, 02138
and
CNPq/Observatório Nacional
Rua José Cristino 77, Rio de Janeiro, RJ, Brasil

ABSTRACT

We review the work we have carried out during the past decade in mapping the galaxy distribution in the southern hemisphere. The ultimate goal of this long-term effort is to complete an all-sky redshift survey to moderate depth, combining our southern data with that being accumulated in the northern hemisphere by the CfA group. Over the years we have measured well over 4000 redshifts. Here we give an overview of the surveys already completed and briefly discuss some of the results obtained. We also present some preliminary results from our ongoing survey of a deeper sample in the southern hemisphere.

1. Introduction

The galaxy redshift surveys conducted over the past decade have had a significant impact in our understanding of the way galaxies are distributed in space, stimulating the current active research on the origin and evolution of the large-scale structures in the universe. We have taken part in this effort carrying out since 1983 an extensive survey of the southern hemisphere. Our ultimate goal is to produce a complete and moderately deep all-sky survey of the nearby universe by combining our database with that of the CfA Redshift Survey[1]. We hope that the combined sample will serve to investigate in detail the clustering properties of galaxies on intermediate scales, to compare the results becoming available from the deep surveys currently in progress[2] and for detailed comparison with model distributions of progressively larger *N-body* simulations[3]. We stress the fact that most of the observations were made utilizing medium-size telescopes in Argentina, Brasil and South Africa, demonstrating that these telescopes can play an important role in observational cosmology. In the subsequent sections we describe in chronological order the surveys we have conducted of different regions of the southern hemisphere.

2. The Southern Sky Redshift Survey (SSRS)

As a first step towards the goal of enlarging the sky coverage of the redshift surveys of galaxies we have extended the original CfA survey[4] (CfA1) to the southern hemisphere, recently completing redshift surveys of different areas of the southern galactic cap. The Southern Sky Redshift Survey[5,6] (SSRS) was the first wide-angle survey of the southern hemisphere to be completed, comprising over 2000 galaxies with diameters greater than $\approx$ 1.3', drawn from the ESO-Uppsala Catalog[7]. It covers an area of 1.73 steradians in the region south of δ = -17.5° and below galactic latitude -30°. The diameter limit was chosen to yield a surface density of galaxies comparable to that of Zwicky galaxies in the CfA1 sample. However, due to the absence of large concentrations of galaxies, such as Virgo, the adopted selection criterion led to a slightly deeper catalog in the south, providing a better definition of the structures to greater distances.

To our knowledge, the SSRS provided the first direct evidence of the existence of extended, wall-like structures stretching to the limits of the survey region[5]. One such structure, the Southern wall, can be seen in the 10° wedge diagram shown in Figure 1, running across the survey volume from 0^h to about 4^h. It can actually be traced over the entire declination range covered by the SSRS. Also very prominent in this declination slice is a 5000 km s^{-1} void, centered at about 23^h and 6000 km s^{-1} (see section 6). Voids this size have since been shown to occur quite frequently, giving the impression that 5000 km s^{-1} is a characteristic scale of the galaxy distribution.

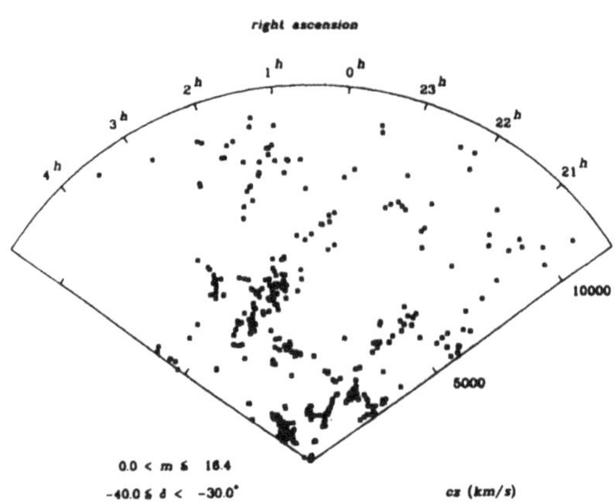

Fig. 1. Redshift map for galaxies in the SSRS.

The SSRS has also been widely used to test the robustness of the statistics used to describe the clustering properties of galaxies, and to confront theoretical models with observations. Sampling a different volume, from that probed by the CfA1, the SSRS may be viewed as an independent realization of the observed universe. Therefore, it can be used to test the universality of the characteristics of the galaxy distribution described by the statistical indicators. Volume-limited samples of the SSRS and CfA1 have been examined and compared by several authors[8,9,10]. These studies have shown that the clustering properties of galaxies on small-scales ($< 10h^{-1}$ Mpc, $H_o = 100$ km s^{-1}/Mpc), are reproducible and can be considered as well established. Unfortunately, there is also considerable evidence that the volumes surveyed so far are still small to represent a fair sample on larger scales, pointing out the need of deeper, wide-angle surveys as currently being carried out by the CfA in the northern hemisphere[11] and in the south as described below in section 6.

3. The Equatorial Survey in the Southern Galactic Cap

More recently, we have extended the SSRS to cover the equatorial strip $-17.5° \leq \delta \leq -2.5°$, filling in the gap between the regions covered by the SSRS and CfA1. Radial velocity measurements, which will soon become available[12], were made for about 850 galaxies brighter than m = 14.5 (450 galaxies with $m_{B(0)} \leq 14.5$) in the interval $20^h \leq \alpha \leq 5^h$, selected from the Morphological Catalog of Galaxies[13] (MCG). We estimate that in the Zwicky system the above magnitude limit corresponds roughly to $m_{B(0)} = 15$. We have used the MCG catalog, despite its incompleteness and inaccuracies, because it is currently the only available catalog of galaxies in the region of interest. We hope that in the near future the situation will improve thanks to the current efforts of the APM group to produce a deep galaxy sample in the region.

In Figure 2 we present a redshift map of the galaxy distribution in a 7.5° slice in the declination range indicated in the figure. Two points are worth noting: first, the extension of the wall identified in the SSRS (see Figure 1), running across the survey volume from about 3^h and $v \approx 4000$ km s^{-1} to 0^h and $v \approx 7000$ km s^{-1}, and confirming it to be an extended feature; second, the presence of a large void at $\alpha \approx 4^h$ and $v \approx 7500$ km s^{-1} with a diameter of about 5000 km s^{-1}, typical of the largest observed voids.

The new data, for the subsample $m_{B(0)} \leq 14.5$, were also combined with those of the CfA1 and SSRS by Pellegrini *et al.* [14] to construct a roughly homogeneous magnitude-limited sample of the entire southern galactic cap ($b \leq -30°$), providing a continuous coverage of 3.13 steradians. The large sky coverage has been used to investigate the extent and geometry of some of the features previously identified within the SSRS region. The sample was particularly useful in demonstrating the planar nature and the large extent of the Southern wall[14]. It has also been used, in conjuction with the CfA1 survey, to calculate the peculiar velocity induced to the Local Group by the surrounding density fluctuations[15].

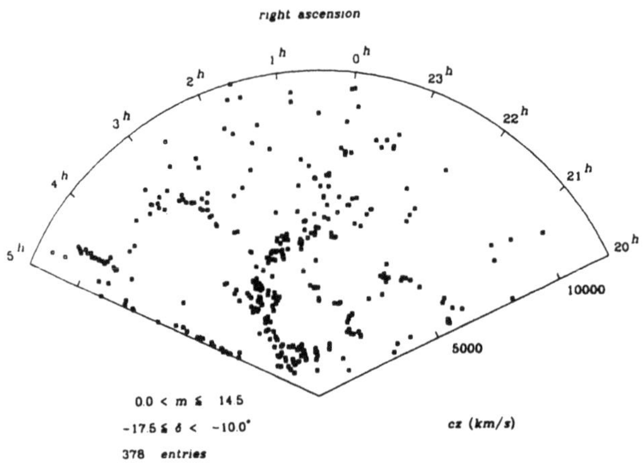

Fig. 2. Redshift map for galaxies brighter than m = 14.5 in the equatorial region covered by the MCG survey.

4. The Hydra-Centaurus Survey

As part of our general effort to complete the coverage of the southern hemisphere we have for many years measured galaxy redshifts in the direction of the complex of Hydra-Centaurus (HC). Our interest in this region was originally driven by the fact that, apart from Virgo, the most prominent structures in the projected distribution of galaxies are located in this general direction. The slow progress of the observations, however, forced us to concentrate our earlier efforts in selected areas, chosen[16,17] to confirm that the concentration of galaxies in this region reflected real structures in space.

The study of the Hydra-Centaurus complex became of greater interest after it was recognized[18] that in order to reconcile the motion of the Local Group with respect to the Microwave Background, with a low infall velocity towards Virgo, one must require the existence of a density perturbation in the direction of the HC complex. This idea was reinforced by the work of Lynden-Bell *et al.* [19] (7S), which showed that in order to account for the observed peculiar motion of elliptical galaxies within 5000 km s^{-1} from the Local Group, the existence of a large concentration of galaxies, the Great Attractor (GA), is required at low galactic latitudes.

This striking result motivated additional interest in systematic redshift surveys of the region. One of these is the recently completed SPS survey conducted by Dressler[20,21] at low galactic latitudes, aimed at confirming the Great Attractor model and thus centered around its presumed position. We have, instead, chosen to study its immediate surroundings, pursuing our goal of completing a magnitude-limited sample covering both galactic caps. Like in the case of the southern galactic cap, we constructed a magnitude-limited sample in the region $b \geq 30°$ and $\delta \leq 0°$, drawing galaxies from the ESO, whenever available, or the MCG catalogs. In order to better probe the GA region we have also included in our sample galaxies at lower latitudes ($b \geq 15°$) for $\delta \leq$-17.5°.

In contrast with our earlier work, the HC sample was constructed using the recently published photometric catalog of Lauberts and Valentjin[22], applying appropriate corrections to the listed magnitudes in order to produce a homogeneous sample, consistent with the Zwicky system of magnitudes. The redshifts available in the literature[21,23,24] were compiled and the remaining galaxies observed at SAAO and CASLEO (Argentina).

In Figure 3 we present a sequence of redshift slabs of the region covered by the HC survey, showing the projected distribution of galaxies on the sky within the velocity range indicated in the figure: in the upper panel we see the southern extension of the Local Supercluster (LSC) at around 13^h between -20° $\leq \delta \leq$ 0°; in the middle panel we see that the LSC connects at about (13^h, -20°) to a feature that extends southwards to the Centaurus cluster (12^h 45^m, -42°); this feature persists in the lower panel and suggests that the LSC extends further to the south, forming an large wall-like structure similar to those observed in the SSRS. The most prominent features in this panel are the Centaurus cluster, the Centaurus Concentration ($13^h \leq \alpha \leq 14^h$, -40° $\leq \delta \leq$-30°), and the Hydra cluster (10^h 30^m, -29°).

The spatial distribution of galaxies in the region can also be examined from the wedge diagrams shown in Figure 4. In this figure we present four slices in the declination range -40° $\leq \delta \leq$ 0°. We should emphasize that one must exercise caution interpreting these redshift maps as obscuration and large peculiar motions are important in this region. In contrast with the region covered by the SSRS, the volume surveyed is densely populated by groups and clusters, and the associated virial distortions make it difficult to describe the underlying galaxy distribution. Nevertheless, thin two-dimensional structures can still be identified such as that described above, between 13^h and 14^h, stretching from about 1000 km s^{-1} to at least 6000 km s^{-1}, clearly visible in the upper two panels. Large empty regions, 5000 km s^{-1} in diameter, can also be identified like the one located between 10^h and 12^h in the declination range -40° $\leq \delta \leq$ -20° centered at $\approx$ 6500 km s^{-1}, demonstrating once again that these large voids are quite common. Also note in the lower-right panel the structure running across the volume at roughly 4000 km s^{-1}, presumably associated to the GA complex.

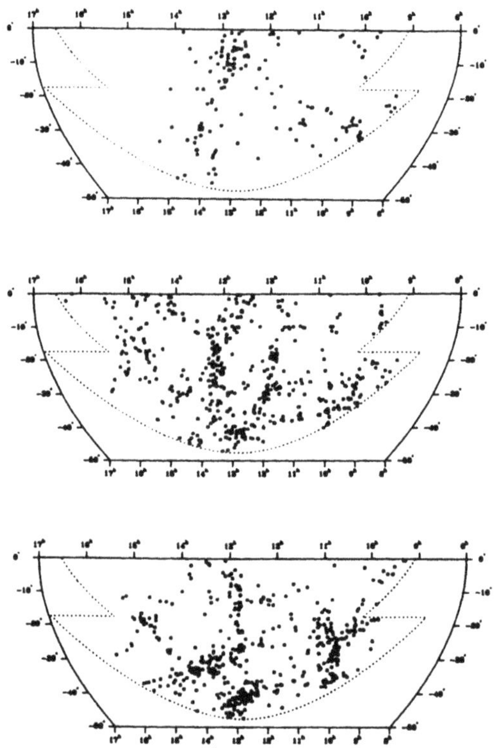

Fig. 3. Distribution of galaxies in the Hydra-Cenataurus region with radial velocities in the range; $0 \leq v < 1500$ km s^{-1} (upper panel); $1500 \leq v < 3000$ km s^{-1} (middle panel); $3000 \leq v < 4500$ km s^{-1} (lower panel). In this plot we include all galaxies with available redshifts with $b \geq 30°$ and $-17.5° \leq \delta \leq 0°$, and $b \geq 15°$ south of $\delta = -17.5°$.

5. The Nearby Universe

The completion of the MCG and the HC surveys, described in the previous sections, provides for the first time the opportunity of studying in detail the distribution of galaxies over $\approx$ 50% of the sky, using a magnitude-limited sample ($m_{B(0)} \leq$ 14.5) with essentially complete redshift information. For the southern galactic cap (b $\leq$-30°) we utilize the sample constructed by Pellegrini *et al.* [14], which combines data from the CfA1, SSRS and MCG survey. For the northern galactic cap (b $\geq$ 40°) we utilize data from the CfA1 and HC surveys. To this well-defined all-sky sample we have also added galaxies with $m_{B(0)} \leq$ 14.5 in the region $\delta \leq$-17.5° and $b \geq 15°$, and galaxies from the diameter-limited SPS in the region $-30° \leq b \leq 15°$ and $290° \leq l \leq 350°$. These latter subsamples have been included to allow us probe

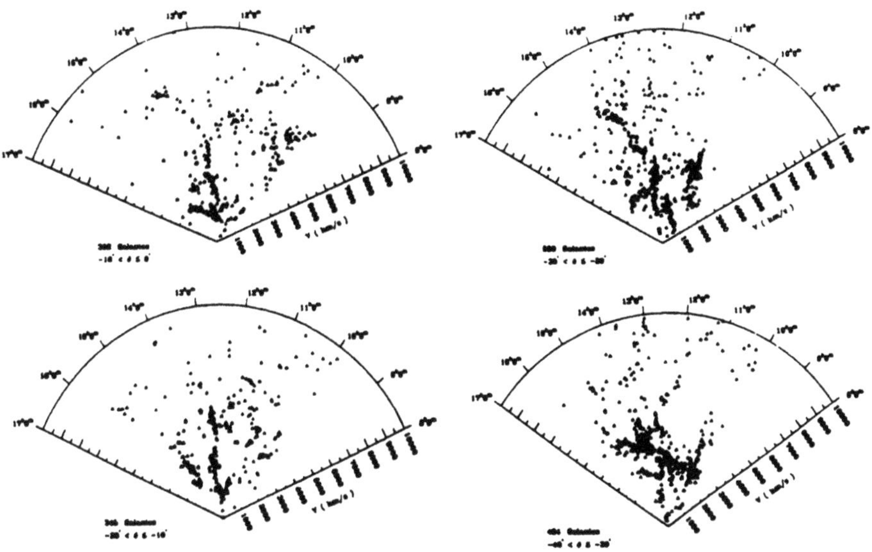

Fig. 4. Redshift maps of adjacent 10° slices in the Hydra-Centaurus Region.

the Great Attractor region at low galactic latitutes, and investigate its nature and relation to other structures in the nearby universe.

The three-dimensional distribution of our all-sky sample is presented in Figure 5, where we have adopted a cartesian coordinate system with the x-y plane coincident with the supergalactic plane. The x-axis is in the direction $L = 0°$ and the y-axis points nearly in the direction of the north galactic pole. Our representation is the same as that utilized by Bertschinger *et al.* [25]. In Figure 5 we display in each row three parallel and non-overlapping slabs, 3000 km s^{-1} thick, with the middle slab centered at the origin. In this coordinate system the approximate location of some of the most well-known clusters are: Virgo (-320, 1150, -60); Ursa Major (610, 1150, 70); Centaurus (-2880, 1200, -690); Hydra (-2150, 1700, -2180); Fornax (-100, -830, -720). Different rows represent orthogonal views of the distribution as indicated in each panel.

A full discussion of the distribution will be presented elsewhere[26]. Here it is worth mentioning only the following points. First, the LSC seems to extend all the way to the Centaurus region, forming a thin and extended two-dimensional structure. This can be seen in the last two panels of the first row and the middle panel of the second row. Second, in the southern hemisphere two wall-like structures can be seen in the first two panels of the third row; these are the Southern wall and the Telescopium-Pavo-Indus (TPI) wall, first identified in the SSRS. Third, the remarkable fact that the direction of the GA coincides with the intersecting line of the LSC and the TPI walls, where a local overdensity of galaxies is expected. We speculate that this intersection, the most prominent in the vicinity of the Local Group, is the source of attraction responsible for the motion of the Local Group. In other words, the Great Attractor is simply the intersection of two walls. If this is correct, one should expect the scale for coherent motions to be of the order of 50–60h^{-1} Mpc, corresponding to the scale of the observed voids in the galaxy distribution.

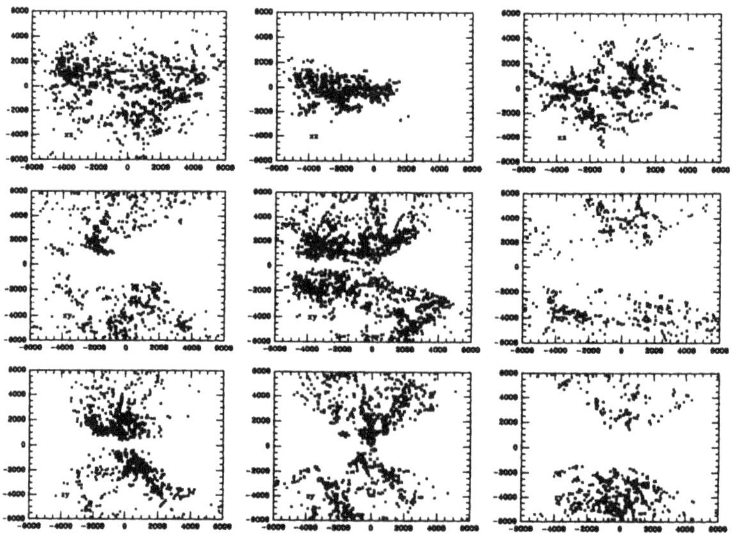

Fig. 5. Three-dimensional distribution of galaxies in supergalactic coordinates, as described in the text. The second panel of the first row is a projection onto the galactic plane, while the second panel of the second row is a projection onto the superglactic plane.

6. The Southern Slices

The indication that the surveys limited at $m_{B(0)}$= 14.5 are too shallow has motivated the ongoing surveys to fainter magnitudes. As the results of the CfA2 survey have demonstrated, there is a significant gain of information in probing galaxies down to $m_{B(0)}$= 15.5. An extraordinary example of this is the discovery of the Great Wall, barely discernible with the shallower CfA1 survey. In extending the SSRS to $m_{B(0)}$=15.5, we have followed the CfA strategy and for the past three years we have been conducting surveys of 10° slices, as described below.

For many years a major difficulty in working in the southern hemisphere was the lack of an adequate photometric catalog from which to extract samples to a well-defined magnitude limit. This was the case in the early-phases of this work when only a sample of early-type galaxies was considered[27]. Since then the situation has dramatically changed with the publication of the photometric catalog of Lauberts and Valentjin[22] and machine-generated catalogs like the APM catalog[28] and STScI Guide Star Catalog[29] (GSC).

In our current work galaxy samples are generated using as our primary list the non-stellar objects of GSC. The position of these objects are compared to those in the APM catalog (limited at $\approx$ 16.5) and objects in common are accepted as galaxies. The objects without an APM match are examined by eye and galaxies included in the final sample. The raw magnitudes of the GSC were calibrated using galaxies in common with the ESO catalog, in one of our strips. We have found that there is a linear relation between GSC and ESO magnitudes down to approximately B_T = 16, with a relatively small scatter ($\approx$ 0.3 mag). The B_T magnitudes, obtained from this relation for the GSC objects, are converted to the B(0) system by adding a mean correction of 0.26. The above procedure has been used to construct $m_{B(0)}$=15.5 magnitude-limited galaxy samples in the strip -40° $\leq \delta \leq$ -30°, which appear to be consistent with those extracted from the Zwicky catalog in the northern hemisphere. We point out that we have also carried out CCD observations which will allow an independent calibration of our magnitudes.

The first slice to be completed is located in the declination range -40° $\leq \delta \leq$ 30°. This slice was chosen because most of the prominent features present in the SSRS, like the Southern wall and a 5000 km s^{-1} void, were very well defined in this declination range. Furthermore, the corresponding strip in the northern hemisphere had also been completed.

In Figure 6 we present a panoramic view of the galaxy distribution showing side-by-side the opposite southern and northern slices. The similarity between the two is striking. Galaxies lie preferentially in thin structures which delineate large empty regions, that appear to be the most prominent feature of the galaxy distribution. Like the Great Wall in the north, at least one long structure extending over 100h^{-1} Mpc is easily detected running across the survey region from 23^h to 4^h 30^m. It is interesting to note that despite the different orientations of the Southern Wall and the Great Wall their thickness appear to be roughly the same, suggesting that it is an intrinsic property of the structure rather than a distortion of redshift maps induced by peculiar motions. Another similar structure is present at about 10000 km s^{-1} extending from 21^h to 2^h, which at that distance corresponds to a linear size of 130h^{-1} Mpc, although breaches as wide as 10h^{-1} Mpc are visible, between 23^h and 0^h. Qualitatively a fair description of the observed distribution is one of closely-packed voids of 5000-6000 km s^{-1} in diameter.

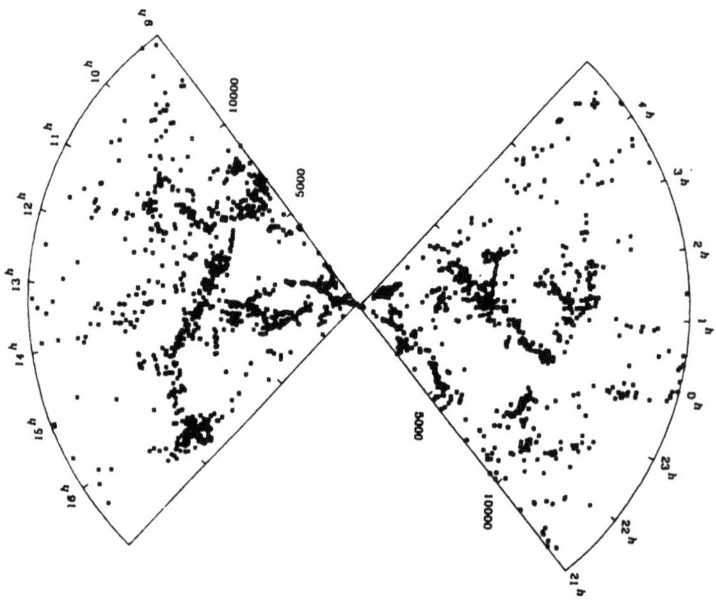

Fig. 6. Panoramic view of the galaxy distribution. Shown are opposite wedge diagrams for the northern (left) and southern (right) hemispheres.

As the size of the observed inhomogeneities are comparable to the survey depth the question arises about how close to a fair sample the available $m_{B(0)}$=15.5 surveys are. One simple way of estimating that is to compare the mean density of galaxies derived using the northern and southern samples. In Figure 7 we show this comparison plotting the variation of the mean density computed for the north, south and combined samples as a function of the survey depth.

Our estimate of the mean density is based on the minimum variance estimator discussed by Davis and Huchra[30], and separate selection functions were derived for each semi-volume limited sample. In the calculation only galaxies brighter than $M_{B(0)}$= -18.5 (left) and $M_{B(0)}$= -19.5 (right) are included. From the figure we see that in the north the density of galaxies is small nearby, rising steadily beyond the depth of the Great Wall ($\approx$ 8000 km s^{-1}) and falling thereafter to about 12 x 10$^{-3}h^{-3}$Mpc^{-3}. In the south the density rises more slowly reaching a constant value beyond 10000 km s^{-1}. The combined sample has an intermediate behavior, as

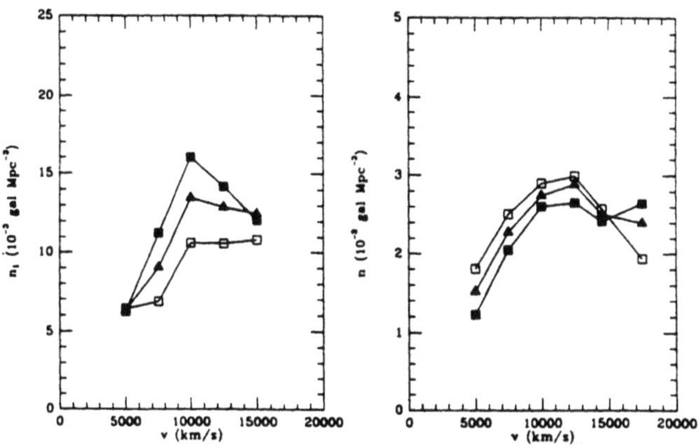

Fig. 7. Mean density of the northern slice (filled square), southern slice (open square) and combined sample (triangles), as a function of the depth of the sample.

expected. The interesting result is that despite the large peak associated with the Great Wall, which sits near the limit of the survey, the mean densities derived for each hemisphere disagree by about 10%. Note that the uncertainties in the comparison of the mean density reflect not only the intrinsic properties of the galaxy distribution but any relative errors in the magnitude system adopted in the north and south. For brighter galaxies the density at 15000 km s^{-1} is roughly the same; beyond that distance the estimates are unreliable as the number of galaxies falls rapidly.

We have also used the new data in the south to compare the two-point correlation function. This comparison is shown in Figure 8 where we plot the correlation function as determined separately for the two semi-volume limited samples. In the left panel we plot log ξ to compare the clustering properties on scales $< 10h^{-1}$ Mpc, while in the right panel we plot the quantity $1+\xi$.

From these plots we see that out to about $20h^{-1}$ Mpc the correlation functions in the north and south are almost indistinguishable, with the first zero-crossing occurring at the same point. For larger separations the correlation is negative, becoming positive again in the range 50-$60h^{-1}$ Mpc. We note that this behavior is consistent with the observed distribution where voids of 2000-6000 km s^{-1} are visible. The correlation presents a maximum in the range 70-$80h^{-1}$ Mpc, with an amplitude of about 0.15 in the north and 0.05 in the south. A secondary peak with the same amplitude is also observed at about 100-$110h^{-1}$ Mpc, when ξ is plotted linearly with r. This result appears to be similar in nature to that obtained by Broadhurst *et al.* [2] although, in our case, the characteristic scale is about half of that claimed by the latter authors. A more detailed analysis of this result is

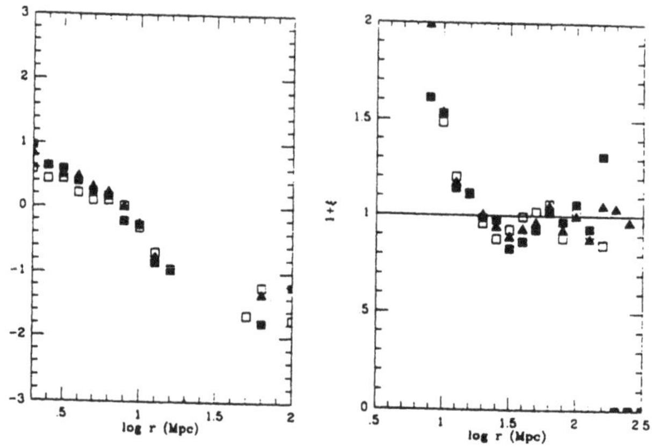

Fig. 8. Comparison of the correlation functions as derived for semi-volume limited samples in the northern (open squares), southern (filled squares) and combined sample (triangles).

currently in progress.

Finally, we point out that we have recently completed a second 10° slice in the range -20° < δ < -10°. We are hopeful that by the end of 1991 we will have completed three adjacent slices covering the declination the range -40° < δ < -10° , yielding a sample comparable to that presently available in the northern hemisphere.

7. Final Remarks

We have presented a brief review of the work carried out in mapping the galaxy distribution in the southern hemisphere. After a decade of effort we have essentially complete redshift information for galaxies brighter than $m_{B(0)}$= 14.5 over 50% of the celestial sphere, approaching the limit of what can be done using optically selected samples. Enormous progress has also been achieved in extending the surveys to $m_{B(0)}$= 15.5 in both hemispheres. These data, on the nearby universe, nicely complement the information from the various deep redshift surveys currently in progress. The combination of these new data sets should provide, in the next few years, a better understanding of the properties of the galaxy distribution over a range of scales.

An effort similar to that invested in redshift surveys should also be made in extending the database of redshift-distance measurements in order to allow a more detailed study of the peculiar velocity field and a better comparison of the matter density field with the observed galaxy distribution. Results from these studies should greatly improve our understanding on galaxy formation and the origin of the large scale structures in the universe.

A major contribution to these efforts could come from new optical and radio telescopes currently under construction or in the planning phases. Among these we call attention for the Large Southern Radio Telescope (LSRT), recently presented to the Astronomy and Astrophysics Survey Committee. The LSRT project envisions the construction of a large collecting area radio telescope, similar to that of the Arecibo Observatory, in southern Brazil. However, the new design being considered would allow for a much larger declination coverage, essentially the entire southern hemisphere. An instrument like that would have an enormous impact in observational cosmology, especially on the programs discussed here.

8. Acknowledgments

I would like to thank D. Altschuler for his hospitality, and the organizing committee of the Winter School for its pleasant atmosphere. I would like to emphasize that some of the programs described here are still in progress and represent the effort of several people. I would like to acknowledge in alphabetical order the following people that have participated in the redshift surveys mentioned here: J. Calderon, A. Fairall, P. Focardi, M. Geller, J. Huchra, M. Kurtz, D. Latham, S. Lipari, P. S. Pellegrini, G. Vettolani and C. Willmer. This work was supported in part by the Smithsonian Institution Scholarly Studies Program.

9. References

1. J. P. Huchra and M. J. Geller, *in Large-Scale Structures and Peculiar Motions*, ed. D. W. Latham & L. N. da Costa, (Brigham Young University, Provo,1991), p. 143.
2. T. J. Broadhurst, R. S. Ellis, D. C. Koo and A. S. Szalay, *Nature*, **343**, (1990), 726.
3. C. Park, *Princeton PhD Thesis*, 1990.
4. J. P. Huchra, M. Davis, D. W. Latham and J. Tonry, *Ap. J. (Suppl. Ser.)*, **52**, (1983), 89.
5. L. N. da Costa, P. S. Pellegrini, W. L. W. Sargent, J. Tonry, M. Davis, A. Meiksin, D. W. Latham, J. W. Menzies and I. A. Coulson , *Ap. J.*, **327**, (1988), 544.
6. L. N. da Costa, P. S. Pellegrini, M. Davis, A. Meiksin, W. L. W. Sargent and J. Tonry, *Ap. J. (Suppl. Ser.)*, **76**, (1991), 935.
7. A. Lauberts, *The ESO/Uppsala Survey of the ESO(B) Atlas*, (European Southern Observatory,Garching, 1982).
8. M. Davis, A. Meiksin, M. Strauss, L. N. da Costa and A. Yahil, *Ap. J. (Letters)*, **333**, (1988), L9.
9. P. S. Pellegrini, C. Willmer, L. N. da Costa and B. X. Santiago, *Ap. J.*, **350**,

(1990), 95.
10. M. Lachièze-Rey, L. N. da Costa and S. Maurogordato, *in preparation*, (1991).
11. M. J. Geller, and J. P. Huchra, *Science*, **246**, (1990), 897.
12. J. P. Huchra *et al.* , *in preparation*, (1991).
13. B. A.Vorontsov-Velyaminov and V. P. Arkhipova, *Morphological Catalogue of Galaxies*, (Moscow University,Moscow, 1963-1968).
14. P. S. Pellegrini, L. N. da Costa, J. P. Huchra, D. W. Latham, and C. Willmer, *A. J.*, **99**, (1990), 751.
15. P. S. Pellegrini and L. N. da Costa, *Ap. J.*, **357**, (1990), 408.
16. L. N. da Costa, M. A. Nunes, P. S. Pellegrini, C. Willmer, G. Chincarini and J. J. Cowan, *A. J.*, **91**, (1986), 6.
17. L. N. da Costa, C. Willmer, P. S. Pellegrini and G. Chincarini, *A. J.*, **93**, (1987), 1338.
18. G. A. Tammann and A. Sandage, *Ap. J.*, **294**, (1985), 81.
19. D. Lynden-Bell, S. M. Faber, D, Burstein, R. L. Davies, A. Dressler, R. J. Terlevich and G. Wegner, *Ap. J.*, **326**, (1988), 19.
20. A. Dressler, *Ap. J.*, **329**, (1988), 519.
21 A. Dressler, *Ap. J. (Suppl. Ser.)*, **75**, (1991), 241.
22. A. Lauberts and E. A. Valentjin, *The Surface Photometry Catalogue of the ESO-Uppsala Galaxies*, (European Southern Observatory,Garching, 1989).
23. J. P. Huchra, private communication.
24. A. Fairall and A. Jones, *Pub. Astr. Dept. University of Capetown*, (University of Capetown,Capetown, 1988).
25. E. Bertschinger, A. Dekel, S. M. Faber, A. Dressler and D. Burstein, *Ap. J.*, **364**, (1990), 370.
26. C. Willmer *et al.* , *in preparation*, (1991).
27. L. N. da Costa, P. S. Pellegrini, C. Willmer and D. W. Latham, *Ap. J.*, **344**, (1989), 20.
28. S. J. Maddox, W. J. Sutherland, G. Efstathiou and J. Loveday, *M.N.R.A.S.*, **243**, (1990), 692.
29. B. M. Lasker, R. S. Conrad, B. J. Mclean, J. L. Russel, H. Jenkner and M. M. Shara, *A. J.*, **99**, (1990), 2019.
30. M. Davis and J. P. Huchra, *Ap. J.*, **254**, (1982), 425.

DON BACKER

PULSARS: THE NEW CELESTIAL CLOCKS

D. C. BACKER
Astronomy Department and Radio Astronomy Laboratory
University of California, Berkeley, CA 94720, USA

ABSTRACT

Pulsating radio sources, or pulsars, are rapidly rotating, highly magnetized neutron stars. They emit a beam of radio emission along their dipole field axis. The star's rotation sweeps the beam past the Earth to create a pulse if the axis is tilted with respect to the rotation axis. New classes of pulsars – including ones with millisecond rotation periods – have been discovered in the last decade.

The rotations of pulsars are extremely stable. This stability allows pulsars to be used as probes of the interstellar plasma, for tests of General Relativity in the case of binary pulsars, and for detection of a cosmic background of gravitational radiation from the early universe.

A discussion of the pular phenomenon is followed by a tutorial on pulsar arrival time measurement and analysis. The use of an array of millisecond period pulsars as a fundamental reference frame is then described.

1.0. Pulsars

1.1. A Brief History of Neutron Stars

Pulsars are firmly associated with highly magnetized, rapidly rotating neutron stars. Neutron stars were first "invented" as a theoretical entity shortly after the discovery of the neutron. In 1934 the eccentric astrophysicist, Zwicky, and his colleague Baade suggested that a neutron star might be produced in the gravitational collapse of massive stars after they run out of fuel, the supernova event. Calculations by Oppenheimer and his student Volkoff in the late 1930's, while inventing the atom bomb, indicated that a one solar mass neutron star would have a 10-km radius and a central density around 10^{15} gm cm^{-3}. Only theoretical progress was made on understanding the nature of neutron stars until the late 1960's. They were not expected to be observable. Their small size would lead to small fluxes of radiation. Furthermore their surface temperature was unknown. One assumed that they were very hot, and therefore their black body spectrum shape would further prevent their detection in the optical channel where most observations were conducted. If they were cold, then detection at longer wavelengths also was not possible since the flux of a blackbody varies with temperature T as T^4. A further clue about their radiative properties would have come from consideration that the gravitational potential of matter on the surface is a tenth of the rest mass, but that gets ahead of our story.

One of the most prescient remarks about neutron stars came in the year prior to the pulsar discovery. Pacini (1967) suggested that a rapidly rotating, highly magnetized neutron star with a misaligned dipole moment would radiate copious energy in the form of magnetic dipole radiation – taking the formula out of Landau & Lifschitz *Classical Theory of Fields*, §9. Pacini proposed that this radiation would provide the continuing energy source that was needed to explain the energetics of the Crab Nebula.

Pulsars were discovered accidentally during an investigation of extragalactic sources by Hewish, Bell and colleagues (1967). Shortly thereafter other pulsating radio sources were found. The most important other object was the pulsar in the Crab Nebula because it provided the observational evidence for the direct link, proposed decades earlier, between neutron stars and supernovae. Pulsars in general are confined to the galactic plane. The spatial distribution on the sky of 450 of the nearly 500 presently known pulsars shown in Fig. 1 does not overwhelmingly show a concentration in the galactic plane owing to the fact that the known objects are rather near the solar system. Pulsars are discussed in books by Manchester & Taylor (1977) and Michel (1991) and a review by Taylor & Stinebring (1986).

Following the pulsar discovery galactic, point, Xray sources were found that were soon identified with binary systems in which mass is being transferred onto the neutron star from the companion star. In these systems the source of energy is gravitational potential energy of the accretted matter. In pulsars the energy source is rotational kinetic energy. In the past decade or so a new class of pulsars – those with binary companions, and those associated with globular clusters, have been discovered (Backer & Kulkarni, 1990). As the field has progressed over the past two decades the study of accretion powered Xray sources and rotation powered pulsars has merged to become comprehensive study of the death of stars (see review by Bhattacharya & van den Heuvel 1991).

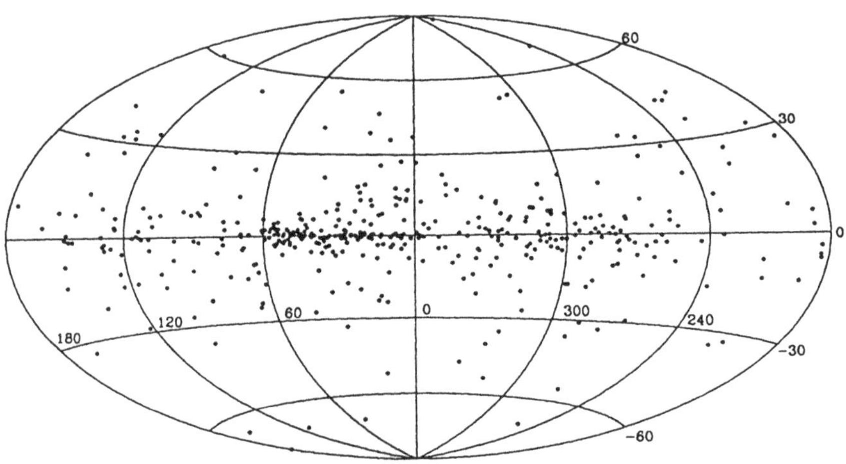

Figure 1 Distribution of 450 pulsars in galactic coordinates: latitude l and latitude b. The concentration of objects near $b = 0°$ and $0° < l < 90°$ is the result of high sensitivity surveys. (From Backer 1988.)

1.2. Standard Model of Pulsars

Observations of pulsars have led to a number of highly developed theories about the structure and dynamics of pulsars. The extremely high matter density, the ultra strong magnetic field, and the rapid rotation of pulsars has pushed astrophysics into many extreme directions. I will discuss the main ideas about pulsars working from the inside out. The internal structure of a neutron star has a number of uncertainties owing to the extreme density of matter that is beyond the limits of laboratory measurements. Condensed matter physicists divide the neutron star into a core, which is a superconducting, superfluid of $^3P_2$ neutrons and $^1S_0$ protons, an inner crust of $^1S_0$ neutrons, and an outer crust of heavy nuclei. The core covers 70% of the stellar radius and 90% of the mass. The inner and outer crusts cover 25% and 5% of the radius, respectively, and the inner crust contains the most of the remaining 10% of the mass. The quantitative values presented here are for a stiff equation of state that would give a 1.4 $M_\odot$ star a 16-km radius.

In the rotating superfluid neutron star the rotation is carried by quantized vortices with a density of $6300/P$ cm^{-2} (P is period in seconds). As the star slows down, which is the consequence of torques applied via the magnetic field to the outer crust, the density of vortices must fall. The vortices move outward and disappear. The inner crust is differentially rotating owing to the pinning of vortices to the lattice structure, a phenomenon seen in the laboratory in He^3. This leaves a piece of the star unconnected to the rotating body that we observe. In §2.7 I discuss instabilities in the smooth rotation and spin down of pulsars that are attributed to sudden or erratic events in the pinning and unpinning of vortices in the inner crust. For further information on neutron star interiors read the tutorial lectures by Sauls and by Alpar & Pines in the NATO ASI volume edited by Ögelman & van den Heuvel (1989) and the textbook by Shapiro & Teukolsky (1983).

Many observable phenomena in pulsars and in accreting Xray binaries lead us to conclude that neutron stars have intense magnetic fields that are dominated at large radii by a dipole component. Furthermore the dipole axis is required to be misaligned with the rotation axis. The lifetimes of pulsars require lifetimes for the currents that sustain the fields in excess of 10^6 years, and perhaps much longer. No one is sure that the crustal conductivity either allows or disallows such stability. The properties of the pulsed radio emission lend the strongest support to the off-axis, dipole model. The radio emission is probably beamed along the dipole axis, and we observe a pulse as the star swings its beam past our line of sight (Fig. 2). Perhaps there are symmetric beams emanating from both poles that can be seen if the axes are orthogonal.

All isolated pulsars have period derivatives that indicate the presence of a spindown torque. A few objects, which I will discuss in §1.4, display apparent spinups, but these are almost certainly the result of a small intrinsic spindown rate coupled with an acceleration in the gravitational potential of a globular cluster star system. A number of young pulsars, with ages determined by their spindown rate, are surrounded by a nebula of energetic particles that one easily concludes are the result of energy emitted by the pulsar. The presence of a torque to spindown the pulsar, and the nebulae around other very young and rapidly slowing pulsars are consistent with Pacini's original idea that magnetic dipole radiation is responsible for both phenomena. If pulsars spindown in this manner, then we can estimate their energy loss rate with:

$$\dot{E} = I\Omega\dot{\Omega} = 4 \times 10^{31} \text{ erg s}^{-1} I_{45} \dot{P}_{-15}/P^2, \tag{1}$$

where I_{45} is the moment of inertia in units of 10^{45} g cm^2, $\dot{P}_{-15}$ is the period derivative in units of 10^{-15} s s^{-1}, and $P \equiv 2\pi/\Omega$ is the period in seconds. Magnetic dipole radiation leads to the relation $\dot{E} \propto \mu_{d,\perp}^2 \Omega^4/c^3$, where $\mu_{d,\perp}$ is the equatorial component of the magnetic dipole moment. An estimate of the surface magnetic field is then

$$B_{d,\perp} = 10^{12} \text{ G } (P\dot{P}_{-15})^{0.5}, \tag{2}$$

for conventional neutron star parameters. The actual surface field strength can be significantly larger as a result of the aligned component of the dipole and higher multipole moments.

Articles about the structure of the magnetosphere surrounding pulsars and the nature of the resultant torque make a clear case for the insufficiency of the vaccuum dipole model (Goldreich and Julian 1969; Gunn and Ostriker 1969; Arons 1979; Michel 1991). Nevertheless the energy loss rate and magnetic field calculated above seem to give values that are supported by other observations: the power required to sustain the Crab Nebula and the magnetic fields inferred from cyclotron lines in the spectra of accreting Xray binaries.

Goldreich and Julian discuss the formation of a charge separated magnetosphere driven by rotationally induced electric fields. A portion of this magnetosphere will corotate with the star as long as the field lines loop out to axial radii less than c/Ω, the light cylinder radius (Fig. 2). The remaining portion of the magnetosphere must reach out into the surrounding medium with field lines that sweep back to form a transverse, strong field wave far from the star. The sweep back requires a current that is driven by the rotation. This current can provide an additional braking torque on the star. While this description is certainly more complex than the vacuum dipole model, estimates of the resultant braking torque and magnetic field are comparable in magnitude and have similar scaling with neutron star parameters. Some models conclude that the current related torque is largely independent of dipole inclination angle. There are no exact solutions for a charged, rotating magnetosphere that are relevant to the pulsar case (see Michel 1991, Arons 1991).

The above discussion summarizes the main properties of a rotating, magnetized neutron star. But how does it emit beamed radio signals? Good question. The single idea that has held the attention of most investigators is that the emission arises in the relativistic current along the dipole axis. The large and rapid sweep of linear polarization that is observed as the pulse swings by our line of sight led early investigators to this conclusion (Radhakrishnan and Cooke, 1969). Nothing has swayed us from this in the intervening years. The qualitative idea is that the relativistic current is ducted along the intense magnetic field lines in the narrow zone of open field lines. Curvature of the field lines leads to an acceleration whose vector is in the plane of curvature. Curvatures ρ between that of the stellar radius and that of the light cylinder, when coupled with the expected radio emission frequency $\gamma^3 c/\rho$, suggest relativistic factors of $\gamma \simeq 10^2$. The voltage drop along these polar field lines would allow much larger factors, and indeed other models for the formation of radio emission exist.

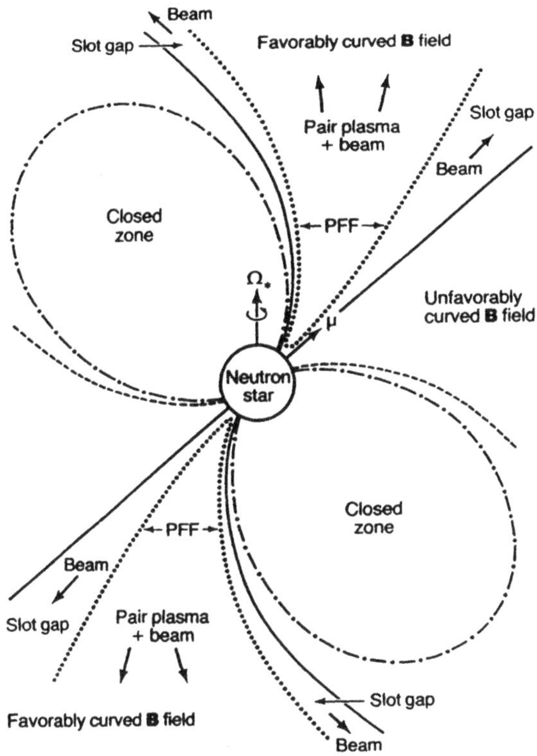

Figure 2 Pulsars are modeled as rotating neutron stars with an off-axis dipole magnetic field. The closed zone rotates with the star and extends to a cylindrical radius of Ω/c where corotation speeds approach c. An e^+e^- plasma is created by a particle beam accelerated off the surface above a pair formation front (PFF) and along favorably curved field lines. Beamed radio emission is generated at low altitudes in the pair plasma. (From Arons 1983.)

The intensity of the radio emission, when coupled with the small size of the emitting region which we determine either from this emission model or from the temporal fluctuations, indicate brightness temperatures of the emission as high as 10^{30} K. The extreme thermodynamic condition of high brightness temperature and low energy (very low) photons require the collective motion of many charges – sometimes called coherent radiation. Many charges moving in unison lead to the coherent addition of emitted voltages rather than the usual addition of incoherent powers. In summary, we typically view the radio emission as the result of turbulence convected relativistically along polar field lines, and carefully arranged to prevent any degrees of freedom for upconverting the copious number of low energy photons to a more balanced thermodynamic state.

1.3. Origin and Evolution of Isolated Neutron Stars

Once a class of objects is isolated in astronomical studies attention turns to understanding their origin and evolution. In a snapshot of the universe, one must make sense out of the different groups of related objects by connecting one class as the progenitors, for example, of another class. Sometimes there are even missing links. If a snapshot is taken of a steady process, then the number of objects in various evolutionary stages will be proportional to the lifetime of the objects in that stage. Students in school are a trivial example. Consider 'classes' as the groups. There are nearly equal numbers in each since the lifetime of all classes is the same, one year. Attrition of students from school will form an independent group and cause the class size to diminish with school year. On the other hand grouping according to primary school, junior high school, and high school, leads to relative lifetimes in the ratio of 6:2:4, for a typical US town. The relative numbers in each group will follow this ratio, apart from the attrition factor.

One of the principal large goals of the study of neutron star evolution is connecting the pulsars we see with the entire population of neutron stars, with the supernova progenitors, and with the massive star population which is continually replenished by star formation in the interstellar medium. This is a tall order. The more restricted task of understanding the life cycle of the pulsar phase of the isolated neutron star is where some progress has been made.

We are fortunate with pulsars in that their signals provide an evolutionary clock in their period derivatives. The ratio of the period to the period derivative gives an age, an age for doubling the period. Furthermore we can take the model presented above and create a more specific age after one or more assumptions are made. If the original period of the pulsar is P_o, and if the braking mechanism evolves with constant perpendicular magnetic moment, then the true age of the star is

$$t_c = \frac{P}{2\dot{P}}\left[1-\left(\frac{P_0}{P}\right)^2\right]. \tag{3}$$

Isochrones based on Eq. 3 are drawn on the $P\dot{P}$ plot of pulsars in Fig. 3; the logarithm of the age in years of each isochrone is noted to the right of each. The typical age of pulsars by this means is a few million years. The boundaries of the isochrones delineate high and low magnetic field extremes. The standard model allows us to interpret the coefficient in terms of the magnetic moment, and with further assumptions, the dipole component of the surface magnetic field (Eq. 2). Another age estimate comes from the location of the pulsar with respect to the presumed layer in the galaxy of the massive stellar progenitors. In many cases we even have the transverse motion of pulsars. Distance estimates required for these parameters are fortunately available from the column density of electrons that disperse pulsar signals (§2.5).

What emerges from this abundance of parameters is that most pulsars are born in the layer that is associated with present day massive star formation. They live as pulsars for an interval of about 10 million years. Their "death" as radio emitting pulsars is the result of a combination of a slow decrease of luminosity combined with an abrupt turnoff (the e^+e^- limit in Fig. 3). During this time they travel away from the galactic plane to form a layer many times thicker than their progenitor stars. Also the torque on the pulsars seems to decay faster than expected based on the simple dipole spindown theory discussed earlier. This rapid decay has been attributed to a reduction of the dipole magnetic field by 1 or 2 orders of magnitude over 10 million years. This is shown in Fig. 3 by bends in the evolutionary paths of low and high field objects. These ideas are consistent with the observations, but there is a range of opinion at present. The data are not overwhelmingly in support of a single synthesis of ideas on the evolution of pulsars during their first 10 million years.

On the origin side, the association of at least six young pulsars with supernova remnants is conclusive about the birth of neutron stars in the cataclysmic event of the death of a massive star. We know that a supernova remnant indicates a massive star's final act by totaling the mass and the kinetic energy of the present day remains. Significant questions such as, "Are all neutron stars formed this way?", remain unanswered. Larger numbers of associations with supernova remnants and more proper motions may be available in the future. In the interim, large monte carlo simulations are being conducted to find a consistent model for the pulsar population origin and evolution.

1.4. Binary, Millisecond and Globular Cluster Pulsars

Most stars one sees in the night sky are members of a binary system. If supernova events do not unbind binary systems, then a significant fraction of neutron stars will also be members of binary systems. The Xray binaries provide firm evidence that at least the first supernova event in a binary, *i.e.*, the first neutron star production event, does not always unbind binaries. Hulse and Taylor (1974) discovered the first binary pulsar, PSR 1913+16. In §2.8 I will show that pulse arrival time measurements of a binary pulsar give us the mass function that allows an estimate of the binary companion mass if one assumes the pulsar mass is 1.4 $M_\odot$. For PSR 1913+16 and a few others the companion mass is a similar 1.4 $M_\odot$. These are the high mass binary pulsars which have progenitors first in the high mass Xray binaries, and earlier in high mass OB binary star systems. They evidently have survived two violent neutron star formation events. However their numbers are small, and the conclusion is that survival is improbable. The spindown lifetimes of these pulsars are an order of magnitude larger than those of most isolated pulsars. The production rate, which is obtained from the number divided by the lifetime, is then much lower than that of ordinary pulsars.

In 1982 an isolated pulsar PSR 1937+21, which is spinning at 642 Hz, was discovered (Backer *et al.* 1982). The evolution of this millisecond period pulsar is a great puzzle. However shortly after this discovery a second millisecond pulsar was found PSR 1953+29 (Boriakoff *et al.* 1983). This object was in orbit around an inferred low mass companion which presumably is a white dwarf. A swarm of models were generated at this time with the leading idea being that these millisecond pulsars had their progenitors in the low mass Xray binaries which, in turn, arose from main sequence systems with one high mass and one low mass star. The division between high mass and low mass is around 8 $M_\odot$, which is the nominal dividing line between stars that form neutron stars and those that evolve to white dwarfs. The models had no answer to how such a system could evolve to an isolated millisecond pulsar like PSR 1937+21. One distinguishing characteristic of these very short period pulsars is that their spindown rates are many orders of magnitude smaller than those of "normal" pulsars. Small spindown rates suggest low magnetic fields (Eq. 2). Low magnetic fields are also required in most models of the low-mass Xray binaries. The age of these systems places a firm limit on how much decay of pulsar magnetic fields can take place. Certainly the fields cannot decay exponentially with a time scale of 10^{6-7} years as many authors had suggested in the 1970's.

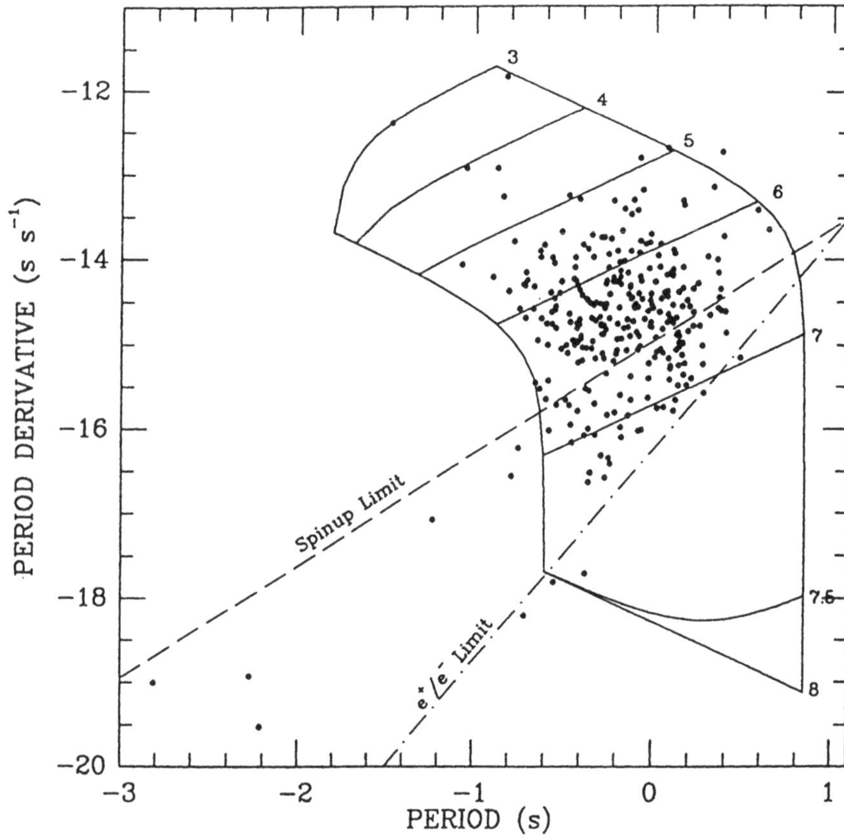

Figure 3 Distribution of rotation periods and period derivatives of pulsars. The standard model discussed in the text leads to the various evolutionary tracks, or limits, in the diagram. (From Backer 1988.)

The role of the magnetic field is critical to the binary formation models. A low magnetic field on an old neutron star is required to allow mass accretion into the neutron star magnetosphere at small enough radii so that an accretion torque can spin up the old, and presumably slowly rotating, neutron star to a millisecond period. The process is limited along a line shown in Fig. 3.

The next episode in pulsar discovery in the 1980's is the detection of pulsars in globular clusters. The clusters are self gravitating systems of 10^{5-6} stars. They form a spherical population hovering around the center of our galaxy and were created very early in the evolution of the galaxy. As a self gravitating system, they have negative heat capacity and will collapse endlessly unless there is a source of heat to stir up the stars. One source of heat is the formation of binary systems in the core where the density is a million times higher than that in the solar neighborhood. Imagine what the night sky would look like there! One type of binary will be a neutron star and a low mass companion. These binary systems are made harder by further collisions, and eventually will evolve first into Xray binaries and then into millisecond pulsar binaries. The tree of possibilities is much richer than just outlined. In the cluster 47 Tucanae there are no less than 10 millisecond pulsars.

Fruchter *et al.* (1988) reported the discovery of a millisecond pulsar in an binary system PSR 1957+20 that is eclipsed once per orbit by a halo of material surrounding the companion. This system had the properties of a system that Ruderman *et al.* (1988) had considered theoretically – that of a close binary where a portion of the rotational energy lost by the pulsar is deposited on the companion and ablates material from the companion's surface. They were considering scenarios by which low mass binary pulsars could eventually become single. A second system undergoing eclipses was found in the globular cluster Terzan 5. In this second case the "weather" in the ablation and subsequent wind formation processes is such that occasionally eclipses last for several orbits. These objects are very exciting. They have whittled their companion down to just 0.02 $M_\odot$, and provide a new constraint on the form of energy and momentum loss by the pulsar as a consequence of its rotation.

In the summer after our winter school Bailes *et al.* (1991) announced yet another pulsar first: evidence for a planetary mass companion. The timing residuals (§2.0) give clear sign of the weak Doppler variations that indicate a companion of less than Jupiter's mass, 0.001 $M_\odot$. And a second pulsar discovered by Wolszczan (1991a) is now showing signs of two planetary mass companions. Wolszczan's object is a millisecond pulsar, and one is drawn to making a connection between this object and the eclipsing systems. These startling new observations are too recent to come to any firm conclusion about their origin.

Scientists are often swayed into believing, and stretching, a strong, comprehensive idea or set of ideas. What you have just read may well be such a case. There is still no solution to the formation of an isolated, low field millisecond pulsar in the binary evolution model described. Now we have objects that, while not isolated, only have planetary mass companions. Other ideas may need strong consideration. One is that single millisecond period pulsars can be formed *ab initio* with low magnetic fields. Another is that single pulsars may be formed from the coalescense of either white dwarf or neutron star binaries. Neutron stars can also form in binaries where both systems were low mass as normal stars. The formation of one white dwarf can be followed by mass accretion that supercedes the limit for a white dwarf, 1.4 $M_\odot$, and leads to neutron star formation. The collapse to a neutron star may be accompanied by a residual disk within which planetary mass coagulation can occur in a manner similar to that proposed for the solar nebula.

In conclusion let me mention that a neutron star spinning 642 times per second has an equatorial speed of 0.15 c. At rotation rates only slightly larger, and here slightly is dependent quantitatively on the equation of state for the stellar interior, the star will be unstable to the growth of a triaxial figure that would emit gravitational radiation and tend to slow the star down. The fastest neutron stars may be very close to their rotational limit. Future pulsar surveys with good sensitivity for 1.0 millisecond rotation periods are required to support this contention.

TABLE 1

Binary, Millisecond and Globular Cluster Pulsars

NAME	ROTATION PERIOD (ms)	PERIOD DERIVATIVE (10**-18 s/s)	DISPERSION MEASURE (pc/cm**-3)	ORBITAL PERIOD (d)	MASS FUNCTION (Msun)	ECCEN-TRICITY	ASSOCIATION	REFERENCE
0021-72C	5.757	-40+/-50	24.4	--	--	--	47 TUC (NGC 104)	89 IAU 4892
0021-72D	5.357	..	24.7	--	--	--	47 TUC (NGC 104)	90 workshop
0021-72E	3.536	..	24.2	..	..	..	47 TUC (NGC 104)	90 NATO ASI
0021-72F	2.624	..	24.4	--	--	--	47 TUC (NGC 104)	90 NATO ASI
0021-72G	4.040	..	24.2	--	--	--	47 TUC (NGC 104)	90 NATO ASI
0021-72H	3.211	..	24.3	..	..	..	47 TUC (NGC 104)	90 NATO ASI
0021-72I	3.485*	..	23.7	..	..	..	47 TUC (NGC 104)	90 NATO ASI
0021-72J	2.101	..	24.6	..	..	..	47 TUC (NGC 104)	90 NATO ASI
0021-72K	1.786	..	24.9	..	..	..	47 TUC (NGC 104)	90 NATO ASI
0021-72L	4.346	..	24.5	..	..	..	47 TUC (NGC 104)	
0021-72M	3.677	..	24.4	..	..	..	47 TUC (NGC 104)	
0655+64	195.671	0.68	8.7	1.03	0.0712	<.00005	field	82 APJ 253 L57
0820+02	864.873	103.9	23.7	1232.47	0.0030	0.01187	field	80 APJ 236 L25
1257+12	6.218	..	10.16	--	--	--	field	90 IAU 5073
1310+18A	33.163	..	25.0	255.84	..	<0.01	M53	89 IAU 4853
1516+02A	5.553	..	29.5	--	--	--	M5	89 IAU 4880
1516+02B	7.947	..	29.5	6.85	0.00065	0.126	M5	89 IAU 4880
1534+12	37.904	..	11.6	0.421	0.32	0.27	field	90 IAU 5073
1620-26A	11.076	0.82	62.9	191.44	0.0080	0.02532	M4 (NGC 6121)	88 NAT 332 45
1639+36A	10.378	<0.05	31.	--	--	--	M13 (NGC 6205)	89 IAU 4819
1745-20	288.60	..	210	--	--	--	?NGC6440	89 IAU 4905
1744-24A	11.563	-.05+/-0.09	242.2	.0708	0.00032	<0.003	TER 5	90 IAU 4974
1744-24B	442.84	..	205				TER 5?	90 IAU 4974
1802-07A	23.10	..	187	2.62	0.0097	0.22	NGC 6539	90 IAU 5013
1820-11	279.828	1378.	428.4	357.76	0.068	0.79462	field	89 preprint NAT
1820-30A	5.440	..	86.0+/-0.3	--	--	--	NGC 6624	90 IAU 4988
1820-30B	378.59	..	86+/-3	--	--	--	NGC 6624	90 IAU 4988
1821-24A	3.054	1.62	120	--	--	--	M28 (NGC 6626)	87 IAU 4401
1831-00	520.947	14.3	88.3	1.81	1.2E-4	0.0001	field	87 NAT 328 399
1855+09	5.362	0.017	13.3	12.33	0.0056	0.00002	field	86 NAT 322 714
1908+00A	3.6	..	200	..	..	..	NGC 6760	90 IAU 5010
1913+16	59.030	8.64	171.6	0.32	0.1323	0.61713	field	75 APJ 201 L55
1937+21	1.558	0.11	71.0	--	--	--	4C 21.53	82 NAT
1953+29	6.133	0.030	104.6	117.35	0.0024	0.00033	2CG 095	83 NAT
1957+20	1.607	0.016	29.1	0.38	5.2E-6	0.000	field	88 NAT 333 237
2127+11A	110.665	-20.0	67.25	--	--	--	M15 (NGC 7078)	89 NAT 337 531
2127+11B	56.133	>0	67.25	--	--	--	M15	89 IAU 4762
2127+11C	30.529	+4.99	67.14+/-0.02	0.335	0.153	0.6814	M15	89 IAU 4772
2127+11D	4.803	<0	67.	--	--	--	M15	90 workshop
2127+11E	4.651	~0	67.	--	--	--	M15	90 workshop
2303+46	1066.371	569.3	61.	12.340	0.2455	0.65838	field	85 NAT 317 787

2. Precise Pulsar Timing

2.1. Space-Time Events

The standard model of pulsars presented above in §1.2 has one or more beams of radiation continuously emitted in a radial direction. The space-time locus of this radiation is a spiral winding away from the neutron star like the stream of water from a lawn sprinkler. Pulses are detected at the observatory as radiation from successive stellar rotations cross the observer's path. The observer records the time of these crossing events as registered on a comoving clock. More precise definitions are given below. The first stage of pulsar timing anlaysis is a transformation of the space-time reception event of the pulse front crossing the observatory to the corresponding transmission event of the launching of this pulse at the neutron star. The transmission event is fictitious and covers our ignorance of the emission process. I will not deal with issues of aberration and retardation of the radiators and the radiation in the pulsar magnetosphere. I am also not interested in the exact orientation of the neutron star, but only its relative orientation between successsive rotations. The second stage of pulsar timing is an analysis of events in the reference frame of the neutron star: do the transmission events observed follow a simple model for the rotation that involves an initial epoch, period, and period derivative to within the noise of the measurement (and precision of the transformation)? Precise pulsar timing has been discussed by Backer & Hellings (1986), Hellings (1986), Backer (1989), Taylor (1989) and Doroshenko & Kopejkin (1990). References for timing binary pulsars will be given later.

2.2. Arrival Time Measurement

In the ideal case one time tags the arrival of individual pulses. In practice individual pulses vary considerably within a well-defined window of rotational longitude, and may be too weak to detect individually. Pulsar timing then requires averaging over many pulses, often thousands, to obtain a stable emission pattern, or profile of flux density, along with a sufficient signal to noise ratio. One of the remarkable features of the pulsar phenomenon is that stable profiles are obtained in spite of the dramatic pulse to pulse fluctuations. Helfand *et al.* (1975) present a quantitative study of the stabilization of profiles as a function of integration time. Similar studies are needed for a wider set of objects to validate this assumed property. Experimentally the samples of flux density are often taken with a time base that is synchronous with the apparent pulse period as estimated from an existing model. This allows the summing of successive blocks of 1024 samples, which is a typical number per period, into an averaging array. The space-time event recorded is the time of the first sample of the first block. All successive samples are taken at deterministic times relative to the first sample. Examples of several pulse profile integrations are shown in Fig. 4.

Pulsar signals travel through the ionized interstellar medium and are dispersed – lower frequencies arrive later than higher frequencies. This dispersion places an upper limit on the bandwidth that can be used for pulse detection observations. The dispersion within the band is typically matched to the sample interval, $0.001P$ in the present example. The dispersion time τ_d within a bandwidth b at a center frequency ν is

$$\tau_d = 8.3\mu s \frac{DM(\text{pc cm}^{-3})\, b(\text{MHz})}{\nu(\text{GHz})^3}, \tag{4}$$

where DM is the column density of electrons along the path in astronomer's favorite unit, pc cm^{-3}, which is 3×10^{18} electrons cm^{-2}. DM's range from 3 to 1000. One can see that $DM = 100$ leads to almost a millisecond of dispersion, for $b = \nu = 1$, and this exceeds the sample spacing of $0.001P$ for many pulsars. Narrow bandwidths are required for short period pulsars. A limit to bandwidth narrowing occurs when τ_d is equal to the maximum sample rate $1/2b$. The dispersion within a band of signals is a frequency dependent phase in the electric field domain, and can be removed prior to power detection. This process, called coherent dispersion removal, is used for precise millisecond pulsar timing. Dispersion removal techniques are discussed by Hankins and Rickett (1975).

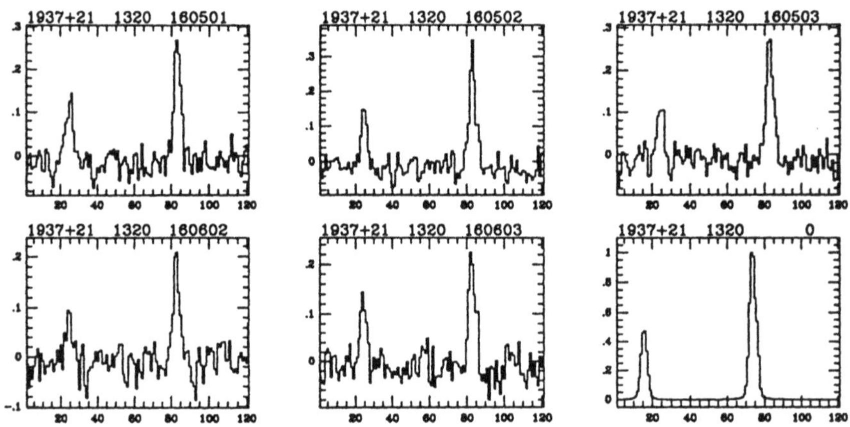

Figure 4 Pulsar timing data profiles. The first five plots contain successive 2-min integrations of PSR 1937+21 at the NRAO 140ft telescope using 20 MHz of bandwidth at 1320 MHz; scale is flux density (Jy) *vs* bin number (1-121). In the lower right is a standard template which is convolved with the data to obtain relative arrival times.

The sensitivity of radio astronomical observations – the ability to detect signal power against a background of unwanted noise coming from the amplifier and other unavoidable processes – is determined by the inverse of the square root of the product of the bandwidth and the integration time. In the present example the signal to noise ratio for each sample of the profile is proportional to $1/\sqrt{bNP/1024}$, where N is the number of pulses averaged. Dispersion limits b, but we can increase our sensitivity by using many adjacent channels of width b. Signal processors that define many channels and then detect and average the results, are a continuing engineering challenge.

Pulsar signals are highly polarized. Most timing observations accumulate averages of orthogonally polarized signals. These are calibrated to a common flux density scale and added together to form the total intensity profile. Careful calibration is required to prevent distortion of the profile from polarization effects.

The next step in pulsar timing is the comparison of a pulse average with a template pulse. The template pulse typically comes from a long term average of the data itself. Alternatively an analytic model can be derived from the data and then used as a template. See example in Figure 4. The observed pulse is convolved with the template pulse to obtain the relative pulse arrival time from the peak of the correlation function. Other approaches such as use of Fourier techniques have also been developed. The relative pulse arrival time is added to the first sample start time to obtain the observatory, or topocentric, time of arrival or TOA. The precision of determination of a TOA will depend on the source average flux density, the width of the pulse, and the instrumental factor discussed above. Short period pulsars can be timed more precisely than longer period objects with equal flux density owing to the narrower pulse width in the shorter period pulsars.

In standard practice the constant time interval in the template from its first sample to a selected fiducial point of the pulse, such as the peak, is added to the TOA. Also, the time interval in a many period integration from the first sample of the first pulse down to the first sample of a pulse in the middle of the integration is added to the TOA. This mid-integration correction reduces the sensitivity of the TOA to systematic drifts that will occur if there is an error in the assumed model. An additional correction of the TOA for propagation from the sampling hardware back to the fiducial point that corresponds to the observatory coordinates in the Earth reference frame should be done, but is not done in present work. This latter effect is essential only if absolute TOAs are compared between observatories.

2.3. Time Correction

Anyone who has some familiarity with astronomical observations knows that there are many, too many, time scales – sidereal time, universal time, atomic time, Julian Date, Gregorian Date, *et al.* See Jesperson & Hanson (1991) for a general discussion of time and frequency issues. We need to record our space-time pulse arrival event on a uniformly running time scale in the observer's reference frame. This time scale, generically, is atomic time. The standard atomic time scale, International Atomic Time - TAI, is based on an assumed value for the frequency of a transition in the Cesium atom (see §3.2), and a particular, but arbitrary epoch for the origin of the scale. The principal calendar scale is Julian Date which increments the day number after 24 hours of TAI and has no year; the day boundary is at noon. 1991 March 27 noon (in PR) corresponds to JD 2448343.167.

Observatory clocks keep universal time, a 24-hour scale that is also known as Greenwich Mean Time. Coordinated Universal Time, UTC, is the most accurate universal time scale which is coordinated by Bureau International des Poids et Mesures in Paris (BIPM). UTC is decimated by various means so that the observatories can keep approximate UTC. The UTC scale now runs at a rate identical to that of TAI. However, the goal of UTC is to assist in predicting the meridian transit of the Sun at any observatory. This means that it is a solar, or Earth rotation, time scale, and must be locked to the somewhat erratic rotation of the Earth. There is an offset in phase, or epoch, between UTC and TAI. This offset is adjusted by leap seconds that are inserted at midnight on December 31, or more often if needed. Observations of the orientation of the Earth are required to determine when leap seconds are needed. As an aside, the Earth's rotation was the basis of civilization's time scale for centuries. After electronic crystal clocks were developed in the 1930's the erratic behavior of the Earth's rotation was observed, and laboratory clocks took over the job of precise time keeping. Sidereal time gives the orientation of the Earth with respect to the stars, and is derived from UTC by a step in epoch called (UT1-UTC), and then a change in rate that reflects rotation with respect to the stars rather than the sun.

There are several means of transfering the epoch of a time scale from a standards laboratory to a remote astronomical observatory. The crudest is the long wavelength radio signals of WWV at 5, 10, 15, 20, and 25 MHz that are broadcast from Fort Collins, CO, and travel large distances around the globe by reflection off the ionosphere. Similar signals are broadcast by the time services of various nations. The accuracy is around 1 ms for distances of a few thousand kilometers. The US maintains a system of 100-kHz transmitters for radio navigation called Loran C. These signals travel as a ground wave and provide microsecond level time transfer for distances up to 1500 km. The most accurate time transfer is now done using the Global Positioning System of satellites. Signals are transmitted in bands between 1300 and 1700 MHz. The Cesium atomic clocks on board on these satellites are monitored from the ground, and a prediction ephemeris is sent back to them to allow real time transfer at the level of 100 ns. Even better transfer can be obtained by real-time tracking of the same satellites at the observatory and at the national standards laboratory – the common view technique. A transfer accuracy of a few ns has been obtained in an experiment where time was transferred between laboratories and satellites in three hops around the globe. (Ashby & Allan 1984). Careful applications of relativity theory are required to conduct this experiment in the rotating, non-inertial reference frame of the laboratories.

The time transfer discussed above makes the best atomic time scale available to the remote observatory. This time scale currently is called Terrestrial Dynamical Time, TDT. TDT is computed at the BIPM by averaging over the International Atomic Time scales, TAI(x), as realized at various standards laboratories identified by x. TDT is proper time for an observer at sea level on the Earth, and is not a uniform time scale in the reference frame of the pulsar. In the pulsar frame the Earth clock is affected by the variable gravitational redshift that is the result of the influence of the Sun and the planetary masses in the solar system, and by the variable time dilation along the more or less elliptical orbit of the Earth. The ratio of the rate of a TDT clock (t_i) to the rate of a clock in the solar system barycenter (τ) in the absence of any solar system bodies is given by

$$\frac{dt_i}{d\tau} = \Sigma_p \frac{Gm_p}{r_{ip}c^2} + (\frac{v_i}{c})^2, \tag{5}$$

where m_p is the mass of solar system body p, r_{ip} is the distance from the clock to the body, and v_i is the velocity of the clock. These two effects, which are the result of the equivalence principle and special relativity, respectively, are inseperable in their time signature. The principle periodic term in the difference between the two clocks can be approximated by an analysis of a clock in a pure elliptical orbit around the Sun,

$$\Delta t_R \equiv t - \tau = 1.66 \text{ ms } [(1 - \frac{1}{8}e^2)\sin M + \frac{1}{2}e \sin 2M + \frac{3}{8}e^2 \sin 3M], \tag{6}$$

where e is the eccentricity and M is the mean anomaly (roughly phase within year) of the Earth's orbit (Clemence & Szebehely 1967; Blandford & Teukolsky 1976).

If we are to conduct pulsar timing observations with submicrosecond precision, then this relativistic clock correction must be done much better than is possible by simple analytic expressions. Several groups who develop detailed models of the solar system dynamics have provided algorithms and ephemerides that have sufficient precision. The barycentric time scale τ is not what is obtained by these means. By international agreement the corrected time scale is declared to have only the periodic terms in $dt/d\tau$ removed. This modified scale is called Barycentric Dynamical Time, TDB.

In summary UTC(observatory) is converted to UTC(standards laboratory) by a satellite time transfer technique, and then to UTC(BIPM) according to an algorithm. UTC(BIPM) is, apart from a constant that is incremented slowly by integer seconds, atomic time TAI(BIPM). TAI(BIPM) is converted to proper time on the geoid TDT by a second algorithm, and then to coordinate time for solar system dynamics TDB through use of a relativistic clock transformation ephemeris.

2.4. Space Correction

The space correction relocates the arrival time from the observatory to the origin of an inertial reference frame, the solar system barycenter. The Earth moves in a nearly circular orbit of radius 1.5×10^{13} cm, or 500 light seconds. The accuracy of the transfer of arrival that is required for the highest precision pulsar timing is 0.1 μs which is 2×10^{-10} smaller than the orbit size.

The space correction is typically effected in two steps (Fig.5). First the arrival time is transferred from the observatory to the Earth center using a simple model for the Earth's orientation in an inertial frame and the star's celestial coordinates. This step is similar to the calculations that are required to perform Very Long Baseline Interferometry (VLBI) at radio wavelengths, although the accuracy requirements for VLBI are much more stringent.

The transformation from the Earth center to the solar system barycenter requires an accurate ephemeris, or prediction table, of the Earth's position as a function of date. There are two principal groups involved in this solar system dynamics work. They assemble all measurements of solar system bodies – optical eclipses and occultations, radar range measurements, planetary probe results – and feed this data into a comprehensive model fitting program. Parameters, masses and orbital elements, are adjusted to obtain the best fit to the data. The resulting model of the solar system can then be extrapolated into the future for uses such as spacecraft tracking, observations of planets and other bodies, and pulsar timing. The ephemeris computer file typically consists of values evaluated every four days. Intermediate results are obtained by interpolation.

2.5. Propagation Correction

2.5.1. Dispersion Delay

Pulsar signals traverse an ionized, dilute plasma that delays low radio frequencies relative to high radio frequencies. This plasma is located in the general interstellar medium, in localized

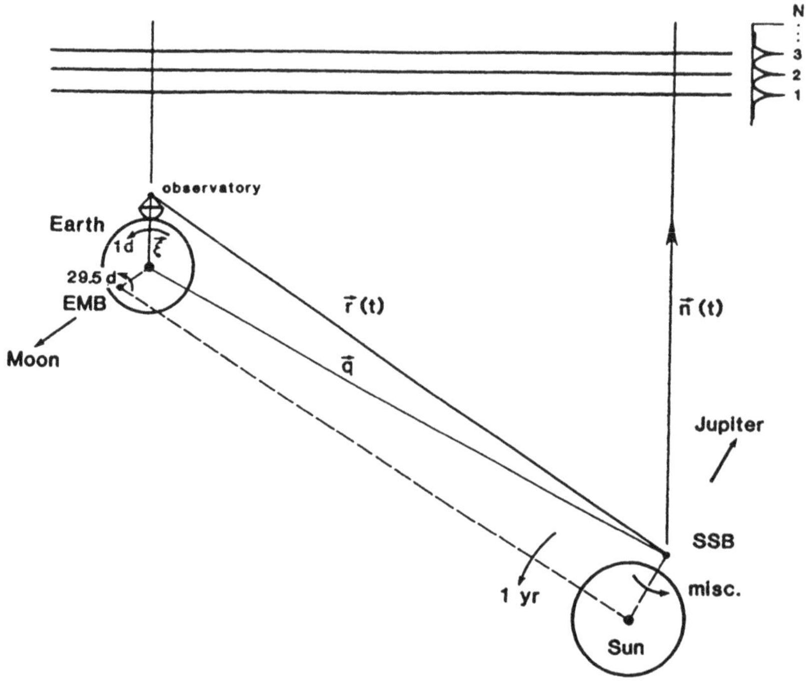

Figure 5 Schematic of space transformation from the observatory to the solar system barycenter. While the relative sizes and locations of objects are inaccurate, the locations of the Earth-Moon mass center with respect to the Earth size, and the solar system barycenter with respect to the Sun size are nearly correct. Pulsar signals are represented as "pulse" fronts numbered by pulse number with normal $\hat{\mathbf{n}}$. (From Backer and Hellings 1986).

ionized clouds around luminous stars, in the ionized solar wind emanating from our Sun, and in the ionosphere of the Earth. If this medium were fixed, then we could measure a dispersion constant once, and apply this to all data recorded at various radio frequencies. However, all of the media mentioned are fully turbulent with a broad spectrum of length scales. Precise pulsar timing requires continuous determination of the dispersive effects.

The dispersive delay is calculated from the path integral of the inverse of the group velocity which is determined using the cold, unmagnetized plasma index of refraction, n. The magnetic field imbedded in the interstellar plasma does not lead to a significant birefringent time delay.

$$n^2 = 1 - \frac{\nu_p^2}{\nu^2}, \quad \text{and} \tag{8}$$

$$\Delta t_D = \int_P^E \frac{ds}{v_g} = 0.00415 \text{ s } DM(\text{pc cm}^{-3})\nu(\text{GHz})^{-2}, \tag{9}$$

where ν is the radio frequency, ν_p is the plasma frequency - 8980 Hz$/\sqrt{N_e(\text{pc cm}^{-3})}$, t_D is the dispersion group delay, v_g is the group velocity derived from the dispersion relation, and DM is the column density of electrons or dispersion measure. The integral is taken from the pulsar P to the Earth E. A DM of 25 leads to an excess delay of 1 second for signals at meter wavelengths. The interstellar density is around 0.03 cm^{-3}. A DM of 25 then indicates a distance of some 750 pc.

Measurement of dispersion requires timing of the pulse arrival at two or more radio frequencies. This is straightforward until one reaches the precision where one must deal with the fact that the pulse shape changes with radio frequency. Then some fiducial point will scale with the inverse of the frequency squared, but other parts may disappear owing to spectral differences, or may be offset early or late. One approach to dealing with this complexity is to make a frequency dependent template of the pulse shape to compare with the data.

2.5.2. Scattering

Turbulence in the interstellar plasma is driven by injection of energy into a gas on a characteristic length scale, and the spread of turbulent cells to other scales through viscous interaction and dissipation. Turbulence in the ionized interstellar medium means that the radio emission wavefront is crinkled in proportion to the path integral of the corresponding index of refraction fluctuations along the unperturbed signal paths. If the phase of the wavefront is perturbed by more than a radian within a Fresnel zone, then diffraction develops. That is, across a plane perpendicular to the path at the location of the observer there will be bright intensity maxima and minima where the signal goes to zero. The Fresnel zone size for a kpc path (3×10^{21} cm) at decimeter wavelengths (30 cm) is 3×10^{11} cm, 4 solar radii for comparison. The instantaneous diffraction pattern will be very frequency dependent if there are many random turns of phase within the Fresnel zone owing to the quadratic dependence of the index of refraction on frequency. Typical transverse scales for a radian of phase change are around 10^9 cm. Eventually, at some high frequency, the phase perturbation within the Fresnel zone will drop below a radian and the diffractive effects will subside.

The Earth observer, the pulsar, and even the medium are all moving with respect to each other. This produces a time variable diffraction pattern. The signal intensity goes up and down with wide swings of intensity when the bandwidth is small enough. Astronomers refer to this as scintillation. At meter wavelengths and for DMs around 25 the bandwidth of scintillation is a fraction of 1 MHz and the scintillation time scale is minutes. Larger radio sources, quasars and radio galaxies, show no such effects owing to their size which blurs the diffraction pattern - stars twinkle and planets do not in the Earth's atmosphere. In summary the scattering observables are: the time scale for scintillation; the frequency scale for "spectral scintillation"; the radio frequency scaling law of these two parameters; and the size scale of the diffraction pattern (when it is smaller than the Earth so that one can observe it). From these parameters one can assess the strength of the turbulence along a given path, and the transverse velocity of the diffraction pattern.

Diffraction requires ray paths that, while formerly diverging from the source, converge on the observer. The path lengths of the various rays that reach the observer differ. Thus, an impulse spit out by the pulsar will now be spread out by a geometric factor,

$$\tau_b = \frac{d\theta_s^2}{2c}, \tag{10}$$

where d is the distance to the turbulent "screen", and θ_s is the "scattering" angle. Let me make a semantic point here. The microphysical phenomenon that does the damage to the wavefront is scattering of the electrical field by electron interactions. So the phenomenon is called ***interstellar scattering***, and not interstellar scintillation which is just one of the effects of interstellar scattering. Eq. 10 is for a source at infinity. An excellent exercise is to develop the scale factor for a source at a finite distance with the phase perturbing screen at an arbitrary fraction of the source-observer distance; simple point to point ray path geometry will do. Note that I have repeatedly referred to a collapse of the phase perturbing medium into a single screen. This thin-screen approximation is crucial to a simple analysis of the effects, a Born approximation approach. The scintillation bandwidth is the conjugate parameter to this broadening time since it is the differential delays between paths that cause the constructive and destructive interference as a function of frequency.

Now one might ask if there isn't also turbulence on scales larger than the Fresnel zone? Yes, and these produce refraction. The source position is shifted by gradients of phase, and quadratic phase terms lead to amplification or dimunition of the intensity. Even more severe perturbations can lead to caustics, and double images which interfere coherently just like an interferometer. These are fascinating areas of current research.

2.5.3. Solar Wind and Ionosphere

Pulsar signals also traverse the solar wind flowing out from the Sun and the ionosphere that surrounds the Earth. At a radio frequency of 1 GHz typical delays are around 0.1 microsecond for both of these media with the solar elongation of 90°. These plasma delays also scale with the inverse square of the radio frequency. Both media are highly variable on a day to day and season to season basis. The solar wind delay is very dependent on the solar elongation owing to the sharp gradient in the density of the outflowing plasma. The Earth's troposphere adds a rather small additional nondispersive delay.

2.5.4. Relativistic Delay in Solar System Potential

Photons from the pulsar will follow null trajectories of the geometry. To first order, the line element is

$$d\sigma^2 = (1 - 2U)c^2dt^2 - (1 + 2\gamma U)(dx^2 + dy^2 + dz^2) = 0. \tag{11}$$

where U is the gravitational potential from all solar system bodies, and γ is a one parameter in the Parametrized Post-Newtonian coordinate system of Will & Nordvedt (1972). An expression for the time of flight extra delay owing to this effect is then

$$\Delta t_S = (1+\gamma)\Sigma_p \frac{Gm_p}{c^3} \ln \left| \frac{\mathbf{n}\cdot\mathbf{r}_p + r_p}{\mathbf{n}\cdot\mathbf{R}_p + R_p} \right|, \tag{12}$$

where $\mathbf{r}_p$ is the position of the observatory relative to the pth solar system body, and $\mathbf{R}_p$ is the pulsar's position relative to the body p. This delay, which was first pointed out by Shapiro (1964), is the *longitudinal* counterpart of the more familiar *transverse* effect that was identified by Einstein as one of his three tests of General Relativity, the bending of light by the Sun. The factor $2Gm/c^2$ is the gravitational radius of an object, and is about 3 km for the Sun. The corresponding delay is then 10 microseconds times the logarithmic factor. This delay, which is 240 microseconds at the limb of the Sun, is very difficult to measure in the presence of very large and variable plasma delays. The most precise measurements of γ come from dual frequency radio interferometry. The best current measurement is $\gamma = 1.0002 \pm 0.0020$ by Robertson & Carter (1991).

There will also be large contributions to the relativistic delay as the signal traverses the interstellar gravitational potential. However, these are so nearly constant that their effect on pulsar timing will not be measureable. This is particularly true when one considers the stellar parameter estimation process that is the subject of the next section.

2.6. Pulsar Parameter Estimation

In the above sections I have stepped through the various terms that are removed in making the space-time transformation from the observer's reference frame to that of the neutron star. This provides a set of times of emission of pulses that can be matched to a model of the rotation of the star. The process, however, is not as linear as just implied. The celestial coordinates of the pulsar are required to remove the effects of the Earth's motion, and these, at the required level of precision, must be derived from the data. The DM must also be derived from the data, and the multi-frequency observations that allow a DM determination may not be simultaneous. What must be done is to iterate from starting parameters that allow connection of data without rotation number ambiguities over limited intervals of time, and then work toward model parameters that are without ambiguity over longer intervals.

The rotation of the star, if it acts as a simple clock, can be described by a polynomial in pulse number,

$$t_N = t_0 + N\ P_0 + N^2\ P_0\ \dot{P}_0/2 + \ \ldots, \tag{13}$$

where t_0, P_0, and $\dot{P}_0$ are the initial epoch, period and period derivative, and t_N is the epoch of the Nth pulse; epoch and arrival time are synonymous. Higher derivatives may also be present. However, if the star's spindown, which is described by $\dot{P}_0$, is the result of magnetic dipole radiation, then the next derivative can be estimated as

$$\ddot{P}_M = -\frac{\dot{P}^2}{P}. \tag{14}$$

In most stars this higher derivative produces unmeasureable effects. The rotation parameters are the strongest effects since they arise on the shortest time scales. After obtaining parameters that are suitable first for hours then for days, one can begin to include the astrometric parameters that have a time scale of one year. Observations of binary pulsars are further complicated by the required estimation of Keplerian parameters of the binary orbit as well (§2.8).

Figure 6 illustrates some of the steps that I have just discussed. In all cases the data displayed are timing residuals R – the difference between observations and a preliminary model. Note that in all cases the residuals are much less than the star's rotation period. This is extremely important since all we know is pulse phase, modulo one period. If the residuals are not small with respect to one period, then the model will have the pulse number, N, in error, and the residuals will not provide the correct data for new parameter estimation. At first data with imperfect estimates of the spin parameters are shown. Next data that extends beyond the region of a previous fit are shown. This illustrates the poor extrapolation capability of a fit over a restricted range; this is expected and is no methodological shortcoming. Finally data for an entire year are displayed that has a small position error, and for several years that shows the effects of proper motion.

2.7. Rotation Noise

When proper account is taken of all the effects mentioned above, then one might expect the residuals to display the character of white noise which would result from uncorrelated errors. However, in many stars, particularly those younger than 10^5 y, the residuals are distinctly not white. Terms like "white" are used to indicate a frequency indendent spectrum of fluctuation, and "red" to indicate a spectrum where the low frequencies (*i.e.*, red) have larger amplitude variations then those at high frequencies. Qualitatively, a red spectrum of variations will appear as a slow meandering of arrival time, or phase, with an amplitude of a fraction of the pulsar period, 1-10%, and occasional sudden events, called glitches.

The glitches are most dramatic. In a matter of seconds to minutes the pulse period may jump by parts per million. The period event is accompanied by a similarly sudden increase in the spindown rate. The net effect of the subsequent evolution is that the excess spin is nearly removed over an interval of weeks to months, and the spindown rate returns to its preglitch magnitude. There is an extensive literature built up around these glitches that discusses the evolution of the internal structure of neutron stars from a condensed matter viewpoint. Briefly, the standard model here is that a thin, inner crust is normally decoupled from the rotation we observe. This inner crust is differential rotating with a specific angular momentum in excess of that in the rest of the star. Some catastrophe occurs that leads to an abrupt coupling of this inner crust to the rest of the star. Its extra momentum leads to the spinup. However, the star now carries an extra moment of inertia, and will spindown more rapidly. Eventually the inner crust will decouple again. The mechanism of coupling and decoupling has to do with the dynamics of the quantized rotation vortices which

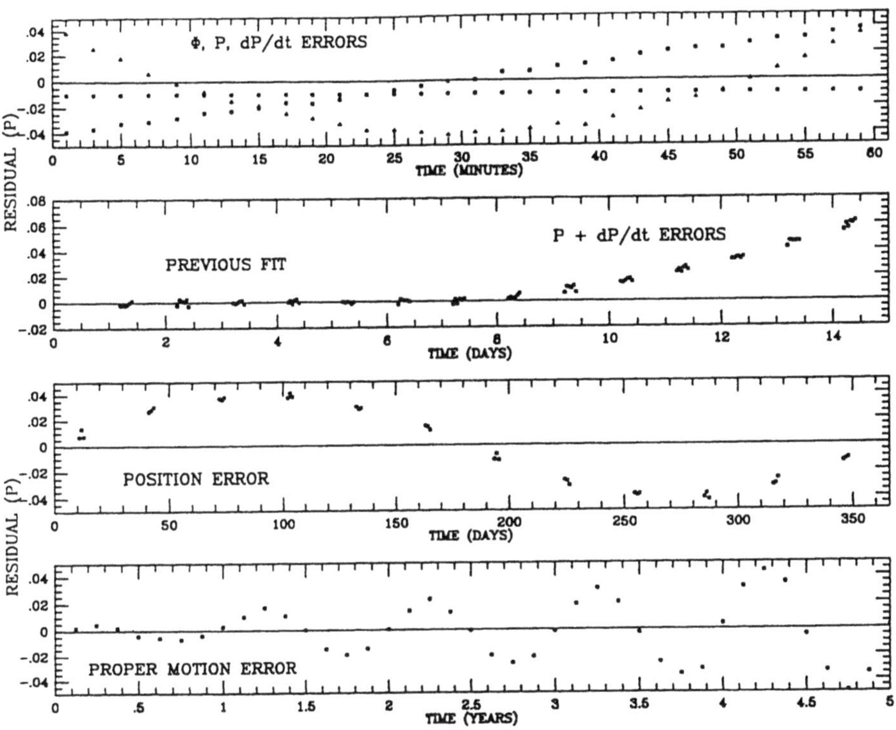

Figure 6 Pulsar parameter estimation steps. (a) Illustration of residuals from model for errors in phase (filled squares), period (open squares) and period derivative (filled triangles) with data sampled every two minutes. (b) Demonstration of imperfect extrapolation of model that was fit over 7 days (4.5 hours each day) to next 7 days. (c) Sinusoidal residuals with one year period from position error. (d) Growing residuals from error in proper motion.

carry the angular momentum in neutron star interiors and the lattice structure of the crust (*e.g.*, see Sauls 1989).

The random noise behavior in pulsar timing residuals is also connected to the evolution of the neutron star figure. The neutron star crust has sufficient strength so that the crust can carry the equilibium figure of the star in a prior state of rotation – in other words, the equatorial bulge of the crust will be larger than one would calculate for an equilibrium, self-gravitating fluid. Eventually gravitational and centrifugal forces will need to balance at the appropriate radius. The transition from the old figure to the new may be marked by a glitch, or by slow meandering.

Investigators have quantified the degree of "red" rotation noise by a logarithmic activity index which is the variance of the timing residuals measured at a selected time interval relative to the variance in the somewhat noisy Crab pulsar. A background level of white noise, which is assumed to result from instrument sensitivity and not the pulsar, is removed. Activity indices vary from +1 for a few pulsars with variances ten times larger than the Crab's, down to upper limits of −4 for some some of the millisecond pulsars. The millisecond pulsars thus have the exciting combination of both short periods, which means short pulse widths and therefore high timing precision capability, and low activity which means they are very stable clocks.

2.8. Keplerian Binary Pulsars

If the neutron star under observation is a member of a binary system, then one must transform the arrival time from the barycenter of the binary system to the neutron star before proceeding with analysis of the spin parameters of the star. A classical description of this transformation is based on Kepler's laws of motion. Figure 7 is an aid to understanding the geometry of a distant binary system. The center of the sphere is both the center of mass of the binary, and one focus of the pulsar's elliptical orbit. The plane of the orbit is inclined by an angle i with respect to the plane of the sky. The intersection of the two planes defines the line of nodes, which has a position angle on the sky χ. The periastron point in the orbit, the point of closest approach of the two bodies, is at an angle ω measured from the line of nodes. The masses of the pulsar and its companion are m_1 and m_2, respectively; $m \equiv m_1 + m_2$.

The location of the pulsar in the orbit with respect to the line through the binary barycenter and the periastron point is given by the true anomaly ν. This line, which is along the major axis of the ellipse, is the line of apsides. The mean anomaly is a uniformly increasing phase angle $M = \Omega_b(T - T_o)$, where T_o is the epoch of a periastron passage, $\Omega_b = 2\pi/P_b$, and P_b is the binary orbital period. The eccentric anomaly E is an angle defined on a circle that just encompasses the orbital ellipse. The origin of E is the line of apsides, and the projection of the radius of the circle at angle E onto the line of apsides is identical to the similar projection of the pulsar location on the ellipse. Kepler's equation relates the mean and eccentric anomalies:

$$M = \Omega_b(T - T_o) = E - e \sin E, \tag{15}$$

where e is the eccentricity of the orbit. Further trigonometry relates the eccentric anomaly E to the true anomaly ν, so that the plane polar coordinates (R_1, ν) of the pulsar relative to the binary barycenter are determined by

$$R_1 = a_1(1 - e \cos E), \quad \text{and} \tag{16}$$

$$\sin \nu = \frac{(1 - e^2)^{0.5} \sin E}{1 - e \cos E}, \quad \text{and} \quad \cos \nu = \frac{\cos E - e}{1 - e \cos E}. \tag{17}$$

The component of this motion along the line of sight is the one that affects the arrival time,

$$\delta t_b = a_1 \sin i (1 - e \cos E) \sin(\nu + \omega). \tag{18}$$

For an approximation one can substitute M into the above equation for both ν and E to obtain an idea about the variations of δt_b. There are five Keplerian parameters that must be determined for an elliptical orbit: $P_b, a_1 \sin i, e, \omega$, and T_o. These provide a constraint on the mass of the companion which is known as the mass function:

$$f(m_1) = \Omega_b^2 (a_1 \sin i)^3 / G = (m_2 \sin i)^3 / m^2. \tag{19}$$

The timing of a binary must proceed from an initial determination of its orbit using Doppler data alone, apparent periods, to a more precise orbit which is obtained by phase connecting the data which uses the above formulation for δt_b.

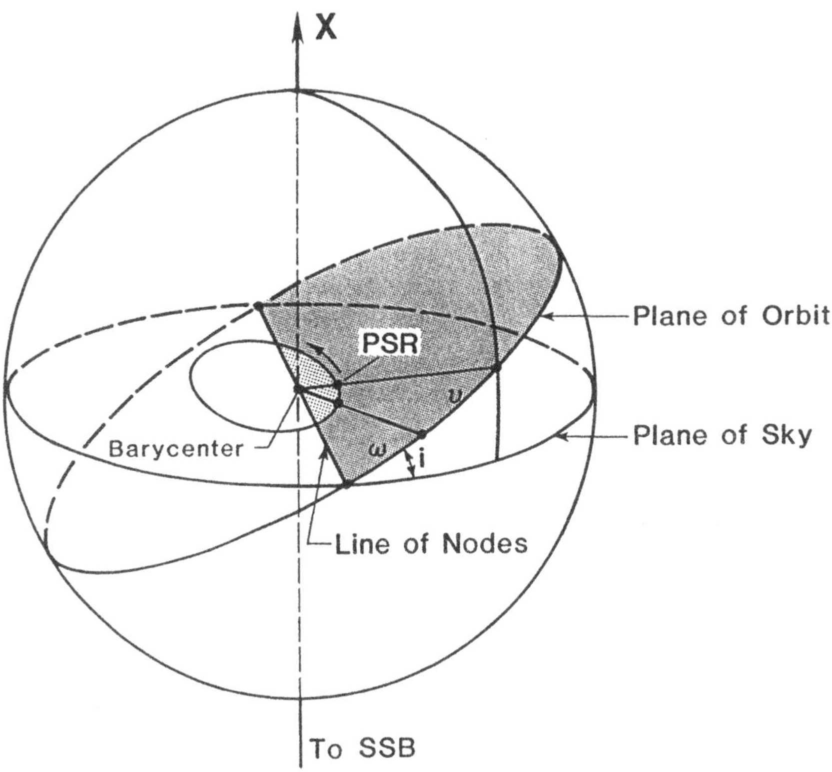

Figure 7 Schematic of orbital motion of binary pulsar about the binary barycenter relative to the line of sight and plane of sky. (From Backer and Hellings 1986.)

2.9. Relativistic Binary Pulsars

2.9.1. Parameterized Post-Keplerian Parameters

A pulsar in a short orbital period binary with a massive companion will require post-Keplerian corrections to its timing model. The largest effect is the secular advance of the line of apsides in an inertial reference frame. The corresponding effect in the solar system is the advance of perhelion of the planet Mercury. A second relativistic effect is the counterpart of the relativistic clock correction for an Earth clock which was presented in §2.3. The neutron star rotation is a clock, and we must transform from coordinate time in the binary barycenter to proper time in the orbital frame of the neutron star. These relativistic parameters depend on orbital phase with independent signatures so that the masses and the inclination of the system can be determined. A fundamental development in pulsar research is the detection of the orbital period decay in PSR 1913+16 that is most likely the result of a back reaction to gravitational radiation by the binary mass system (Taylor *et al.* 1979; Damour & Taylor 1991a). The over determination of the elements of the system that is provided by the three effects mentioned above allow one to actually predict the gravitational wave orbital decay and hence compare the observations to theory. Binary pulsars provide the only experimental evidence for the existence of gravitational radiation and the first strong field tests.

The light travel time of signals from the pulsar through the gravitational potential of its companion provides a further effect, just as it does for propagation past our Sun, which is detectable in several binary pulsar systems. Damour & Taylor (1991b) have developed a post-Keplerian parameterization of the weak and strong field effects that are, or may be, detected in binary pulsar signals.

The most prominent relativistic effect is apsidal motion of the orbit. This is so large, and the measurements of PSR 1913+16 by Taylor and his colleagues have been so precise, that proper modeling of the data must take into account that the motion is not uniform along the elliptical orbit. The time averaged rate of apsidal motion can be calculated from the usual Einsteinian formula:

$$\dot{\omega} = \frac{3(Gm)^{2/3}(\Omega_b)^{5/3}}{(1-e^2)c^2}, \quad \text{or} \tag{20}$$

$$\dot{\omega} = 0.20^\circ\ \mathrm{y}^{-1} \left(\frac{P_b}{\mathrm{days}}\right)^{-5/3} \left(\frac{m}{\mathrm{M}_\odot}\right)^{2/3} (1-e^2)^{-1}. \tag{21}$$

Parameters for the three relativistic binaries, 1534+12, 1913+16 and 2127+11, are given, or can be estimated, from the data in Table 1.

The relativistic clock transformation leads to a periodic correction Δt_{Rb} which is similar to that discussed in §2.3 for the Earth clock.

$$\Delta t_{Rb} = \frac{G^{2/3}}{c^2}\frac{m_2(1+m_2/m)}{(m\Omega_b)^{1/3}} e \sin E, \quad \text{or} \tag{22}$$

$$\Delta t_{Rb} = 0.0069 \text{ s} \left(\frac{P_b}{\text{days}}\right)^{1/3} \left(\frac{m_2}{M_\odot}\right) \left(\frac{m}{M_\odot}\right)^{-1/3} \left(1 + \frac{m_2}{m}\right) e \sin E. \tag{23}$$

The fourth relativistic effect is the decay of the orbit that results from the back reaction to emission of gravitational radiation. The field equations of general relativity for a dynamic system reduce to a wave equation for the perturbation of the spacetime metric about its flat space form in the limit of small perturbations, or weak fields and slow speeds. While the perturbation of the spacetime metric in the immediate vicinity of a neutron star is not small (*i.e.*, does produce strong field effects), the binary system is much larger and weak field calculations are probably sufficient. The question of constraints on strong field parameters from binary pulsar timing observations is now under consideration (Damour & Taylor 1991b; Taylor *et al.* 1991). The orbital decay rate $\dot{\Omega}_b$ for two point masses in General Relativity is given by

$$\dot{\Omega}_b = \frac{96}{5} \frac{G^{5/3}}{c^5} \frac{m_1 m_2^5}{m^{1/3}} \frac{(1 + \frac{73}{24}e^2 + \frac{37}{96}e^4)}{(1 - e^2)^{7/2}} \Omega_b^{11/3}. \tag{25}$$

$$T_b \equiv \frac{\Omega_b}{\dot{\Omega}_b} = 1.3 \times 10^{11} \text{ y} \left(\frac{m_1}{M_\odot}\right)^{-1} \left(\frac{m_2}{M_\odot}\right)^{-2/3} \left(1 + \frac{m_1}{m_2}\right)^{1/3} \left(\frac{P_b}{\text{days}}\right)^{8/3} f(e). \tag{26}$$

In Eq. 26 the two masses are taken to be 1.4 $M_\odot$ neutron stars and $f(e)$ is the ellipticity factor in Eq. 25. For the dual neutron star binaries with orbital periods of 0.3 days and modest eccentricities gravitational radiation will lead to a coalescence, or explosion, of the two stars on a time scale of 100 million years. This means that there are many events like this in the lifetime of our galaxy, and that these are moderately common events in the manifold of galaxies.

2.9.2. Precession

If the spin angular momentum of a pulsar in a tight orbit around a companion neutron star is misaligned with the orbital angular momentum of the system, then several torques arise in general relativity that will lead to a slow precession of the spin axis of the observed pulsar. The microwave beam that allows us to detect a pulsar is thought to be circular, or mildly elliptical. On time scales short with respect to the precession period we would expect secular variations of the pulse shape. Eventually we might "lose contact" if the precession amplitude were sufficiently large.

The theoretical tree of precession effects is thoroughly covered in the black hole text by Thorne, Price, and Macdonald (1986). An isolated neutron star will undergo free precession if its figure axis and its angular momentum axis are misaligned. A binary system with two stars each with arbitrary spin angular momenta and quadrupole mass moments is a complex affair with classical and general relativistic effects. The first of three effects is that of Lense-Thirring which is referred to as the gravitomagnetic (GM) effect by Thorne *et al.*: angular momentum produces a GM field which can produce a torque on a body moving in this field. The E&M analog is the $\mathbf{v} \times \mathbf{B}$ body force on a charge moving in a magnetic field. Matter circulation is like a current which in E&M produces $\mathbf{B}$. If the pulsar has a misaligned quadrupole moment, which would arise naturally from a misaligned spin, then its spin axis would precess. In our case the orbital angular momentum is dominant over the spin angular momentum of the companion. Gravity Probe B is designed to measure this as yet unmeasured effect of GR.

The second two effects come under the heading of geodetic precession (GD). One component of GD precession is from spin-orbit coupling – the analog of the precession of a magnet in an external magnetic field, or the spin of an electron as it moves through the electrical field of the nucleus. (There is no Thomas precession analog.) The second term is the result of space-curvature, and has no classical analog: a vector moved along a closed orbit in curved space-time does not return to the same orientation.

In an equal mass binary system these three terms have amplitudes in the ratio of 1:1:2. Specifically the relation for the precession rate is given by

$$\Omega_p = \frac{G^{0.67}\Omega_b^{1.67}(\mu + 3m_2)}{2c^2(1-e^2)} \tag{27}$$

$$\Omega_p = 0.57^\circ \mathrm{y}^{-1}\left(\frac{P_b}{8\mathrm{h}}\right)^{-1.67}\left(\frac{m_1}{\mathrm{M}_\odot}\right)^{0.67}\frac{g(m_2/m_1)}{(1-e^2)}, \quad \text{where} \quad g(x) = \frac{0.63x(1+3x)}{(1+x)^{1.33}}. \tag{28}$$

In this expression $g(x)$ is unity for equal masses. The precession amplitude will depend on the figure of the neutron star.

There is more to this than talk. In recent years observations by two groups have been exploring the secular stability of the radiation pattern of 1913+16. Weisberg, Romani and Taylor (1989) have shown clearly that the ratio of the two components of emission in 1913+16 at a single radio frequency is slowly changing. If the size of the radiation beam in rotational latitude is comparable to that in longitude, then an inferred precession rate would be 0.05° y^{-1}. This is small relative to the value computed from the above formula given the eccentricity of this system, 1.2° y^{-1}. Of course we can be at an unlucky viewing angle in the precession period, and the spin and orbit can be well aligned. Cordes and his colleagues have monitored the polarization signature of 1913+16 and show further evidence for precession of the pulsar.

The recent discovery of a 1534+12 by Wolszczan at the Arecibo Observatory (1991b) provides another object for similar investigation. The strong polarization and wide emission profile suggest strong misalignment between its spin and orbital angular momenta.

3.0. Pulsar Timing Array

3.1. Concept

In §2.0 the methods of pulsar timing and the parameters required for investigations of isolated and binary pulsars were discussed. The data for each pulsar were treated independently using a model of the Earth's space-time coordinate as a function of epoch that is assumed to be more accurate than the precision of the pulsar data. In this section the limits of this model are discussed, and the influence of a stochastic background of gravitational radiation on pulsar arrival times is presented. The section concludes with a description of the use of an array of precisely timed pulsars to solve for *global* parameters that affect all objects in a correlated manner. The ideas presented here were the subject of a recent workshop – ***The Impact of Pulsar Timing on Relativity and Cosmology***, 1990 June, Berkeley (workshop notes are available). Ideas concerning a Pulsar Timing Array are developed by Foster (1990), Foster and Backer (1990) and Romani (1989).

3.2. Time Coordinate

In §2.3 I described the transfer of *time* from a standards laboratory to the observatory. We are, of course, interested not in some fuzzy, metaphysical concept like *time*, but in the very concrete concept of *time interval*, or its inverse, frequency. The time coordinate for us is just the numbered sequence of cycles of a frequency. We are comparing rotations of our star to a fundamental atomic frequency. There are two concepts used in evalation of the quality of an standards lab time scale: accuracy and stability.

The atomic time scale is determined by defining the second, and by identifying an epoch. The second of Terrestrial Time is 9,192,631,770 cycles of the frequency corresponding to a fine structure transition in ground state of the Cesium 133 atom at sea level. (An accurate definition of what we mean by sea level is currently being negotiated.) The epoch when atomic time reads some value, like zero or 2440000.0 days 0 seconds, is not of particular concern here. What does concern us is the accurate realization of this definition. How do we make a device that produces this countable frequency that is stable from now until forever? The answer is that it is not easy. The measurements of transitions of Cesium atoms are affected by temperature, magnetic fields, vibrations of room, interactions with container walls, and so on. Many of these effects can shift the frequency. The task of achieving high accuracy is to identify all such systematic influences and remove them.

If the systematic shift in an atomic clock is constant, then the clock can be very stable, but not accurate. If the shift varies with time, then the clock is both unstable and inaccurate. Time averaging can improve the stability of the clock if the shifts average to zero sufficiently rapidly with increased averaging time. A stability analysis always involves a time scale of measurement. Instability mechanisms have a characteristic time scale and amplitude. For example, if the clock is affected by temperature, then one would expect changes at various rates: (a) an air conditioner cycle; (b) diurnal; (c) annual. Often the amplitude of instabilities increase with an increase of time scale. It is just tougher to stabilize a clock over long scales, than over short time scales. Allan at the National Institutes of Standards & Technology developed a useful algorithm to characterize oscillator stabilities as a function of measurement interval. Consider a series of time measurements, $(t_{ij} - t_{0j})$, of clock i relative to clock 0 with measurement index j. Assume that clock 0 is perfect. Then form the Allan standard deviation:

$$\sigma_i(\tau) = \frac{< [(t_{ij+1} - t_{ij}) - (t_{ij} - t_{ij-1})]^2 >_j^{0.5}}{\sqrt{2}\tau}, \tag{29}$$

where $\tau = t_{j+1} - t_j$ is the interval between measurements. The double difference in this measure removes slopes between clocks since one is interested in stability not accuracy. The measure is also known as the fractional frequency stability since one can write the statistic in terms of fractional frequency differences, $(t_{j+1} - t_j)/\tau$. If the numerator is independent of time, the stability is limited by measurement interval only. This is often true on short time scales where a signal is being measured against a background of white, frequency independent, noise. On longer time scales unwanted, and uncontrollable influences ultimately set a limit on σ. A stability floor is reached. On longer time scales the stability typically degrades even further. This characterization is true for every oscillator ever measured: electronic crystals, atomic clocks, the rotation of the Earth, and my Casio watch (Fig. 8).

The stability floor for the best Cesium atomic time scales is around 10^{-14} on time scales of six months to a year or so (Fig. 8). The PTB scale, from the German standards laboratory, is evaluated as the best in existence. The BIPM in Paris is responsible for the collection and comparison of time scale data from around the globe. They form two time scales: International Atomic Time (TAI) and Terrestrial Time (TT-BIPM); the former is by agreement with all participants, and the latter is their formulation that they claim is the best scale for use in precise pulsar timing (Guinot and Petit 1991). A stability of 10^{-14} corresponds to a numerator of 0.4 μs in Eq. 29, or a typical timing residual of 0.3 μs for time intervals of one year. On longer time scales estimates of stability suggest a modest rise with interval. The atomic time community has made steady progress with accuracy and stability since time keeping began – even back to the Stonehenge era. Keep this limit in mind when we come to discuss the timing of an array of pulsars.

3.3. Space Coordinate

In §2.4 the transformation of the pulse arrival time event at the observatory to the corresponding arrival time event at the solar system barycenter, the local inertial reference frame was discussed. This transformation requires accurate estimation of the barycenter to observatory vector at the time, which is on an inertial clock time scale, of the observatory pulse arrival event. We use an ephemeris of the Earth's motion as a function of time to estimate this vector. The ephemeris is constructed from a detailed model of solar system dynamics that includes the Sun, all planetary systems, minor planets, and a band of matter representing the asteroid belt, and uses an appropriate level of relativistic effects (*e.g.*, Ash *et al.* 1967, Standish 1990). The parameters of this model are adjusted to fit data from *all* observations – from ancient eclipses to recent satellite planet flybys. The table below gives the current estimates of the errors in planetary masses, and the corresponding offset between the estimated barycenter location and the "true" location. These mass errors are large with respect to asteroid and cometary masses, but comparable to masses of the largest minor

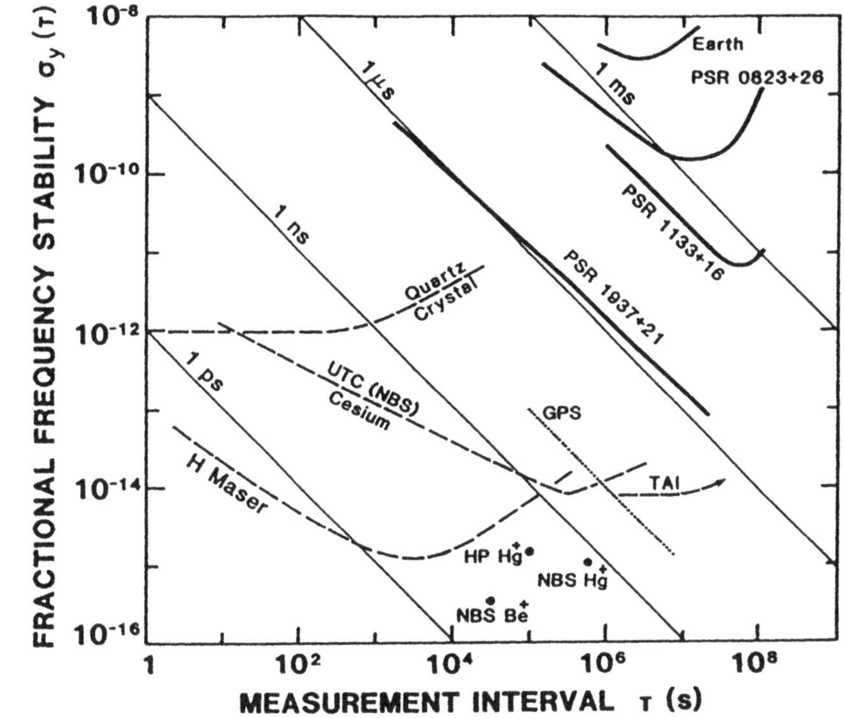

Figure 8 Fractional frequency stability of both natural (solid) and laboratory oscillators (dashed). Dotted line gives satellite time transfer stability. White noise time errors correspond to lines identified with values and slope of -1. (From Backer & Hellings 1986.)

planets and larger moons. The barycenter will also be shifted by errors in the celestial coordinates of the planets, but a list of these errors is not readily available.

We can take these mass errors and simulate their effect on pulsar timing. The arrival time transformation is effected by a scalar product between the observatory vector, which is of order 500 s – the AU – in length, and a unit vector in the direction of the source. This means that the planetary timing error for pulsar j in direction $\mathbf{n_j}$ will be

$$\delta t_{Pj}(t) = \Sigma_p \left(\frac{\Delta m_p}{\mathrm{M}_\odot}\right) \mathbf{r_p(t)} \cdot \mathbf{n_j(t)}, \tag{30}$$

where Δm_p and $\mathbf{r_p}$ are the mass error and position, respectively, of planet p. Figure 9a gives a simulation of this error for one pulsar with the Δm_p's taken as those in Table 2.

The simulation in Figure 9a is for 90 years. If we have only 10 years of data, then a large part of the variations, which are dominated by the orbits of Jupiter and Saturn, will be absorbed in fits for the pulsar epoch, spin, and spindown rate. Figure 9b shows such fits in 10-year segments. Figure 9c shows the residual that one would expect to see from the mass errors in the current ephemeris calculations. An amplitude of a fraction of one microsecond is typical. The residuals for different pulsars will appear different as a result of the dot product in Eq. 30. There will be a slow variation around the sky with a dipole signature in the residuals. The dominant temporal variation will be cubic.

Table 2

Planetary Mass Errors

System	Δm_p (10^{21} gm)	P_b (y)	r_p (AU)	$\Delta r_p/c$ (ns)	Reference
Mercury	1.4	0.25	0.39	1	Mariner 10
Venus	1.5	0.62	0.72	3	Mariner 5/10
Mars	0.2	1.88	1.52	1	Mariner 4
Jupiter	150	11.9	5.20	1950	Voyager 1/2
Saturn	300	29.5	9.52	7140	Voyager 1/2
Uranus	15	84.0	19.2	720	Voyager 2
Neptune	30	165	30.0	2250	Voyager 2

3.4. Gravitational Wave Background

3.4.1. Influence of gravitational radiation on pulsar timing

Following the review of Backer and Hellings (1986) and references therein I start with the metric of a spacetime interval as perturbed by gravitational radiation that is specified by its dimensionless strain amplitude, h:

$$ds^2 = c^2dt^2 - dz^2 - (1 + h\cos 2\psi)dx^2 - (1 - h\cos 2\psi)dy^2 - h\sin 2\psi dx dy. \tag{31}$$

This expression results from a vacuum solution of Einstein's linearized field equations. The solution corresponds to a temporal spectrum of plane gravitational radiation propagating along the $+z$-axis with amplitude $h(t - z/c)$, where h is an arbitrary function of $(t - z/c)$, and ψ is the constant polarization angle between the principal polarization direction and the x-axis. The Earth and the pulsar are assumed to be at rest. We choose the origin of the coordinate system at the Earth, and place the pulsar in the xz-plane at an angle θ from the propagation direction of the gravitational radiation (the $+z$-axis) and at a distance l from the Earth (Fig. 10).

A photon emitted at the pulsar and moving toward the Earth will follow a path in flat space that may be written as

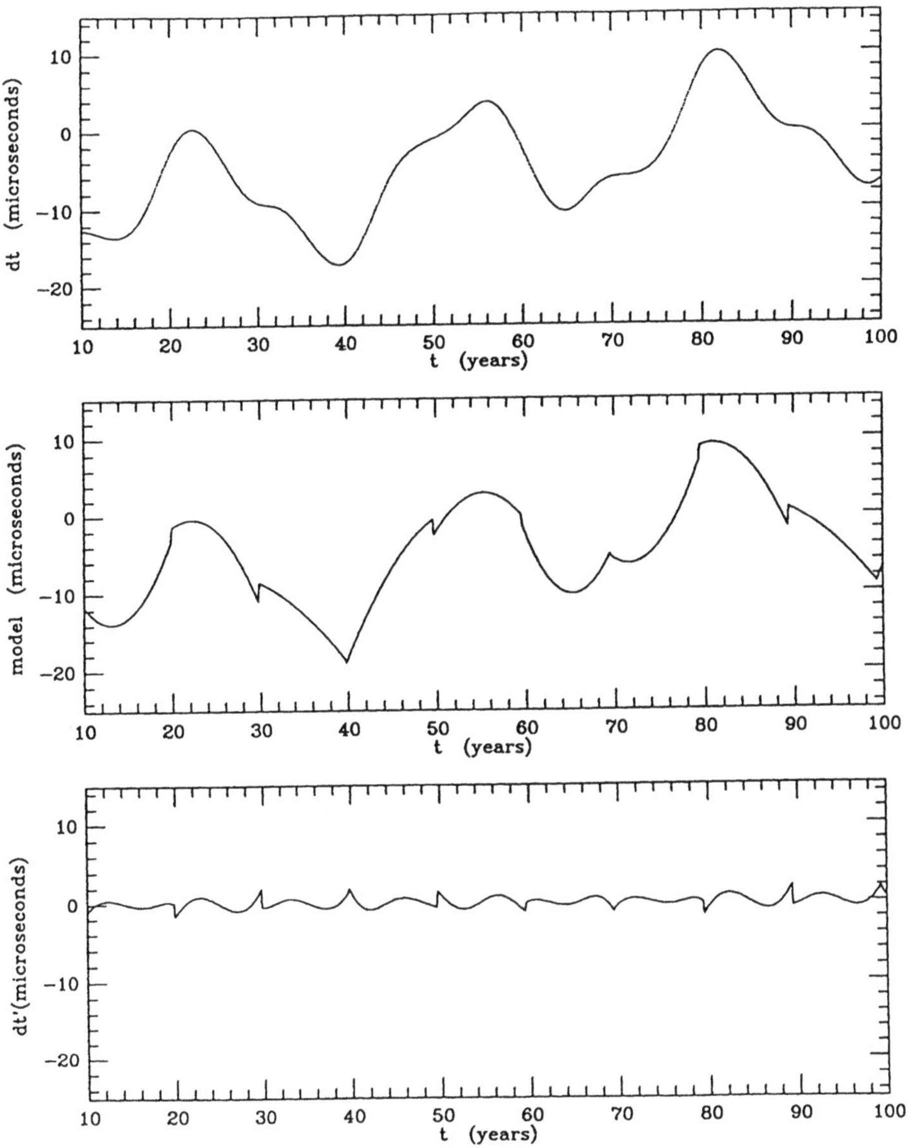

Figure 9 Simulation of timing errors that result from errors in planetary masses for PSR 0021-72. (a) The errors. (b) The results of fits to errors for pulsar spin parameters in 10-y segments. (c) The final residuals that result from subtracting (b) from (a).

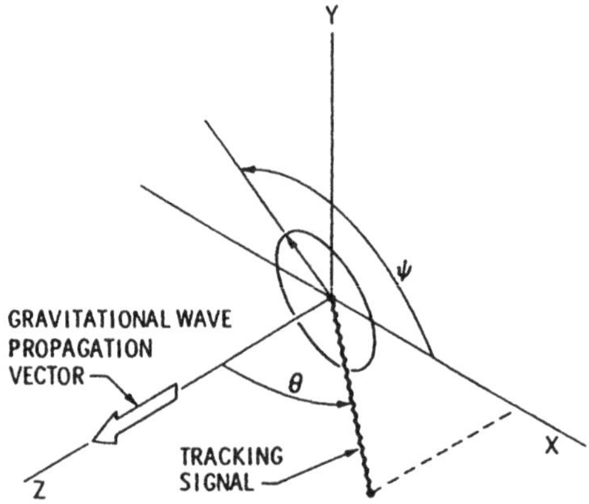

Figure 10 Geometry of observer, pulsar, and gravitational wave propagation vector. (From Backer & Hellings 1986.)

$$x = (l - s)\sin\theta, \qquad z = (l - s)\cos\theta, \tag{32}$$

where s is the distance parameter. The null geodesic may be written as

$$c^2dt^2 = (1 + h\cos 2\psi \sin^2\theta)ds^2. \tag{33}$$

Inspection shows that the influence of the gravitational radiation appears as an index of refraction, albeit one that changes with space and time, and I can immediately see that the photon propagation will be advanced or retarded compared to its vacuum path in flat space. Integration of the above equation from the spacetime event of emission at the pulsar (T, l) to the spacetime event of reception at the telescope $(t, 0)$ gives

$$c(t - T) = l + \frac{1}{2}c(1 - \cos\theta)\cos 2\psi[H(t) - H(T - l\cos\theta/c)], \tag{34}$$

where $H(u)$ is the indefinite integral of $h(u)$.

Consider the plane wave case where $h(t - z/c) = h_o \sin(\omega t - kz)$, where $k = c/\omega$. Substitution leads to

$$c(t - T) = l + \frac{1}{2}\frac{h_o c}{\omega}(1 - \cos\theta)\cos 2\psi[\cos(\omega t) - \cos(\omega t - \omega l/c(1 - \cos\theta))]. \tag{35}$$

Photons from the pulsar arrive early or late according to the product of an amplitude h_o/ω, an angular factor in (θ, ψ), and the sum of two phase factors. The second phase factor is from the incomplete phase of the wave in the vicinity of the emission event, near the pulsar, and the first is from the incomplete phase in the vicinity of the reception event, near the Earth. The first term will be correlated for all pulsars observed, as discussed later, while the second term is uncorrelated amongst the objects for all but very nearby emitters – small $\omega l\theta^2/c$.

The angular factor winds through 4π as the polarization angle moves around the sky transverse to the emission direction which demonstrates the quadrupolar signature of the radiation. As θ goes to zero the residual delay goes to zero, where the photons travel in the reverse direction of the wave. This demonstrates the transverse nature of gravitational radiation. As θ goes to π the absolute value of the residual delay increases toward 2 from the angular factor in θ, but eventually goes to zero as the two phase factors cancel exactly at π where again the photon path is parallel to the wave. For a wave period of one year and a pulsar distance of 1 kpc, this cancellation will occur only for θ much less than $0.4°$.

Pulsar timing has reached the level of 1 μs for timing several pulsars over intervals of a year. This provides a sensitity to gravitational radiation with dimensionless strains of $h_o \simeq 3 \times 10^{-14}$.

3.4.2. The Spectrum of a Stochastic Background of Gravitational Radiation

In the preceding section the effects of a single wave on the pulsar arrival time were considered. In the more general case $h(t)$ will contain a full spectrum of waves. Furthermore there will be waves coming from arbitrary directions. Consider first the spectrum. An arbitrary function $h(t)$ has the Fourier spectrum

$$\tilde{\mathbf{h}}(\omega) = \int_{-\infty}^{\infty} h(t) \exp[jwt] dt. \tag{36}$$

For a stochastic variable one is interested in the power spectral density $P_G(\omega) \equiv |\tilde{\mathbf{h}}(\omega)|^2$. If the statistics of h(t) are stationary, then the integral over the power spectrum is equal to the integral of the variance over the data (Parseval's Theorem).

$$\int_{-\infty}^{\infty} P_G(\omega) d\omega = \int_{-\infty}^{\infty} h(t)^2 dt. \tag{37}$$

The units of h are dimensionless, while those of $\tilde{\mathbf{h}}$ are t, and those of P_G are then t^2. This exercise in transforms and units is necessary if one wishes to take a segment of the spectrum and turn it into an effective wave amplitude. In the case of discretely sampled data, the corresponding relation is:

$$\Sigma_i P_{Gi} \frac{\delta\omega}{\delta t_i} = \Sigma_i h_i^2, \tag{38}$$

where δt is the sample spacing, $\delta\omega$ is the resolution of the spectrum, $1/T$, and T is the duration of the data sampled. $P_G/\delta t$ has "density" units of variance per frequency interval in the spectrum.

One assumption about the gravitational wave background is that it is scale invariant with equal energy density in each octave, or decade, of the spectrum. The energy density of a wave with amplitude h and frequency ω is given by

$$\rho_G = \frac{\dot{h}^2}{8\pi G} = \frac{h^2\omega^2}{8\pi G}. \tag{39}$$

A spectrum of waves then leads to a spectrum $\tilde{\rho}_G(\omega)$. If this spectrum contributes equal energy per octave, then $\rho_G(\omega) = \rho_o\omega_o/\omega$. The constant $\rho_o\omega_o$ is referred to as the logarithmic spectrum because one integrates it with $d\ln\omega$ which is dimensionless.

The results of §3.4.1 can now be combined with this spectrum discussion to convert an arrival time precision over a time interval T into a limit on the gravitational wave background logarithmic energy density spectrum. Consider timing residuals with a variance of 1 μs over ten years. If sampled ten times per year, then these would produce a discrete power spectral density with amplitude 2 μs^2 y, *i.e.*, 2 μs^2 per frequency interval of a cycle per year. This corresponds to a power spectral density $P_G/\delta t$ of about 10^{-19} at one cycle per year frequency. The energy density per cycle per year of the spectrum is then 2.5×10^{-26} gm cm^{-3}. And finally the energy density per logarithmic interval is 2×10^{-35} gm cm^{-3}. This value scales with the amplitude of the residuals squared, which follows directly from Eqs. 35,39. The value also scales with frequency to the fourth power: two factors from Eq. 39, one from the conversion of timing residual into wave amplitude (Eq. 35) and one from expressing the results in terms of a logarithmic spectrum.

The energy density of background radiations are typically expressed in terms of the density that is required to close the universe, $\rho_c = 2 \times 10^{-29}$ gm cm^{-3}:

$$\Omega_G = \frac{\rho_G(\omega)\omega}{\rho_c} \approx 10^{-6}\frac{R(\mu s)^2}{T(y)^4}, \tag{40}$$

where a Hubble constant of 100 km s^{-1} Mpc^{-1} is assumed in the computation of ρ_c, and R is the timing residual defined in §2.6. Present limits on Ω_G from pulsar timing are presented in §3.5.

3.4.3. Sources of the Stochastic Gravitational-Wave Background

In this discussion the focus is on sources from energetic phenomena in the early universe. The origin of the stochastic background can be discussed in a number of ways. One classification of sources is by their temporal character: short bursts from isolated, violent astrophysical events, quasi-periodic sources from rapidly spinning or orbiting bodies, and the broadband, stochastic background. Carr (1980) summarizes the possible contributions according to whether the source is *primordial*, by which he means having its nature related to the structure of the early universe, or *generated*, by which he means some object forms and then radiates gravitational radiation according to the quadrupole formula. A related categorization simply uses the epoch, or temperature, of the universe at the time the stochastic background is formed.

Let's start with the primordial sources and work forward in time through the evolution of the detectable universe. A thermal distribution of gravitons from the Planck era is possible. If this

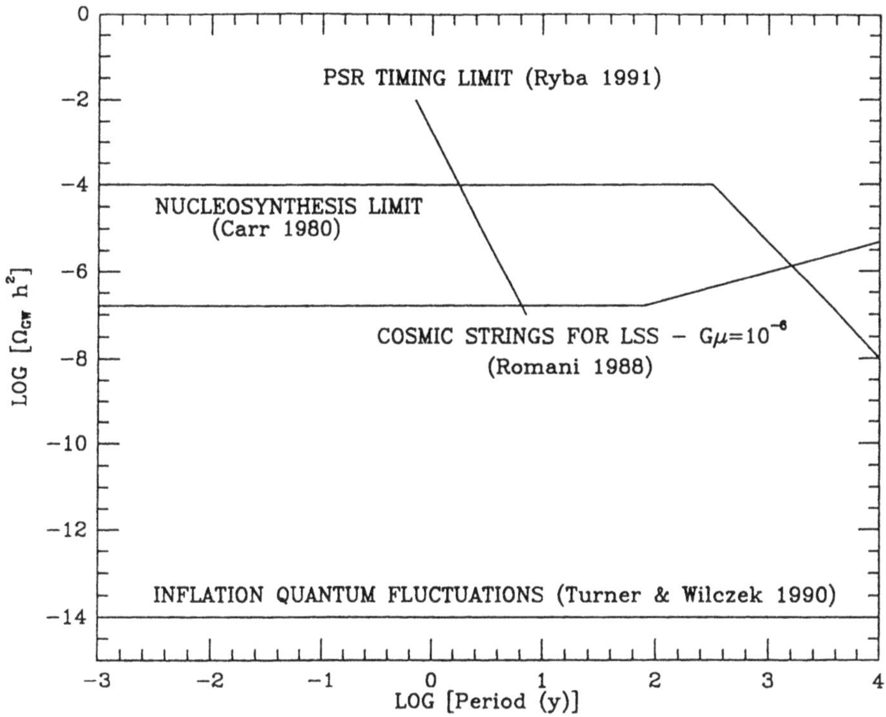

Figure 11 The logarithmic gravitational wave background spectrum. Estimates of the background resulting from dissipation of cosmic strings that might be seeds for large scale structure and galaxy formation and from inflation quantum fluctuations are shown. Limits from nucleosynthesis and pulsar timing are given.

spectrum decouples and evolves into the present era, then it will be redshifted down below the microwave background spectrum to wavelengths of order 3 mm, or a temperature of 1 K. This is not of interest to the pulsar timing measurement. Grishchuk (1977) has given arguments for a nonthermal spectrum. The existence of a stochastic background of gravitational radiation at this early epooch creates an anisotropy which, if sufficiently large, would affect nucleosynthesis of helium and other light elements. Carr (1980) places a limit of $\Omega_g = 10^{-4}$ from this consideration for periods ranging from 1 s to 300 y (Fig. 11). Further limits on the spectrum resulting from early epoch anisotropy can be placed using the isotropy of the microwave background. Gravitational radiation with amplitude h_o, in the present era can induce a quadrupole term in the microwave background comparable to h_o, and radiation at the time of recombination can produce smaller scale anisotropies.

The next class of primordial sources occur during the QCD transition from the false-vacuum phase to the radiation-dominated phase, and are a mix between Carr's primordial and generated sources. Starobinskii (1979) considered the production of gravitational radiation by structural inhomogeneities during this transition interval, and came up with a broadband, scale invariant spectrum. Witten (1984) discusses processes that occur at the beginning of the radiation era where bubbles of low-temperature phase, quark nuggets, amidst the high-temperature phase will collide, and summarizes the principle sources of the stochastic background in an excellent appendix. Turner and Wilczek (1991) have developed the ideas of gravitational-wave production by nucleating and percolating bubbles during ***extended inflation*** era between phases. Their spectrum is not very promising for the pulsar timing experiment (Fig. 11). Other sources have been considered which involve nucleation of the low temperature phase by impurities such as magnetic monopoles, cosmic strings and texture. Furthermore black holes could be produced in this era whose collisions and evaporation could add to the stochastic background.

Vilenkin (1981) developed the spectrum that would be produced by a cosmic strings that come out of the phase transition era and survive to act as seeds for galaxy formation. These topological defects in spacetime form closed oscillating loops and decay be emission of gravitational radiation. Matter falls into the loop and remains after the decay. Vilenkin's spectrum is divided into three parts according to whether the loop forms and decays wholely in the radiation era, or forms in the radiation era and decays in the matter era, or forms and decays wholely in the matter era. Clearly holy stuff! The strings form on the scale of the horizon and then evolve with twists and connections to form loops. Various authors have made detailed models of the evolution of a network of strings with both galaxy formation as one result and observable consequences in the microwave and gravitational wave backgrounds as the constraint (Bouchet and Bennet, 1990).

Finally let me turn to a post-galactic source of generated radiation. The standard model for quasars, and active galactic nuclei in general, is mass accretion onto a massive black hole, $M = 10^{6-9} M_{\odot}$. Now galaxies and presumably quasars live in clusters where tidal interactions and collisions are likely during the lifetime of the system. There is considerable evidence for galaxy mergers, cosmic cannabalism. The existence of massive, binary black holes is a clear possiblity. Sillanpaa *et al.* (1989) have recently discussed the case for a binary black hole driving the nearly periodic outbursts of the radio source OJ287 with interval of 10 y. Begelman *et al.* (1980) have previously looked at the mechanism and time scales for binary black hole production, while Blandford (1979) considered the coalescence rate for massive black holes in the universe as a source of short wavelength gravitational radiation. The new lines of evidence in favor of energetic events in galactic nuclei suggest that the contribution of these objects to the stochastic background needs to be reconsidered. In particular, the accretion of smaller black holes into orbit around a principle black hole is a likely scenario given the observations. A single source, as calculated by Detweiler (1979), will produce a dimensionless strain of

$$h \simeq 2 \times 10^{-17} (\frac{1+z}{n})^{0.67} (\frac{T}{\mathrm{y}})^{-0.67} (\frac{D}{\mathrm{Gpc}})^{-1} (\frac{m_p}{10^8 \mathrm{M}_{\odot}})^{1.67} f(\frac{m_s}{m_p}, n, e), \tag{41}$$

where z is the redshift, n is the harmonic number, T is the observed time scale which is connected back to the binary period using the redshift and harmonic number, D is the distance, m_p is the primary black hole mass, m_s is the secondary black hole mass, and f is a function of order unity. A hole of mass 10^9 would lead to a 30 ns effect over one year period. While this is currently unobservable, a contribution of 100 such objects emitting with random periods and phases would bring the effect up to observable levels.

3.4.4. Basis Function for Pulsar Timing Array Measurements

The program of timing an array of millisecond pulsars for the detection of a background of gravitational radiation requires a set of basis functions which can be used to fit to the data. There are five degrees of freedom in the perturbations of the space time metric that result from an arbitrary form of gravitational radiation. In §3.4.2 a wave along one direction with an arbitrary polarization was analyzed. The effects of three waves can then be superposed to obtain the net effect of an arbitrary signal.

The formulation of Detweiler (1979) is most useful for computation of the angular distribution of the perturbation that each degree of freedom has on the pulsar data. His formulation gives the Doppler shift for a wave incident along the z axis with orthogonal polarization components h_+, $h_\times$ as a function of the direction cosines of the pulsar α, β, γ.

$$z = \frac{(\alpha^2 - \beta^2)h_+ + 2\alpha\beta h_\times}{2(1+\gamma)}. \tag{42}$$

The Doppler shift is just the time derivative of the timing residuals, $R(t)$.

The perturbations of the space-time metric are then formed by permuting Eq. 41 for waves along the x, y, z axes:

$$h_{\mu\nu} = \begin{pmatrix} h_{x+} & h_{x\times} & 0 \\ h_{x\times} & -h_{x+} & 0 \\ 0 & 0 & 0 \end{pmatrix} + \begin{pmatrix} h_{y+} & 0 & h_{y\times} \\ 0 & 0 & 0 \\ h_{y\times} & 0 & -h_{y+} \end{pmatrix} + \begin{pmatrix} 0 & 0 & 0 \\ 0 & h_{z+} & h_{z\times} \\ 0 & h_{z\times} & -h_{z+} \end{pmatrix}, \tag{43}$$

which sums to

$$h_{\mu\nu} = \begin{pmatrix} h_{x+} + h_{y+} & h_{x\times} & h_{y\times} \\ h_{x\times} & h_{z+} - h_{x+} & h_{z\times} \\ h_{y\times} & h_{z\times} & -h_{y+} - h_{z+} \end{pmatrix}. \tag{44}$$

This final matrix is symmetric and trace free. The five degrees of freedom are then h_{11}, h_{12}, h_{13}, h_{22}, and h_{23} with $h_{33} = -h_{11} - h_{22}$. The instantaneous Doppler shift patterns can then be computed for each degree of freedom from the constituent parts that are expressed by Detweiler's equation. Fig. 12 displays the h_{11} pattern. Note that there is considerable angular structure in this pattern beyond the quadrupole. The signature of gravitational radiation on an array of pulsars will be distinct from that of the other effects – the clock which has a monopole angular signature and the earth orbit which has a dipole angular signature.

This discussion has avoided mention of the temporal signature of the various effects. Only time variations of any effect will be detectable owing to the fact that we do not have an absolute measure of the pulsar distance. Furthermore linear and quadratic effects will also be absorbed into the determination of the pulsar rotation parameters which are also unknown *a priori*. Only cubic and higher order temporal modulations will be available for fitting to these global effects on the pulsar timing array (Foster & Backer 1990). In the case of the Earth orbital perturbations the angular and temporal basis function for various solar system parameters will be available because the errors arise in limited precision of known physical effects.

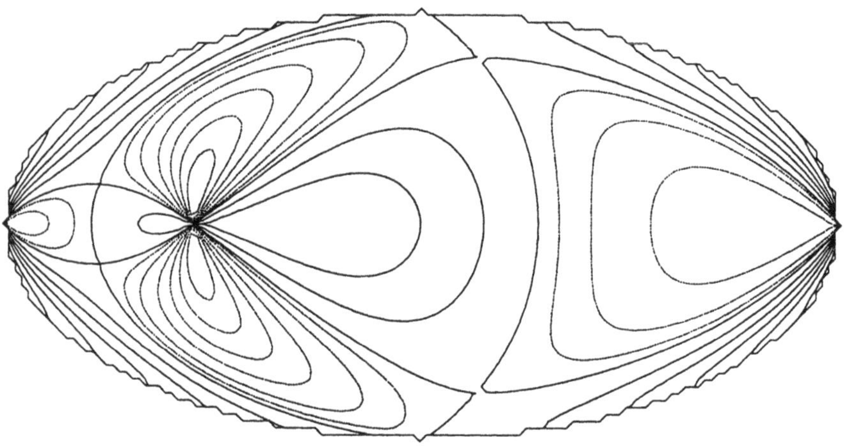

Figure 12 The Doppler shift pattern that results from gravitational wave background component h_{11} in Eq. 44. Celestial coordinate system is arbitrary, but can be thought of as right ascension of 360° horizontally and declination of 180° vertically.

3.5. Pulsar Timing Array Experiments

The goal of timing an array of millisecond period pulsars is just coming within our reach with the experiments in progress. The most detailed study has involved timing the original millisecond pulsar PSR 1937+21 since the installation of improved data acquisition hardware in late 1983 (Stinebring, Ryba, & Taylor 1990; Ryba 1990). Since 1985 data was recorded at two radio frequencies so that dispersion measure variations could be removed. Data from a single pulsar can be used to place a limit on the gravitational wave background spectrum as outlined in §3.4.2. The importance of the results presented by Stinebring *et al.* is that they show a "signal" in that the residuals are not consistent with white noise. The residuals appear to be dominated by a cubic term in phase over the 5 years of dual frequency data. The amplitude of this cubic term is about 1 μs. The origin of this effect is unknown. There are a number of candidates. The atomic time scales available are no more stable than 10^{-14} over durations of a few years (Fig. 8). This corresponds to a residual of 1 μs which is comparable to the observed timing residual. The ephemeris of the Earth's motion as presently provided does not include the latest values of the outer planetary masses. The resultant contribution of the ephemeris to timing residuals is then worse than that presented in Figure 9 and Table 2; a microsecond of ephemeris error is also possible. Stinebring *et al.* attribute the systematic residual to a combination of intrinsic timing noise (§2.7) and interstellar propagation effects (§2.5). Most of their discussion is reserved for the important new limit that they can establish on the gravitational wave background,

$$\Omega_G < 9 \times 10^{-8} \quad \text{for frequencies near} \quad 0.14 \quad y^{-1} \tag{45}$$

with 68% confidence. Ryba (1990) improves on these limits by a factor of four with somewhat extended data span and improved data analysis.

$$\Omega_G < 2.5 \times 10^{-8} \quad \text{for frequencies near} \quad 0.14 \quad y^{-1} \tag{46}$$

with 68% confidence.

Table 3

Millisecond Pulsar Timing Array Programs

Observatory	Investigator	Pulsars
Arecibo	Taylor	1855+09, 1937+21, 1953+29, 1957+20
	Wolszczan	1257+12, 1534+12, 2127+15A
Green Bank	Backer	1257+12, 1620-26, 1821-24, 1855+09, 1937+21
Nançay	Lestrade	1821-24, 1937+21
VLA	Taylor	1937+21
Usada	Hirao	1937+21
Kashima	Imae	1937+21

A number of programs are underway at several observatories to time the best millisecond pulsar "clocks". These are summarized below in Table 3. Other observatories such as Parkes and Jodrell Bank are also involved in millisecond pulsar timing, but with less emphasis on pulsar timing array aspects. I started a pulsar timing array experiment in 1987 shortly after the 3-ms pulsar in the globular cluster M28 was discovered. Our first results have been described by Foster (1990) and Foster & Backer (1990). We have made steady progress in sensitivity improvements, and have started dual frequency observations that allow removal of effects of dispersion measure variations. Timing residuals for three of the pulsars in our program are displayed in Figure 13. We are continuing to explore methods of improving the sensitivity of this experiment by improved receivers and new signal processing hardware. The goal is to have timing residuals at or below one microsecond for a modest number of pulsars distributed across the sky.

The present status is that we are starved for objects distributed around the sky. Pulsars in the globular cluster 47 Tucanae (0021-72) may help in this distribution once their timeability is established. The high latitude pulsar PSR 1257+12 seemed to be an ideal object owing to its location and its flux density. However observations now demonstrate that it has at least two planetary mass companions around it and possibly more. This circumpulsar "space junk" would seem to ruin it as a good clock, while making the astrophysics of the system a delight! Searches are underway to discover more objects throughout the sky. The future is reasonably bright that a significant number of new clocks will be found.

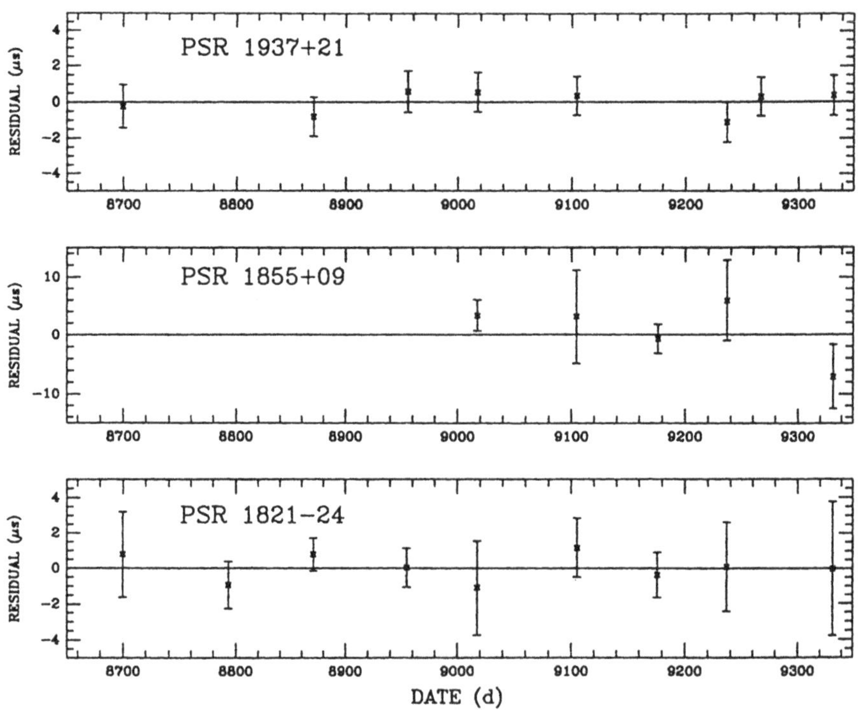

Figure 13 Pulsar timing array data from experiment in progress at the NRAO 140ft telescope. The display contains timing residuals for PSRs 1821-24, 1855+09 and 1937+21. The 6-parameter model that is fit to the PSR 1821-24 data leave 3 degrees of freedom. The PSR 1855+09 data is compared to the model derived by Ryba (1991) with only a relative epoch removed. The PSR 1937+21 data is also compared to a model from Ryba (1991) with a relative epoch and a small period difference removed.

Acknowledgements

This discussion of pulsar timing and related subjects is a summary of the work of many scientists over the past 20 years. I have given credit when specifically needed, but not for the general development of the field.

REFERENCES

J. Arons, *Sp. Sci. Rev.* **24** (1979) 437.

J. Arons, *Ap. J.* **266** (1983) 213.

J. Arons, in *The Magnetospheric Structure and Emission Mechanisms of Radio Pulsars*, ed. T. H. Hankins, J. M. Rankin, and J. A. Gil, (Pedagogical University Press, Zielona Góra, 1991), p. 62.

M. E. Ash, I. I. Shapiro, and W. E. Smith, *Astron. J.* **72** (1967) 338.

N. Ashby and D. W. Allan, *Radio Sci.* **14** (1979) 649.

D. C. Backer, S. R. Kulkarni, C. Heiles, M. M. Davis, and W. M. Goss, *Nature* **300** (1982) 615.

D. C. Backer and R. W. Hellings, *Ann. Rev. Astron. Ap.* **24** (1986) 537.

D. C. Backer, in *Galactic and Extragalactic Radio Astronomy*, ed. K. Kellermann and G. Verschuur, (Springer-Verlag, Berlin, 1988), p. 480.

D. C. Backer, see Ögelman and van den Heuvel (1989), p.3.

D. C. Backer and S. R. Kulkarni, *Physics Today* **43** (1990) 26.

M. Bailes, A. G. Lyne, and S. L. Shemar, *Nature* **352** (1991) 311.

M. C. Begelman, R. D. Blandford, and M. J. Rees, *Nature* **287** (1980) 307.

D. P. Bennet and F. R. Bouchet, *Phys. Rev. D* **41** (1990) 720.

D. Bhattacharya and E. P. J. van den Heuvel, *Phys. Rep.* **203** (1991) 1.

R. D. Blandford, in *Sources of Gravitational Radiation*, L. L. Smarr, ed., (Cambridge, Cambridge, 1979).

R. D. Blandford and S. A. Teukolsky, *Ap. J.* **205** (1976) 580.

V. Boriakoff, R. Buccieri, and F. Fauci, *Nature* **304** (1983) 417.

B. J. Carr, *Astron. Astrophys.* **89** (1980) 6.

G. M. Clemence and V. Szebehely, *Astron. J.* **72** (1967) 1324.

T. Damour and J. H. Taylor, *Ap. J.* **366** (1991a) 501.

T. Damour and J. H. Taylor, 1991b preprint.

S. Detweiler, *Ap. J.* **234** (1979) 1100.

O. V. Doroshenko and S. M. Kopejkin, *Sov. Astron.* **34** (1990) 496.

R. S. Foster, Ph. D. thesis, University of California, Berkeley (1990).

R. S. Foster and D. C. Backer, *Ap. J.* **361** (1990) 300.

A. S. Fruchter, D. R. Stinebring, and J. H. Taylor, *Nature* **333** (1988) 237.

P. Goldreich and W. H. Julian, *Ap. J.* **157** (1969) 869.

L. P. Grishchuk, *Ann. NY Acad. Sci.* **302** (1977) 439.

B. Guinot and G. Petit, *Astron. Astrophys.* **248** (1991) 292.

J. E. Gunn and J. P. Ostriker, *Nature* **221** (1969) 454.

T. H. Hankins, and B. J. Rickett, *Methods in Comp. Physics*, **14** (Academic Press, New York, 1975)

D. J. Helfand, R. N. Manchester, and J. H. Taylor, *Ap. J.* **198** (1975) 661.

R. W. Hellings, *Astron. J.* **91** (1986) 650.

A. Hewish, S. J. Bell, J. D. M. Pilkington, P. F. Scott, and R. A. Collins, *Nature* **217** (1968) 709.

R. A. Hulse and J. H. Taylor, *Ap. J. (Lett.)* **195** (1975) L51.

J. Jesperson and D. W. Hanson, *Proc. IEEE* **79** (1991) No.7.

F. C. Michel, *Theory of Neutron Star Magnetospheres*, (University of Chicago, Chicago, 1991).

R. N. Manchester and J. H. Taylor, *Pulsars*, (W. H. Freeman, San Francisco, 1977).

H. Ögelman, and E. P. J. van den Heuvel, ed., *NATO ASI C262 : Timing Neutron Stars*, (Kluwer, Dordrecht, 1989).

V. Radhakrishnan and D. J. Cooke, *Ap. Lett.* **3** (1969) 225.

D. S. Robertson, W. E. Carter, and W. H. Dillinger, *Nature* **349** (1991) 768.

R. W. Romani, *Phys. Lett. B* **215** (1988) 477.

R. W. Romani, in Ögelman & van den Heuvel (1989), p. 113.

M. Ruderman, J. Shaham, and M. Tavani, *Ap. J.* **336** (1989) 507.

M. F. Ryba, Ph. D. thesis, Princeton University, 1990.

J. Sauls, in Ögelman & van den Heuvel (1989), p. 457.

S. L. Shapiro, and S. A. Teukolsky, *Black Holes, White Dwarfs, and Neutron Stars*, (Wiley, New York, 1983).

I. I. Shapiro, *Phys. Ref. Lett.* **13** (1964) 789.

A. Sillanpää S. Haarala, M. J. Valtonen, B. Sundelius, and G. G. Byrd, *Ap. J.* **325** (1988) 628.

A. A. Starobinskii, *JETP Lett.* **30** (1979) 682.

D. R. Stinebring, M. F. Ryba, J. H. Taylor, and R. W. Romani, *Phys. Rev. Lett.* **65** (1990) 285.

J. H. Taylor and D. R. Stinebring, *Ann. Rev. Astron. Astrophys* **24** (1986) 285.

J. H. Taylor, see Ögelman and van den Heuvel (1989), p. 17.

J. H. Taylor, L. A. Fowler, and P. M. McCulloch, *Nature* **277** (1979) 437.

J. H. Taylor, A. Wolszczan, T. Damour, and J. M. Weisberg, 1991, preprint.

K. S. Thorne, R. H. Price, and D. A. Macdonald, ed., *Black Holes: The Membrane Hypothesis*, (Yale, New Haven, 1986).

M. S. Turner and F. Wilczek, preprint (1990).

A. Vilenkin, *Phys. Lett.* **107B** (1981) 47.

C. M. Will and K. Nordvedt, *Ap. J.* **177** (1972) 757.

J. M. Weisberg, R. W. Romani, and J. H. Taylor, *Ap. J.* **347** (1989) 1030.

E. Witten, *Phys. Rev. D* **30** (1984) 272.

A. Wolszczan, 1991a, in preparation.

A. Wolszczan, *Nature* **350** (1991b) 688.

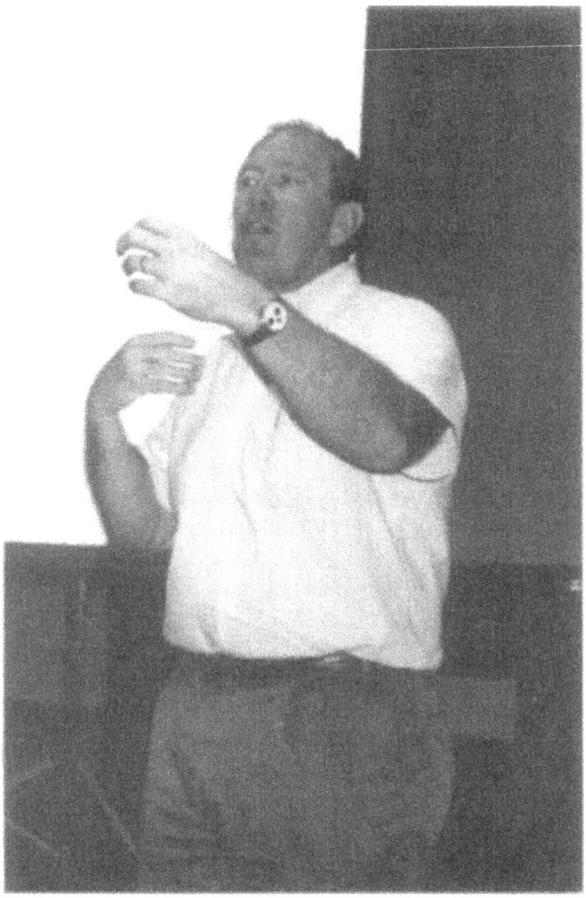

DAVID SCHRAMM

TESTING THE BIG BANG: LIGHT ELEMENTS, NEUTRINOS, DARK MATTER AND LARGE-SCALE STRUCTURE

DAVID N. SCHRAMM
The University of Chicago,
5640 S. Ellis Avenue, Chicago, IL 60637
and
NASA/Fermilab Astrophysics Center,
Fermi National Accelerator Laboratory
Box 500, Batavia, IL 60510-0500

ABSTRACT

In this series of lectures, several experimental and observational tests of the standard cosmological model are examined. In particular, detailed discussion is presented regarding (1) nucleosynthesis, the light element abundances and neutrino counting; (2) the dark matter problems; and (3) the formation of galaxies and large-scale structure. Comments will also be made on the possible implications of the recent solar neutrino experimental results for cosmology. An appendix briefly discusses the 17 keV "thing" and the cosmological and astrophysical constraints on it.

Introduction

These lectures will examine several topics where the modern cosmological model is being experimentally and/or observationally examined. Some specific areas are: (1) nucleosynthesis and neutrino counting; (2) the dark matter problems; and (3) the formation of galaxies and large-scale structure. Comments will also be made on the possible implications of the recent solar neutrino experimental results for cosmology. An appendix will discuss the $17keV$ "thing" and the astrophysical and cosmological constraints upon its properties. However, before going into these specific topics, let us first note the strength of the basic big bang framework.

While Hubble's work in the 1920's established an expanding universe, the establishment of modern physical cosmology begins with the predictions of Gamow and his colleagues.

The basic mathematical space-time framework for the model that is now known as the Big Bang dates to the mid 1920s and the work of Alexander Friedman. However, Chandrasekhar has assured me that the prevailing cosmological picture in the 1930s was that the Hubble expansion started from a quasi-static Lemaitre-like model where galaxies never were squeezed together to form a different phase of matter. In particular, the hot big bang model hinges on two key quantitative observational tests: (1) the microwave background, and (2) big bang nucleosynthesis (BBN) and the light element abundances. This paper will focus on the second of these since that is more directly connected to high energy physics. However, it is worth noting that just as the new COBE[1] results have given renewed confidence in the $3K$ background argument, the LEP collider (along with the SLC) has given us renewed confidence in the BBN arguments. We will return to this point momentarily. Note also that the microwave background probes events at temperatures $\sim 10^4 K$ and times of $\sim 10^5$ years, whereas the light element abundances probe the Universe at temperatures $\sim 10^{10} K$ and times of ~ 1 sec. Thus, it is the nucleosynthesis results that played the most significant role in leading to the particle-cosmology merger that has taken place this past decade.

Since the popular press sometimes presents misleading headlines implying *doubts* about the big bang, it is important to note here that the real concerns referred to in these articles are really in regard to observations related to models of galaxy and structure formation. The basic hot big bang model itself is in fantastic shape with high accuracy confirmations from COBE and, as we will discuss, nucleosynthesis. However, there is admittedly no fully developed model for galaxy and structure formation that fits all of the observations. (But, of course, there is also no fully developed first principles model for star formation either.) That we might not really know exactly how to make galaxies and large-scale structure in no way casts doubt on the hot, dense early universe which we call the big bang. (We also have trouble predicting earthquakes and tornadoes, but that hasn't meant that we question celestial mechanics or a round Earth.) We will return to the problems of galaxy and structure formation towards the end of these lectures.

Before going into the specific argument as to the relationship of BBN to neutrino counting, let us review the history of BBN. This will draw heavily on other recent conference proceedings.[2]

History of Big Bang Nucleosynthesis

It should be noted that there is a symbiotic connection between BBN and the $3K$ background dating back to Gamow and his associates, Alpher and Herman. The initial BBN calculations of Gamow and his associates[3] assumed pure neutrons as an initial condition and thus were not particularly accurate, but their inaccuracies had little effect on the group's predictions for a background radiation.

Once Hayashi recognized in 1950 the role of neutron-proton equilibration, the framework for BBN calculations themselves has not varied significantly. The work of Alpher, Follin and Herman[4] and Taylor and Hoyle,[5] preceeding the discovery of the 3K background, and Peebles[6] and Wagoner, Fowler and Hoyle,[7] immediately following the discovery, and the more recent work of our group of collaborators[8,9,10,11,12] all do essentially the same basic calculation, the results of which are shown in Figure 1. As far as the calculation itself goes, solving the reaction network is relatively simple by the standards of explosive nucleosynthesis calculations in supernovae (c.f. the 1965 calculation of Truran *et al.*),[13] with the changes over the last 25 years being mainly in terms of more recent nuclear reaction rates as input, not as any great calculational insight (although the current Kawano/Walker code[11,12] is somewhat streamlined relative to the earlier Wagoner code[7]). With the possible exception of

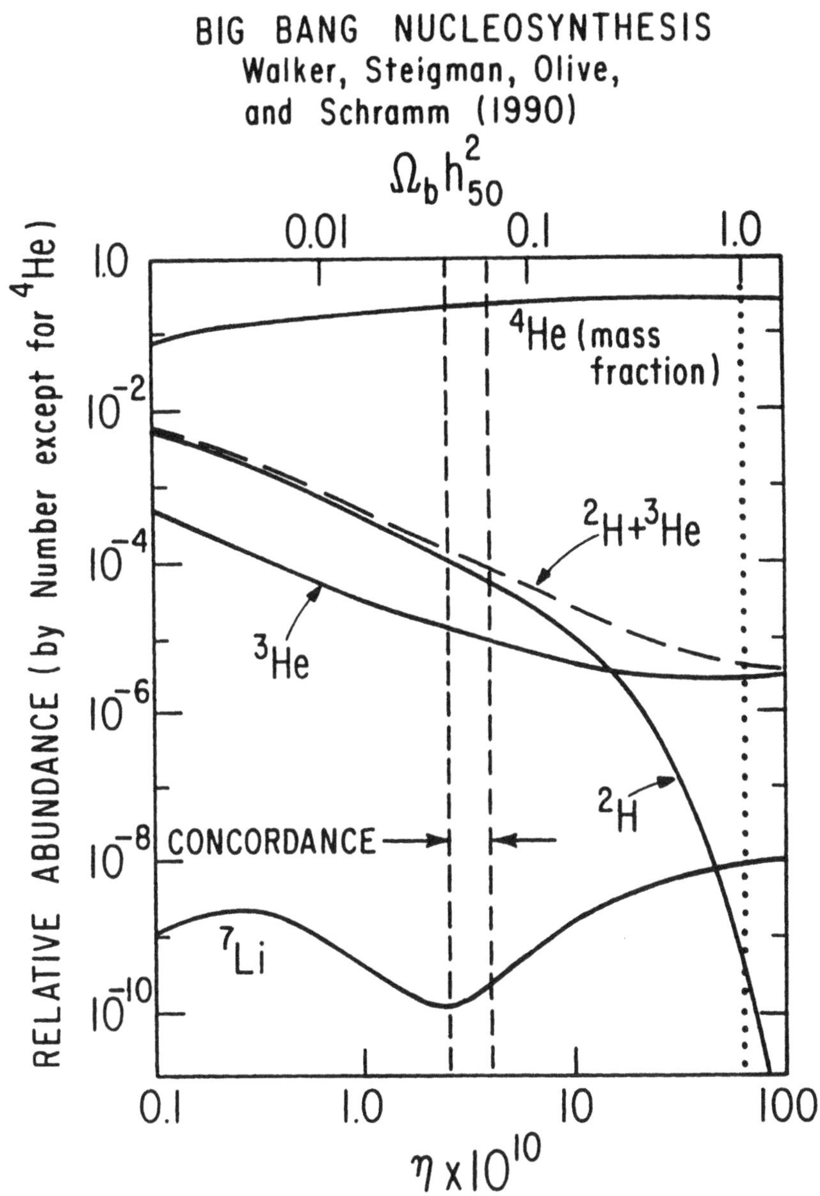

Figure 1. BBN abundances versus the baryon to photon ratio, η, or equivalently the fraction of the critical density, Ω_b.

^7Li yields, the reaction rate changes over the past 25 years have not had any major affect.[9,11,12,13] The one key improved input is a better neutron lifetime determination, a point to which we will also return shortly.

With the exception of the effects of elementary particle assumptions to which we will also return, the real excitement for BBN over the last 25 years has not really been in redoing the basic calculation. Instead, the true action is focused on understanding the evolution of the light element abundances and using that information to make powerful conclusions. In particular, in the 1960's, the main focus was on ^4He which is very insensitive to the baryon density. The agreement between BBN predictions and observations helped support the basic big bang model but gave no significant information at that time with regard to density. In fact, in the mid-1960's, the other light isotopes (which are, in principle, capable of giving density information) were generally assumed to have been made during the T-Tauri phase of stellar evolution,[15] and so, were not then taken to have cosmological significance. It was during the 1970's that BBN fully developed as a tool for probing the universe. This possibility was in part stimulated by Ryter *et al.*[16] who showed that the T-Tauri mechanism for light element synthesis failed. Furthermore, ^2H abundance determinations[17,18] improved significantly with solar wind measurements and the interstellar work from the Copernicus satellite. Reeves, Audouze, Fowler and Schramm[19] argued for cosmological ^2H and were able to place a constraint on the baryon density excluding a universe closed with baryons. Subsequently, the ^2H arguments were cemented when Epstein, Lattimer and Schramm[20] proved that no realistic astrophysical process other than the big bang could produce significant ^2H. It was also interesting that the baryon density implied by BBN was in good agreement with the density implied by the dark galactic halos.[21]

By the late 1970's, a complimentary argument to ^2H had also developed using ^3He. In particular, it was argued[22] that, unlike ^2H, ^3He was made in stars; thus, its abundance would increase with time. Since ^3He like ^2H monotonically decreased with cosmological baryon density, this argument could be used to place a lower limit on the baryon density[23] using ^3He measurements from solar wind[17] or interstellar determinations.[24] Since the bulk of the ^2H was converted in stars to ^3He, the constraint was shown to be quite restrictive.[9] Support for this point[25] also comes from the observation of ^3He in horizontal branch stars which, as processed stars still having ^3He on their surface, indicates the survivability of ^3He.

It was interesting that the lower boundary from ^3He and the upper boundary from ^2H yielded the requirement that ^7Li be near its minimum of $^7Li/H \sim 10^{-10}$, which was verified by the Pop II Li measurements of Spite and Spite,[26] hence, yielding the situation emphasized by Yang *et al.*[9] that the light element abundances are consistent over nine orders of magnitude with BBN, but only if the cosmological baryon density is constrained to be around 6% of the critical value. It is worth noting that ^7Li alone gives both an upper and a lower limit to Ω_b. However, while its derived upper limit is more than competitive with the ^2H limit, the ^7Li lower limit is not nearly as restictive as the $^2H +^3 He$ limit. Claims that big bang nucleosynthesis can yield Ω_b lower than 0.01 must necessarily neglect the $^3He +^2 H$

limit.

The other development of the 70's for BBN was the explicit calculation of Steigman, Schramm and Gunn,[27] showing that the number of neutrino generations, N_ν, had to be small to avoid overproduction of ^4He. This will subsequently be referred to as the SSG limit. (Earlier work had noted a dependency of the ^4He abundance on assumptions about the fraction of the cosmological stress-energy in exotic particles,[28,5] but had not actually made an explicit calculation probing the quantity of interest to particle physicists, N_ν.) To put this in perspective, one should remember that the mid-1970's also saw the discovery of charm, bottom and tau, so that it almost seemed as if each new detector produced new particle discoveries, and yet, cosmology was arguing against this "conventional" wisdom. Over the years the SSG limit on N_ν improved with ^4He abundance measurements, neutron lifetime measurements and with limits on the lower bound to the baryon density; hovering at $N_\nu \lesssim 4$ for most of the 1980's and dropping to slightly lower than 4[29,30,10] just before LEP and SLC turned on.

Big Bang Nucleosynthesis: Ω_b and N_ν

The power of big bang nucleosynthesis comes from the fact that essentially all of the physics input is well determined in the terrestrial laboratory. The appropriate temperature regimes, 0.1 to $1MeV$, are well explored in nuclear physics labs. Thus, what nuclei do under such conditions is not a matter of guesswork, but is precisely known. In fact, it is known for these temperatures far better than it is for the centers of stars like our sun. The center of the sun is only a little over $1keV$, thus, below the energy where nuclear reaction rates yield significant results in laboratory experiments, and only the long times and higher densities available in stars enable anything to take place.

To calculate what happens in the big bang, all one has to do is follow what a gas of baryons with density ρ_b does as the universe expands and cools. As far as nuclear reactions are concerned, the only relevant region is from a little above $1MeV$ ($\sim 10^{10}K$) down to a little below $100keV$ ($\sim 10^9K$). At higher temperatures, no complex nuclei other than free single neutrons and protons can exist, and the ratio of neutrons to protons, n/p, is just determined by

$$n/p = e^{-Q/T},$$

where $Q = (m_n - m_p)c^2 \sim 1.3MeV$.
Equilibrium applies because the weak interaction rates are much faster than the expansion of the universe at temperatures much above $10^{10}K$. At temperatures much below 10^9K, the electrostatic repulsion of nuclei prevents nuclear reactions from proceeding as fast as the cosmological expansion separates the particles.

Because of the equilibrium existing for temperatures much above $10^{10}K$, we don't have to worry about what went on in the universe at higher temperatures. Thus, we can start our calculation at $10MeV$ and not worry about speculative physics like the theory of everything (T.O.E.), or grand unifying theories (GUTs), as long as a gas of neutrons and protons exists in thermal equilibriuim by the time the universe has cooled to $\sim 10MeV$.

After the weak interaction drops out of equilibrium, a little above $10^{10}K$, the ratio of neutrons to protons changes more slowly due to free neutrons decaying to protons, and similar transformations of neutrons to protons via interactions with the ambient leptons. By the time the universe reaches 10^9K $(0.1MeV)$, the ratio is slightly below 1/7. For temperatures above 10^9K, no significant abundance of complex nuclei can exist due to the continued existence of gammas with greater than MeV energies. Note that the high photon to baryon ratio in the universe ($\sim 10^{10}$) enables significant population of the MeV high energy Boltzman tail until $T \lesssim 0.1\ MeV$.

Once the temperature drops to about 10^9K, nuclei can exist in statistical equilibrium through reactions such as $n + p \leftrightarrow {}^2H + \gamma$ and ${}^2H + p \leftrightarrow {}^3He + \gamma$ and ${}^2H + n \leftrightarrow {}^3H + \gamma$, which in turn react to yield ${}^4He$. Since ${}^4He$ is the most tightly bound nucleus in the region, the flow of reactions converts almost all the neutrons that exist at 10^9K into ${}^4He$. The flow essentially stops there because there are no stable nuclei at either mass-5 or mass-8. Since the baryon density at big bang nucleosynthesis is relatively low (much less than $1g/cm^3$) and the time-scale short ($t \lesssim 10^2 sec$), only reactions involving two-particle collisions occur. It can be seen that combining the most abundant nuclei, protons and ${}^4He$ via two body interactions always leads to unstable mass-5. Even when one combines ${}^4He$ with rarer nuclei like ${}^3H$ or ${}^3He$, we still get only to mass-7, which, when hit by a proton, the most abundant nucleus around, yields mass-8. (A loophole around the mass-8 gap can be found if $n/p > 1$, so that excess neutrons exist, but for the standard case $n/p < 1$). Eventually, ${}^3H$ radioactively decays to ${}^3He$, and any mass-7 made radioactively decays to ${}^7Li$. Thus, big bang nucleosynthesis makes ${}^4He$ with traces of ${}^2H$, ${}^3He$, and ${}^7Li$. (Also, all the protons left over that did not capture neutrons remain as hydrogen.) For standard homogeneous BBN, all other chemical elements are made later in stars and in related processes. (Stars jump the mass-5 and -8 instability by having gravity compress the matter to sufficient densities and have much longer times available so that three-body collisions can occur.) With the possible exception of ${}^7Li$,[9,10,11,12,14] the results are rather insensitive to the detailed nuclear reaction rates. This insensitivity was discussed in reference 9 and most recently using a Monte Carlo study by Krauss and Romanelli.[14] An n/p ratio of $\sim 1/7$ yields a ${}^4He$ primordial mass fraction,

$$Y_p = \frac{2n/p}{n/p+1} \approx \frac{1}{4}.$$

The only parameter we can easily vary in such calculations is the density that corresponds to a given temperature. From the thermodynamics of an expanding universe we know that $\rho_b \propto T^3$; thus, we can relate the baryon density at $10^{10}K$ to the baryon density today, when the temperature is about $3K$. The problem is that we don't know today's ρ_b, so the calculation is carried out for a range in ρ_b. Another aspect of the density is that the cosmological expansion rate depends on the total mass-energy density associated with a given temperature. For cosmological temperatures much above $10^4 K$, the energy density of radiation exceeds the mass-energy density of the baryon gas. Thus, during big bang nucleosynthesis, we need the radiation density as well as the baryon density. The baryon density determines the density of the nuclei and thus their interaction rates, and the radiation density controls the expansion rate of the universe at those times. The density of radiation is just proportional to the number of types of radiation. Thus, the density of radiation is not a free parameter if we know how many types of relativistic particles exist when big bang nucleosynthesis occurred.

Assuming that the allowed relativistic particles at $1MeV$ are photons, e, μ, and τ neutrinos (and their antiparticles) and electrons (and positrons), Figure 1 shows the BBN yields for a range in present ρ_b, going from less than that observed in galaxies to greater than that allowed by the observed large-scale dynamics of the universe. The ^4He yield is almost independent of the baryon density, with a very slight rise in the density due to the ability of nuclei to hold together at slightly higher temperatures and at higher densities, thus enabling nucleosynthesis to start slightly earlier, when the baryon to photon ratio is higher. No matter what assumptions one makes about the baryon density, it is clear that ^4He is predicted by big bang nucleosynthesis to be around 1/4 of the mass of the universe.

The SSG Limit - Cosmological Neutrino Counting

Let us now look at the connection to N_ν. Remember that the yield of ^4He is very sensitive to the n/p ratio. The more types of relativistic particles, the greater the energy density at a given temperature, and thus, a faster cosmological expansion. A faster expansion yields the weak-interaction rates being exceeded by the cosmological expansion rate at an earlier, higher temperature; thus, the weak interaction drops out of equilibrium sooner, yielding a higher n/p ratio. It also yields less time between dropping out of equilibrium and nucleosynthesis at $10^9 K$, which gives less time for neutrons to change into protons, thus also increasing the n/p ratio. A higher n/p ratio yields more ^4He. As we will see in the next section, quark-hadron induced variations[31] in the standard model also yield higher ^4He for higher values of Ω_b. Thus, such variants still support the constraint on the number of relativistic species.[32]

In the standard calculation we allowed for photons, electrons, and the three known neutrino species (and their antiparticles). However, following SSG and doing the calculation (see Figure 2) for additional species of neutrinos, we can see when ^4He yields exceed observational limits while still yielding a density consistent with the ρ_b bounds from ^2H, ^3He, and now ^7Li. (The new ^7Li value gives approximately the same constraint on ρ_b as the others, thus strengthening the conclusion.) The bound on ^4He comes from observations of helium in many different objects in the universe. However, since ^4He is not only produced in the big bang but in stars as well, it is important to estimate what part of the helium in some astronomical object is primordial—from the big bang—and what part is due to stellar production after the big bang. The pioneering work of the Peimberts[33] showing that ^4He varies

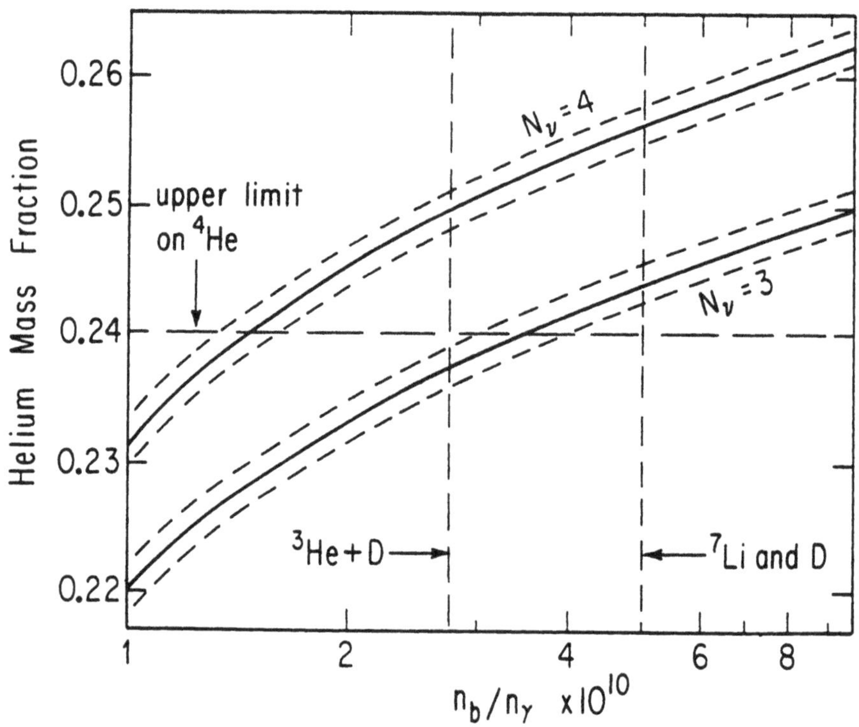

Figure 2. The SSG argument with recent parameter constraints showing the BBN helium mass fraction versus η for $N_\nu = 3$ and 4. Note that 4 is excluded.

with oxygen has now been supplemented by examination of how ^4He varies with nitrogen and carbon. The observations have also been systematically re-examined by Pagel.[34] The conclusions of Pagel,[34] Steigman *et al.*[35] and Walker *et al.*[11] all agree that the ^4He mass fraction, Y_p, extrapolated to zero heavy elements, whether using N, O, or C, is $Y_p \sim 0.23$ with an upper bound allowing for possible systematics of 0.24.

The other major uncertainty in the ^4He production used to be the neutron lifetime. However, the new world average of $\tau_n = 890 \pm 4s(\tau_{1/2} = 10.3\ min)$ is dominated by the dramatic results of Mampe *et al.*[36] using a neutron bottle. This new result is quite consistent with a new counting measurement of Byrne *et al.*[37] and within the errors of the previous world average of $896 \pm 10s$ and is also consistent with the precise C_A/C_V measurements from PERKEO[38] and others. Thus, the old ranges of $10.4 \pm 0.2\ min$, used for the half-life in calculations,[39,9] seem to have converged towards the lower side. The convergence means that, instead of the previous broad bands for each neutrino flavor, we obtain relatively narrow bands (see Figure 2). Note that $N_\nu = 4$ is excluded. In fact, the SSG limit is now$N_\nu < 3.4$.[10,11]

The recent verification of this cosmological standard model prediction by LEP, $N_\nu = 2.98 \pm 0.06$, from the average of ALEPH, DELPHI, L3 and OPAL[40] collaborations as well as the SLC[40] results, thus experimentally confirms our confidence in BBN. (However, we should also remember that LEP and cosmology are sensitive to different things.[41] Cosmology counts all relativistic degrees of freedom for $m_x \lesssim 10MeV$, with LEP and SLC counting particles coupling to the Z^0 with $m_x \lesssim 45GeV$.

While ν_e and ν_μ are obviously counted equally in both situations, a curious loophole exists for ν_τ since the current experimental limit $m_{\nu_\tau} < 35MeV$ could allow it not to contribute as a full neutrino in the cosmology argument.[42] Proposed experiments which push the m_{ν_τ} limit down to less than a few MeV should eliminate this loophole. It might also be noted that if we assume m_{ν_τ} is light so that cosmologically $N_\nu = 3$, we can turn the argument around and use LEP to predict the primordial helium abundance ($\sim 24\%$), or even use limits on ^4He to give an upper limit on Ω_b (also $\lesssim 0.10$). Thus, LEP strengthens the argument that we need non-baryonic dark matter if $\Omega = 1$. In fact, note also that with $N_\nu = 3$, if Y_p is ever proven to be less than ~ 0.235, standard BBN is in difficulty. Similar difficulties occur if Li/H is ever found below $\sim 10^{-10}$. In other words, BBN is a falsifiable theory.

Alternative Proposals

As noted above, BBN yields all agree with observations using only one freely adjustable parameter, ρ_b. Thus, BBN can make strong statements regarding ρ_b if the observed light element abundances cannot be fit with any alternative theory. Before exploring the implications for ρ_b, let us examine alternative proposals which have arisen to try to escape the power of the homogeneous BBN conclusions.

The two alternatives that have recently received interest are:

(1) Decaying particles;[43] and

(2) Quark-hadron transition inspired inhomogeneities.[31]

The first of these notes that if a species of massive particle ($M \gtrsim few\ GeV$) were to decay after traditional BBN, it could redo nucleosynthesis. While previous decaying particle proposals had been made, the new idea[43] emphasizes the importance of the resulting hadron cascade which, they argue, will dominate the yields. While interesting results are obtained, problems with detailed abundance determinations do result. In particular, this class of models seems to predict inevitably that $^6Li/^7Li >> 1$, whereas observations show $^7Li/^6Li \gtrsim 10$. While at first this might seem fatal, it is almost avoidable by noting that ^6Li is much more fragile than ^7Li; thus, it is easy to deplete ^6Li and obtain the observed ratios. However, Brown and Schramm[44] have pointed out that for high surface temperature Pop II stars, the convective zones do not go deep enough to destroy any primordial ^6Li. Pilachowski *et al.*[45] have now looked at those specific stars and indeed find no ^6Li, again seeing $^7Li/^6Li > 10$. Therefore, unless the Brown and Schramm convection argument can be surmounted, ^6Li seems to contrain this model seriously. Steigman, Audouze and others have noted additional problems with this model for ^3He and ^2H ratios.

Let us now look at the quark-hadron inspired inhomogeneity models.[31] While inhomogeneity models had been looked at previously (c.f. reference 9) and were found to make little difference, the quark-hadron inspired models had the added ingredient of variations in n/p ratios.

The initial claim by Applegate *et al.*, followed by a similar argument from Alcock *et al.* that $\Omega_b \sim 1$ might be possible, created tremendous interest. Their argument was that if the quark-hadron transition was a first-order phase transition (as some preliminary lattice gauge calculations implied), then it was possible that large inhomogeneities could develop at $T \gtrsim 100MeV$. The preferential diffusion of neutrons versus protons out of the high density regions could lead to big bang nucleosynthesis occurring under conditions with both density inhomogeneities and variable neutron/proton ratios. In the first round of calculations, it was claimed that such conditions might allow $\Omega_b \sim 1$, while fitting the observed primordial abundances of $^4He, ^2H, ^3He$ with an overproduction of ^7Li. Since ^7Li is the most recent of the cosmological abundance constraints and has a different observed abundace in Pop I stars versus the traditionally more primitive Pop II stars,[26] some argued that perhaps some special depletion process might be going on to reduce the excess ^7Li. Reeves and Audouze each argued against such processes and tried to turn the argument around and use lithium abundances to constrain the quark-hadron transition.

At first it appeared that if the lithium constraint could be surmounted, then the constraints of standard big bang nucleosynthesis might disintegrate. (Although Audouze, Reeves and Schramm emphasized that the number of parameters needed to fit the light elements was somewhat larger for these non-standard models, nonetheless, a non-trivial loophole appeared to be forming.) To further stimulate the flow through the loophole, Mullaney and Fowler showed that, in addition to looking at the diffusion of neutrons out of high density regions, one must also look at the subsequent effect of excess neutrons diffusing back into the high density regions as the nucleosynthesis goes to completion in the low density regions. (The initial calculations treated the two regions separately.) Mullaney and Fowler argued that for certain phase transition parameter values (e.g. nucleation site separations $\sim 10m$ at the time of the transition), this back diffusion could destroy much of the excess lithium. Recent work by Banerjee and Chitre (private communication) suggests that more accurate treatment of the diffusion calculation could reduce the interesting separation distance by several orders of magnitude.

However, Kurki-Suonio, Matzner, Olive and Schramm,[32] the Tokyo group,[46] and the Livermore group[47] have recently argued that in their detailed diffusion models, the back diffusion not only effects ^7Li, but also the other light nuclei as well. They find that for $\Omega_b \sim 1$, ^4He is also overproduced (although it does go to a minimum for similar parameter values as does the lithium). One can understand why these models might tend to overproduce ^4He and ^7Li by remembering that in standard homogeneous big bang nucleosynthesis, high baryon densities lead to excesses in these nuclei. As back diffusion evens out the effects of the initial fluctuation, the averaged result should approach the homogeneous value. Furthermore, it can be argued that any narrow range of parameters, such as those which yield relatively low lithium and helium, are unrealistic since in most realisitic phase transitions there are distributions of parameter values (distribution of nucleation sites, separations, density fluctuations, etc.). Therefore, narrow minima are washed out which would bring the ^7Li and ^4He values back up to their excessive levels for all parameter values with $\Omega \sim 1$. Furthermore, Adams and Freese[48] have argued that the boundary between the two phases may be fractal-like rather than smooth. The large surface area of a fractal-like boundary would allow more interaction between the regions and minimize exotic effects.

Figure 3 shows the results of Kurki-Suonio *et al.*[32] for varying spacing l with the constraints from the different light element abundances. Notice that the Li and even the ^4He constraint do not allow $\Omega_b \sim 1$. (The ^4He abundance constraint used in Kurki-Suonio *et al.* was a generous $Y_p \lesssim 0.25$; for the preferred $Y_p \lesssim 0.24$, the ^4He bound is about as tight as the Pop II Li constraint.) Note also that with the Pop II ^7Li constraint, the results for Ω_b are quite similar to the standard model with a slight excess in Ω_b possible if l is tuned to ~ 10.

Furthermore, initially it looked like quark-hadron inspired models might enable leakage[49] beyond mass-7, thus enabling ^9Be, ^{14}N, or maybe even r-process elements to become probes as whether or not the universe had such a transition (even if $\Omega_b \sim 1$). However, Tarasawa and Sato[46] have shown that when full multizone calculations of the type used by Kurki-Suonio *et al.* are utilized, then no significant leakage occurs.

One possible signature that remains for a first order quark-hadron transition is a slightly larger allowed range for Y_p that is concordant with $N_\nu = 3$ and with the other light element abundances. In particular, if ^4He were ever shown to be definitively $\lesssim 0.23$, it might be evidence for such a quark-hadron induced behavior since the standard homogenous case cannot accomodate such values. Of course, excessively low values for Y_p would still be unallowable.

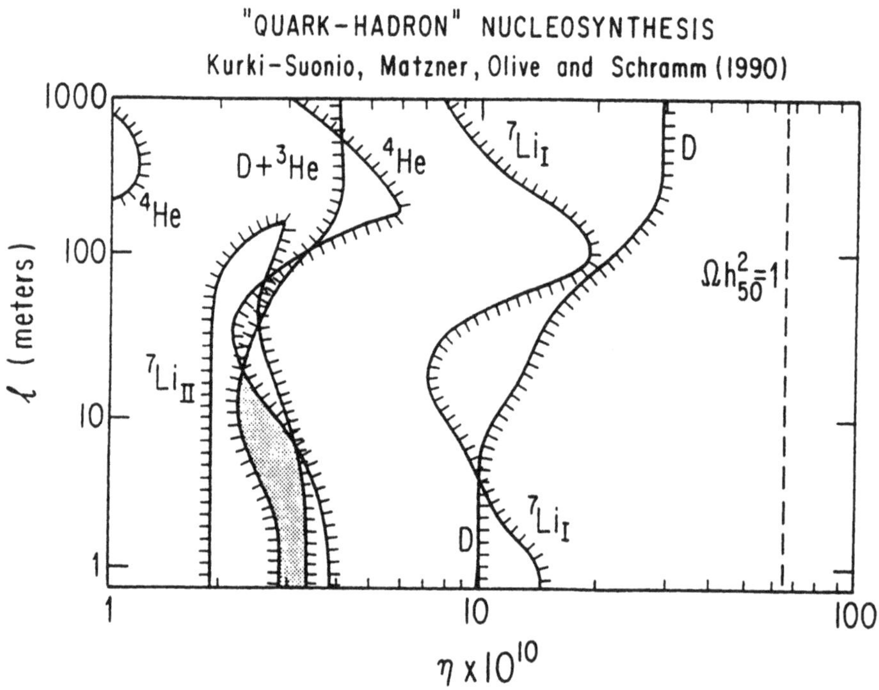

Figure 3. This shows the constraints on η of the various observed abundances in a first-order quark-hadron phase transition with nucleation sites separated by a distance l with density contrast $R \lesssim 10^3$. The Pop II lithium abundance used here is from the compilation of data given by Walker *et al.*[11] and is slightly more restrictive on η than that used in Figure 2 or used in the original Kurki-Suonio *et al.*[32] calculation from which this figure is derived. It should be noted that work by the Tokyo[46] group and by the Livermore group[47] confirms the conclusions on restricting Ω_b to values similar to the standard result even when $R \rightarrow \infty$.

One can conclude from the failure of the attempts to circumvent the standard BBN results that the results are amazingly robust. Even when many new free parameters are added, as in the quark-hadron case, the bottom line, when one requires concordance with the light element abundances, is essentially the same as the standard result. In other words, $\Omega_b \sim 0.06$ (although with fine-tuning the upper bound might be relaxed a bit to ~ 0.2 rather than 0.1).

One loophole which can yield variations in Ω_b outside the above range is to allow for degenerate neutrinos. This possibility has been discussed by many authors over the years with the most recent being Olive *et al.*[121] (also see references therein). The basic point is that creating any significant deviations requires excess lepton number densities, Li, that are comparable to η_γ However, most grand unified/SUSY theories require Lepton number excesses that are comparable to baryon number excess. Thus, $Li \sim \eta_B \ll \eta_\gamma$.

Furthermore, in order to fit the observed abundances, (word missing) requires that $Le \neq L_{\mu,\tau}$. Since again unified theories tend not to produce Le vs. L_μ, τ in the required ratios, but more like $Le \sim L_\mu \sim L_\tau \sim \eta_b$, this loophole appears rather unnatural and requires additional parameters that require artificial tuning.

Limits on Ω_b and Dark Matter Requirements

The narrow range in baryon density for which concordance occurs is very interesting. Let us convert it into units of the critical cosmological density for the allowed range of Hubble expansion rates. For the big bang nucleosynthesis constraints,[9,10,11,12,29,30] the dimensionless baryon density Ω_b, that fraction of the critical density that is in baryons, is less than 0.11 and greater than 0.02 for $0.04 \lesssim h_0 \lesssim 0.7$, where h_0 is the Hubble constant in units of $100 km/sec/Mpc$. The lower bound on h_0 comes from direct observational limits and the upper bound from age of the universe constraints.[49] Note that the constraint on Ω_b means that the universe *cannot be closed with baryonic matter.* If the universe is truly at its critical density, then nonbaryonic matter is required. This argument has led to one of the major areas of research at the particle-cosmology interface, namely, the search for non-baryonic dark matter.

Another important conclusion regarding the allowed range in baryon density is that it is in very good agreement with the density implied from the dynamics of galaxies, *including their dark halos.* An early version of this argument, using only deuterium, was described over fifteen years ago.[21] As time has gone on, the argument has strengthened, and the fact remains that galaxy dynamics and nucleosynthesis agree at about 6% of the critical density. Thus, if the universe is indeed at its critical density, as many of us believe, it requires most matter not to be associated with galaxies and their halos, as well as to be nonbaryonic.

Let us put the nucleosynthetic arguments in context. The arguments requiring some sort of dark matter fall into two separate and quite distinct areas. First are the arguments using Newtonian mechanics applied to various astronomical systems that show that there is more matter present than the amount that is shining. These arguments are summarized in Figure 4. It should be noted that these arguments

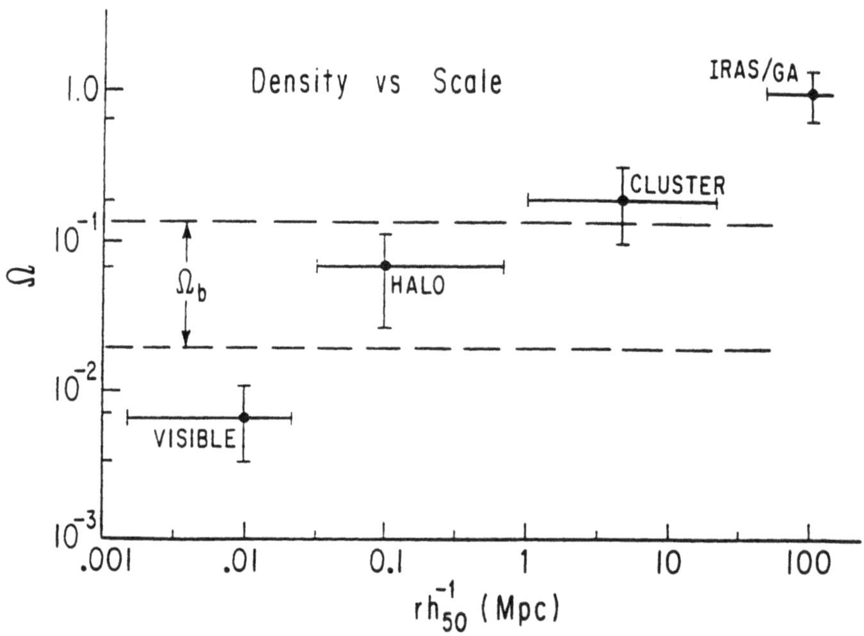

Figure 4. The inferred density in units of the critical density as a function of the scale on which it is "measured." Note the increase in Ω towards unity as larger scales are probed. Note also that Ω_b agrees with densities on the scale of galactic halos and is greater than the amount of visible matter.

reliably demonstrate that galactic halos seem to have a mass $\sim$ 10 times the visible mass.

Note, however, that big bang nucleosynthesis requires that the bulk of the baryons in the universe be dark since $\Omega_{vis} << \Omega_b$. Thus, the dark halos could in principle be baryonic.[21] Recently, arguments on very large scales[51] (bigger than clusters of galaxies) hint that Ω on those scales is indeed greater than Ω_b, thus forcing us to need non-baryonic matter. This is the first observational support for an Ω bigger than what can be accomodated with Ω_b.

Of course, it has been long anticipated, since the only long-lived natural value for Ω is unity, that inflation[52] or something like it provided the early universe with the mechanism to achieve that value and thereby solve the flatness and smoothness problems.

Some baryonic dark matter must exist, since from the $^2H +^3 He$ argument we know that the lower bound from big bang nucleosynthesis is greater than the upper limits on the amount of visible matter in the universe. However, we do not know what form this baryonic dark matter is in. It could be either in condensed objects in the halo, such as brown dwarfs and jupiters (objects with $\lesssim 0.08M_\odot$ so they are not bright shining stars), or in black holes (which at the time of nucleosynthesis would have been baryons). Or, if the baryonic dark matter is not in the halo, it could be in hot intergalactic gas, hot enough not to show absorption lines in the Gunn-Peterson test, but not so hot as to be seen in the x-rays. Evidence for some hot gas is found in clusters of galaxies. However, the amount of gas in clusters would not be enough to make up the entire missing baryonic matter. Another possible hiding place for the dark baryons would be failed galaxies, large clumps of baryons that condense gravitationally but did not produce stars. Such clumps are predicted in galaxy formation scenarios that include large amounts of biasing where only some fraction of the clumps shine.

Hegyi and Olive[53] have argued that dark baryonic halos are unlikely. However, they do allow for the loopholes mentioned above of low mass objects or of massive black holes. It is worth noting, as Schramm[2] points out, that these loopholes are not that unlikely. Furthermore, recent observational evidence[54] seems to show that disk formation is relatively late, occurring at red shifts z $\lesssim$ 1. Thus, the first several billion years of a galaxy's life may have been spent prior to the formation of the disk. In fact, if the first large objects to form are less than galactic mass, as many scenarios imply, then mergers are necessary for eventual galaxy size objects. Mergers stimulate star formation while putting early objects into halos rather than disks. Mathews and Schramm[55] have recently developed a galactic evolution model which does just that and gives a reasonable scenario for chemical evolution. (This scenario also provides a natural explanation for the number-versus-redshift relation of low luminosity galaxies found by Cowie.[56] Thus, while making halos out of exotic material may be more exciting, it is certainly not impossible for the halos to be in the form of dark baryons. One application of William of Ockham's famous razor would be to have us not invoke exotic matter until we are forced to do so.

Non-baryonic matter can be divided following Bond and Szalay[57] into two major categories for cosmological purposes: hot dark matter (HDM) and cold dark matter (CDM). Hot dark matter is matter that is relativistic until just before the epoch of galaxy formation, the best example being low mass neutrinos with $m_\nu \sim 25eV$. (Remember $\Omega_\nu \sim \frac{m_\nu(eV)}{100h_0^2}$).

Cold dark matter is matter that is moving slowly at the epoch of galaxy formation. Because it is moving slowly, it can clump on very small scales, whereas HDM tends to have more difficulty in being confined on small scales. Examples of CDM could be massive neutrino-like particles with masses, M_x, greater than several GeV or the lightest super-symmetric particle which is presumed to be stable and might also have masses of several GeV. Following Michael Turner, all such weakly interacting massive particles are called "WIMPS." Axions, while very light, would also be moving very slowly[58] and, thus, would clump on small scales. Or, one could also go to non-elementary particle candidates, such as planetary mass blackholes or quark nuggets of strange quark matter, possibly produced at the quark-hadron transition.[59] Another possibility would be any sort of massive topological remnant left over from some early phase transition. Table 1 summarizes the matter options. Note that CDM would clump in halos, thus requiring the dark baryonic matter to be out between galaxies, whereas HDM would allow baryonic halos.

When thinking about dark matter candidates, one should remember the basic work of Zeldovich,[60] resurrected by Lee and Weinberg[61] and others,[62] which showed that for a weakly interacting particle, one can obtain closure densities, either if the particle is very light, $\sim 25eV$, or if the particle is very massive, $\sim 3GeV$. This occurs because, if the particle is much lighter than the decoupling temperature, then its number density is the number density of photons (to within spin factors and small corrections), and so the mass density is in direct proportion to the particle mass, since the number density is fixed. However, if the mass of the particle is much greater than the decoupling temperature, then annihilations will deplete the particle number since, as the temperature of the expanding universe drops below the rest mass of the particle, Boltzmann suppression prohibits production while the number is depleted via annihilations until the annihilation reacton freezes out. For normal weakly interacting particles, decoupling occurs at a temperature of $\sim 1MeV$, so higher mass particles are depleted. It should also be noted that the curve of density versus particle mass turns over again (see Figure 5) once the mass of the WIMP exceeds the mass of the coupling boson[63,64,65] so that the annihilation cross section varies as $\frac{1}{M_x^2}$, independent of the mass of the coupling boson. In this latter case, $\Omega = 1$ can be obtained for $M_x \sim 1TeV \sim (3K \times M_{Planck})^{1/2}$, where $3K$ and M_{Planck} are the only energy scales left in the calculation (see Figure 5). A loophole to this argument occurs if there is a matter-antimatter assymmetry as in the case of baryons. However, such particles would have to be Dirac particles and we will see that they are still severely constrained.

A few years ago the preferred candidate particle was probably a few GeV mass WIMP. However, LEP's lack of discovery of any new particle coupling to the Z° with $M_x \lesssim 45GeV$, coupled with underground experiments,[66] clearly eliminates that

Table 1

MATTER

Baryonic $\Omega_b \sim 0.06$

VISIBLE $\Omega_{vis} \lesssim 0.01$

DARK

Halo
Jupiters
Brown Dwarfs
Stellar Black Holes

Intergalactic
Hot gas at $T \sim 10^5 K$
Stillborn Galaxies

Non Baryonic $\Omega_{nb} \sim 0.94$

HOT

$m_{\nu_\tau} \sim 25eV$

COLD

Wimps/Inos $\sim 100Gev$
Axions $\sim 10^{-5}eV$
Planetary Mass Black Holes

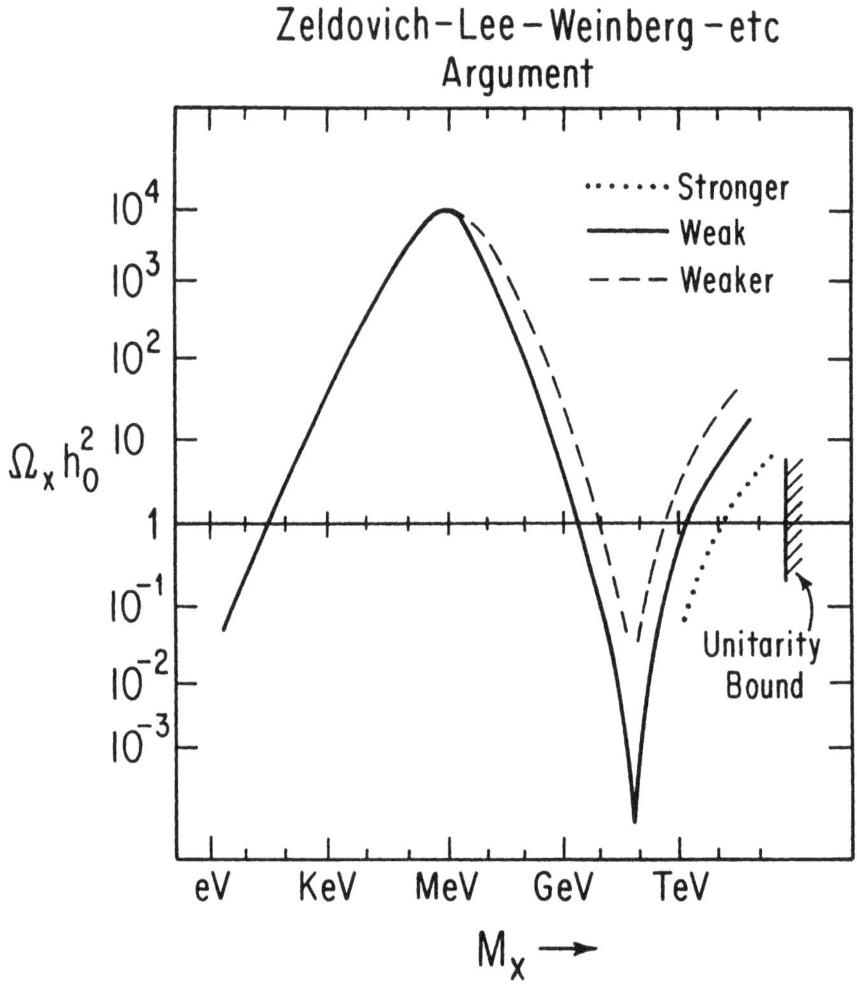

Figure 5. $\Omega_x h_0^2$ versus M_x for weakly interacting particles showing three crossings of $\Omega h_0^2 = 1$. Note also how the curve shifts at high M_x for interactions weaker or stronger than normal weak interaction (where normal weak is that of neutrino coupling through Z^0). Extreme strong couplings reach a unitarity limit at $M_x \sim 340 TeV$.

candidate.[67,68] Constraints for particles not fully coupled to the Z^0 were discussed by Ellis, Nanopoulos, Roskowski and Schramm[68] and are updated and presented in Figures 6a and 6b. (The inclusion of the Kamiokande II results as well as the newer LEP limits yields an important update over the results of Ellis *et al.*[68] since it closes the loophole for Dirac particles near $12GeV$.) Note also that the generic constraints of Figure 6 also apply to other hypothetical particles since CDF and UA2 do not see any squarks, sleptons, W' or Z' up to masses significantly greater than M_{Z^0}. Thus, whatever the coupling boson is, it must be greater than M_{Z^0} which means the effective value for $\sin^2 \phi_Z$ is < 1.

Furthermore, as Krauss[67] has emphasized, scaler particles such as sneutrinos interact like Dirac neutrinos so that the Kamiokande II and ionization experimental limits[66] also apply. Since asymmetric candidates are all Dirac particles, the restricted part of Figure 6b constrains asymmetric candidates where $\Omega = 1$ is no longer required to follow the locus shown. Thus, it seems that whether the particle is matter-antimatter symmetric or not, it is required to have an interaction weaker than weak and/or have a mass greater than $\sim 20GeV$. Future dark matter searches should thus focus on more massive and more weakly interacting particles.

Also, as Dimopoulos[63] has emphasized, the next appealing crossing of $\Omega = 1$ (see Figure 5) is $\gtrsim 1TeV$ (but, in any case, $\lesssim 340TeV$ from the unitarity bound[65]), which can be probed by SSC and LHC as well as by underground detectors. After the correct experimental constraints are taken into account, the favoured CDM particle candidate is now either a $10^{-5}eV$ axion or a gaugino with a mass of many tens of GeV. Of course an HDM ν_τ with $m_{\nu_\tau} \sim 20 \pm 10eV$ is still a fine candidate as long as galaxy formation proceeds by some mechanism other than adiabatic gaussian matter fluctuations.[69,72] This latter candidate becomes particularly attractive if recent hints from the gallium experiment[73] require the solution to the solar neutrino problem to have neutrino mixing with $\nu_e - \nu_\mu$ mass scales of 0.01 to 0.001 eV, making multiple eV mass scales for ν_τ quite plausible from see-saw type models where $m_{\nu_\tau} \sim m_{\nu_\mu}(\frac{M_{f_3}}{M_{f_2}})^2$ and M_{f_i} is an associated fermion mass for the ith generation. For example, if one uses the heavy quark masses, $(\frac{M_t}{M_c})^2 \sim 10^4$, so that ν_τ becomes ideal HDM. Such possibilities also may help late-time phase transition models for producing structure.[72]

The $17keV$ reports are discussed in the appendix and are not dark matter candidates due to the instability requirement.

Structure Formation

Perhaps the most outstanding problem in physical cosmology today is that of the formation of structure. Let us review the basic framework of structure formation in the universe. In particular, let us note that structure formation requires that density fluctuations grow. In order for this to occur, $\rho_{m(atter)}$ must be greater than $\rho_{r(adiation)}$. If we define T_{eq} as the temperature where $\rho_m = \rho_r$, then for an $\Omega = 1$ universe with h_0 equal to 0.5, equality is approximately 10^4 times the present temperature T_o. The horizon mass at T_{eq} is $\sim 5 \times 10^{16}(\frac{0.5}{h_0})^4 M$ which

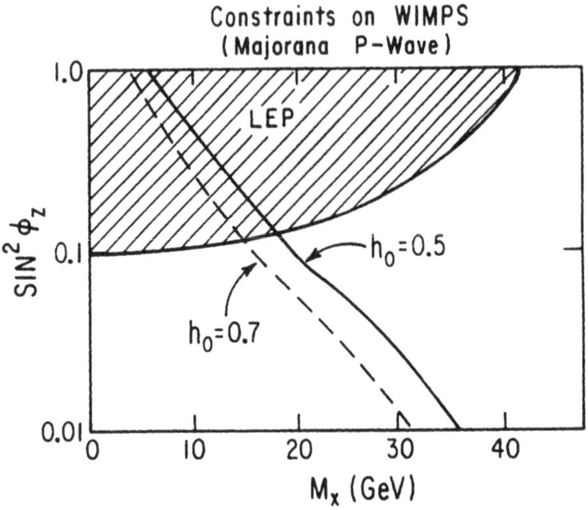

Figure 6a. Constraints on WIMPs of mass M_x versus $\sin^2\theta_z$, the relative coupling to the Z^0. The constraints are shown assuming Majorana particle (p-wave interactons). The diagonal lines show the combinations of M_x and $\sin^2\theta_z$ that yield $\Omega=1$. The cross-hatched region is what is ruled out by the current LEP results. Note that $\Omega = 1$ with $H_0 = 0.5$ is possible only if $M_x \lesssim 20GeV$ and $\sin^2\theta_z < 0.1$. This figure is revised from that of Ref. 68 using latest LEP results.

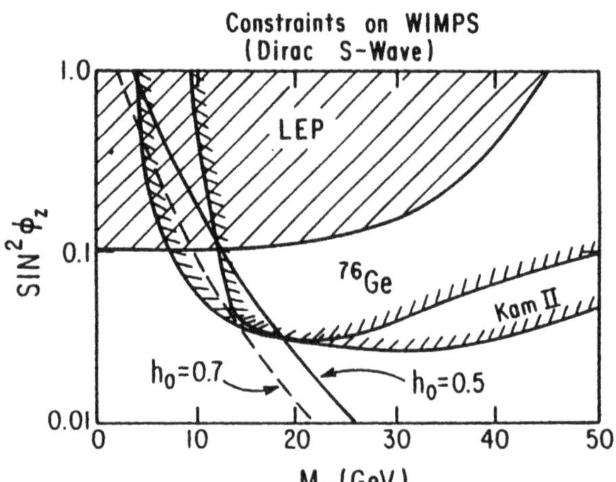

Figure 6b. This is the same as 6a but for Dirac particles (s-wave interactions). The ^{76}Ge region is that ruled out by the Caldwell *et al.* double-β decay style experiments. This figure is revised from that of Ref. 68 using latest LEP results and using new Kamiokande limits which closed a possible loophole near $M_x \sim 10GeV$. The current results require $M_x \gtrsim 20GeV$ and $\sin^2\theta_x \lesssim 0.03$ for matter-antimatter symmetric particles and also exclude the entire cross-hatched region for asymmetric particle candidates.

SO DO YOU THINK $\Omega = 1$, CARLO?

gives a present comoving scale of $\sim 60(\frac{0.5}{h_0})^2 Mpc$. The recombination epoch T_{rec} for an $\Omega = 1$ universe occurs slightly after matter domination. At $T_{rec} \sim 1100 T_o$ baryon fluctuations begin to grow after recombination and the horizon mass at recombination is about $10^{18}(\frac{0.5}{h_0})M_\odot$ with a comoving scale of $200(\frac{0.5}{h_0})Mpc$. We also know that the fluctuations in the microwave background temperature at the time of recombination are less than a few parts in 10^5. [74] Thus, in traditional models with primordial fluctuations existing prior to matter domination, growth begins at matter domination with the limits from $\frac{\delta T}{T}$ forcing $\frac{\delta\rho}{\rho}$ to be less than the order of 10^{-4} since

$$\frac{\delta \rho_m}{\rho} \lesssim 3\frac{\delta T}{T} \lesssim 10^{-4}.$$

Since small fluctuation $\delta\rho$ grows linearly with $1 + z$, this would mean that fluctuations could reach the order of unity only at the present epoch. Non-linear growth, and thus true structure formation, does not begin until $\frac{\delta\rho}{\rho}$ has reached unity (see Figure 7). Thus, in the standard model, the existence of objects at $z > 1$ (see for example Gunn, Schneider, and Schmidt[75]) requires that there be fluctuations far larger than the average in order that these objects currently exist. As Efstathiou and Rees[76] point out, the gaussian fluctuation model for primordial fluctuations would not allow a large number of quasar-like objects to form at $z \gtrsim 5$.

All models for structure formation require at least two basic ingredients for that structure:

(1) the matter,
(2) the seeds.

In traditional models, the seeds are random fluctuations in the density field generated at the end of the GUT phase transition, presumably accompanying inflation.[77,78]

As mentioned in the previous section, the matter in any model of galaxy formation with $\Omega = 1$ consists of normal baryonic matter with Ω the order of 0.06 and some non-baryonic matter, either hot or cold, with Ω the order of 0.94.[79]

The seeds which clump the matter to form objects may be divided into two broad categories (see Table 2) which can further be subdivided. The two broad categories would be (1) random gaussian seeds, presumably induced by quantum fluctuations at the end of a phase transition, and (2) topological defects produced in a vacuum phase transition. For the random gaussian seeds, the traditional assumption has been that the phase transition is the one associated with inflation.[77,78] However, it has been shown that similar kinds of fluctuations can also be generated in late-time phase transitions.[15,16] Similarly, for the topological defects, they could be formed either at the end of a GUT phase transition ($\sim 10^{15} GeV$) or in some late-time transition.[73,82,83] In some sense this current division of random versus topological replaces the old division of adiabatic versus isothermal (or isocurvature). In fact, the current "random gaussian" are indeed "adiabatic" and the topological are isothermal and isocurvature. However, the latter have the new added feature of

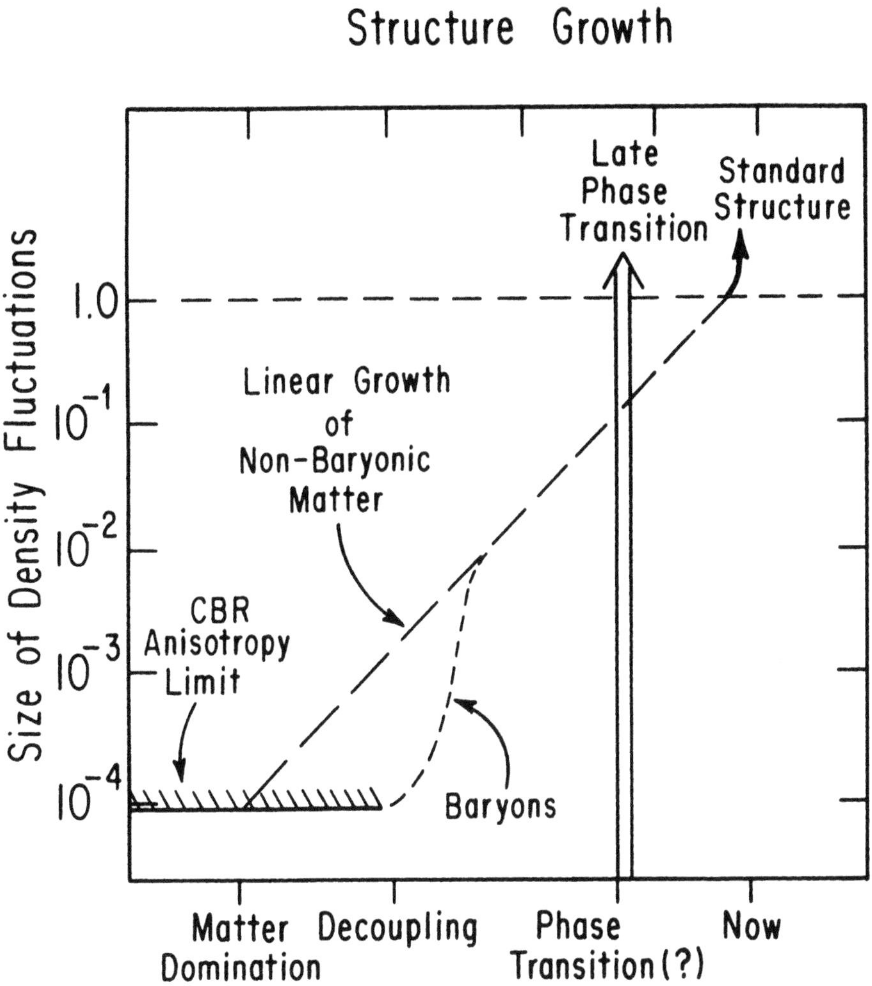

Figure 7. The growth of density fluctuations with the expansion of the universe. Note that LTPT can yield non-linear growth and thus structure at epochs much earlier than standard primordial models.

Table 2

SEEDS

I	RANDOM GAUSSIAN, Quantum Fluctuations
	A. End of Inflation
	B. LTPT
II	TOPOLOGICAL DEFECTS
	A. GUT
	B. LTPT

also being non-gaussian.

Let's note that all models for galaxy formation require new fundamental physics beyond the current particle standard model

$$SU_3 \times SU_2 \times U_1.$$

In particular, all non-baryonic dark matter, whether hot or cold, requires new physics, and similarly, all seeds, whether GUT scale or late-time and whether random gaussian or topological, require vacuum phase transitions. No model exists that does not invoke new physics. In fact, the existence of structure in the universe is one of the most important clues to the existence of physics beyond the standard model.

We should also note that not all combinations of seeds and matter are possible. For example, if one uses random gaussian seeds, then the non-baryonic matter must be cold, whereas if one uses topological seeds, the non-baryonic matter can be either hot or cold. One should also note that baryonic halos would require hot dark matter and hence topological seeds. Thus, searches for the dark baryons will also help constrain the non-baryonic candidates.

All current seed models require some form of vacuum phase transition. Thus, let us explore what possible phase transitions might occur (see Table 3). It should be noted in looking at Table 3 that of the three general classifications of cosmological phase transitions—the early, intermediate and late—the only ones that we absolutely know must have occurred are in the intermediate category when there is a horizon problem, namely that the horizon at the time of that transition is too small to generate galactic sized structure, and yet, the transition is not accompanied by significant inflation. The traditional early transitions have been used in the past because, while their horizon is small, inflation can amplify the effects to large scales. The other option is that of a late-time transition, where the universe waits until the horizon is sufficiently large that the physics of the phase transition directly yields the structures without having to use inflation to avoid the horizon problem.

Potential Observations to be Explained

In the last couple of years there have been a number of observations affecting galaxy formation and large-scale structure that have been a potential problem for traditional models which invoked early random gaussian fluctuations. However, because each of these observations is new and has not stood the test of time, in this discussion we refer to these as potential observations. In particular, many of the advocates of gaussian fluctuations and cold dark matter have tried to argue that these observations are statistical flukes that have yet to be established. Obviously, if these potential observations continue to hold up and are verified and are shown to be ubiquitous rather than statistical rareties, then the traditional models are in serious trouble. Table 4 summarizes these potential observations. Perhaps the most potentially damning would be observations of microwave anisotropies $\frac{\Delta T}{T}$ at levels significantly below 10^{-5}. However, at the present time, observations of small

Table 3

VACUUM PHASE TRANSITIONS

EARLY	(Small horizon but inflation)
	$\sim 10^{19} GeV$ - T.O.E.
	$\sim 10^{16} GeV$ - GUT
INTERMEDIATE	(Known to occur but horizon problem)
	$\sim 10^{2} GeV$ - Electroweak
	$\sim 1 GeV$ - QCD
LATE	(Horizon large)
	$\sim 10^{-2} eV$ - Family symmetries, etc.

Table 4

POTENTIAL OBSERVATIONS

1. $\frac{\delta T}{T} \lesssim 10^{-5}$
2. Structures $\gtrsim 100 Mpc$
3. Large coherent velocity flows
4. Objects existing at $z \gtrsim 5$
5. Large cluster - cluster correlations

scale anisotropy are at the level of a couple times 10^{-5}. Observations on angular scales of degrees or more are also approaching a few 10^{-5}. As this paper is being written, the measurements have not yet reached the point of ruling out the model of random fluctuations. However, as noted by Smoot,[74] within the not too distant future, COBE may be able to achieve limits as low as 3×10^{-6} on scales of a few degrees and larger, and antarctic studies may also push to similar levels on somewhat smaller scales, as might the baloon studies of Meyer *et al.* at MIT.

The next observation that can be a potential problem for traditional models is the existence of structures with scales greater than the order of $100Mpc$. In particular, the great wall observed by Geller and Huchra[84] shows that there is at least one such wall in the universe. The observations of Broadhurst *et al.*[85,86] show evidence for a multiplicity of such great walls with the characteristic spacing comparable to the size of the Geller-Huchra wall itself. While much debate has been made about whether or not the multiple walls of Broadhurst *et al.* are periodic or quasi-periodic, it does seem clear from their observations, as well as the work reported by Szalay,[87] that there is significant structure in the universe on scales of $\sim 100Mpc$. This is thoroughly supported by the large coherent velocity flows where the Seven Samurai[88] and others have found evidence for the existence of an object they call the "Great Attractor" towards which the Virgo cluster and the Hydro-Centaurus cluster all seem to be flowing with a velocity $\sim 600km/sec$. This again seems to indicate evidence of structures on the scales of at least $60Mpc$.

Perhaps most constraining of the traditional astronomical measurements is the existence of objects at very large redshifts. In particular, Schneider, Schmidt and Gunn[75] have found a quasar with a redshift of 4.73 (and they have privately reported one at 4.9). As Efstathiou and Rees[76] have noted, if such objects are ubiquitous, this would be fatal for primordial gaussian fluctuation models. Similarly, if one ever finds a quasar-type object at much larger redshifts, that would also be fatal.

Another potentially fatal observation for gaussian fluctuation models comes from the work of Bahcall and Soneira,[89] and Klypin and Khlopov[90] where they find that clusters of galaxies seem to be more strongly correlated with each other than galaxies are correlated with each other. While Primack and Dekel[91] have warned of the dangers of projection effects on such observations, it seems difficult to understand how projection effects would give the fractal-like behavior.[92] Furthermore, the southern hemisphere work of Huchra[93] also seems to support high cluster correlations. Most recently Vandenburg and West[94] have also found similar correlations for the CD galaxies observed at cluster centers. These CD's should not have the projection effect problems because redshifts are known. Even Primack and Dekel now acknowledge that there seems to be some excess in cluster correlations. If such large correlations turn out to be real, they too cannot be easily explained in the gaussian model, and, as Szalay and Schramm[93] note, they seem to be best fit by some sort of fractal-like pattern, as one might get from topological defects induced by a phase transition.

Late-Time Transitions

By late-time transition we will mean any non-linear growth occurring shortly after recombination. As mentioned above, such non-linear growth can be related either to a gaussian pattern or to a topological pattern such as walls, strings or textures. It is also possible that some normal random gaussian pattern from the very early universe could be triggered to undergo non-linear growth by some sort of phase transition or related phenomenon occurring after recombination. An example of this latter case would be the neutrino flypaper model of Fuller and Schramm.[95]

In general we will see that these late-time transitions can give the smallest possible $\frac{\Delta T}{T}$ for a given size structure. They can produce non-gaussian structural patterns, fractal-like with large velocity flows. It might be noted that the co-moving horizon at the time of the transition is not too different than the scale associated with the largest structures observed. No model of primordial fluctuations naturally imbeds this horizon scale onto the structural pattern. If some non-linear growth is associated with the patterns, the horizon scale can be imposed on the structure.

Another very dramatic advantage of late-time transitions, illustrated in Figure 7, is that it can produce structure with $\frac{\delta\rho}{\rho} \gtrsim 1$ at $z \gtrsim 10$. Thus, one could have significant structure and a significant number of objects at high redshift, which is a problem in any normal model with the seeds forming prior to recombination.

Let us now explore the possible physics that might give rise to a late-time transition, that is, a transition with a critical temperature between $0.001eV$ and $1eV$. It might be noted that in some sense it is a "hierarchy" rather than a "fine-tuning" problem to obtain a transition in this temperature range. We are trying to find a small mass scale somewhat analogous to how one would like to find the mass scale of the electron, or, for that matter, the Z^0 boson, when the natural mass scales to the problem are closer to $10^{19}GeV$, as in superstring models, or to 0. The hierarchy problem of trying to find the intermediate scale of the electroweak interaction of somewhere between the quark-lepton scale and the GUT or Planck scale has traditionally been approached with either a supersymmetric solution or a dynamical solution ("technicolor"). This supersymmetric solution, in some sense, is analogous to the model proposed in the appendix of Hill, Schramm and Fry,[72] denoted as HSF, which is an adaptation of the Hill-Ross[96] mechanism. A dynamical solution which has been proposed by Dimopoulos[97] involves a shadow $SU3$. The scale of a physics that might be associated with an HSF mechanism was relating to the MSW mixing solution to the solar neutrino problem.

The MSW[98,99] mixing solution to the solar neutrino problem is achieved if the neutrino mass difference squared, δm^2, is of the order of 10^{-4} to $10^{-7}eV^2$, or, in other words, neutrino masses of the order of a fraction of an electron volt. If we assume, following HSF, that the neutrino masses are generated by a pseudo-Nambu-Goldstone boson mechanism with mass

$$m_\phi \sim \frac{m_\nu^2}{f}$$

and with a transition occurring at $T_{crit} \sim m_\nu$, and if we further assume that the coupling f is related to the GUT scale, since we want to imbed this in some sort of unified theory, then the Compton wavelength $\lambda_\phi \sim 1Mpc$, in other words, a galactic scale. The density of the ϕ field at the time of the transition is the order of the cosmological density, in other words,

$$\frac{\rho_\phi}{\rho} \sim 1.$$

(Note that this is natural for phase transitions, whereas the requirement for primordial transitions to have small fluctuations, as inflation requires, is a fine tuning requirement.) Furthermore, the average spacing of the nucleation sights, L, can be estimated from Coleman's theory on spontaneous nucleation to yield spacings today that are interesting:

$$\frac{R_H}{L} \sim Log(\frac{M_p}{T_{crit}})$$

$$L_{co} \equiv L(1+z_{crit}) = \frac{6000}{Log(\frac{M_p}{T_{crit}})}(\frac{0.5}{h_0})(1+z_c)^{-1/2}$$

where $z_{crit} = z_c = (T_{crit}/T_0 \; - 1)$,

R_H is the horizon radius at z_c' and $M_p \sim 10^{19}GeV$. This yields for $T_{crit} \sim 10^{-2}eV\text{to}10^{-3}eV/L_{co(moving)} \sim 40\text{to}140Mpc$.

As we mentioned previously, recent impetus for new physics at this energy scale has come from the SAGE experiment which detects neutrinos from the PP chain in the sun. The previous solar neutrino experiments, the chlorine and the Kamiokande experiments, are mainly sensitive to the rare ^8B branch of the solar energy generating reactions. It is well established that the ^8B experiments have seen fluxes at levels somewhat below theoretical predictions.[100] However, there has always been the worry that the ^8B channel may be supressed due to astrophysical effects since its yield is very temperature sensitive. However, the PP chain that produces the neutrinos to be detected by SAGE must work if the sun is burning by fussion. Thus, the report[101] of no significant counts above background after five months of running the gallium experiment when they expected nineteen counts for the standard model implies that something is happening to the neutrinos on their way between emission and arrival at earth. (Or, that something is wrong with the detector, such as ^{71}Ge produced by ν-capture because it starts as an ion may have different chemistry than neutral ^{71}Ge.) Of course, the present results are very preliminary. Questions with regard to estimates of background, counting efficiencies, systematics, statistics, etc., remain, but the tantalizing hint that the ν_e's mixed into some other species of neutrino on their way out of the sun is certainly exciting. The final state of this experiment will not be known for several years. The similar gallium experiment operated by the GALLEX collaboration in the Grand Sasso Tunnel in Italy is also beginning to run but so far has had some background problems. The GALLEX chemistry may be somewhat cleaner and we will thus have an independent check on SAGE. Furthermore, both of these gallium experiments will be callibrated using ^{51}Cr sources of MeV neutrinos. Thus, one will have a true check of their counting efficiencies, etc., and both of these experiments will run for a long-enough time that the statistics will reach significant levels. If the neutrinos really are mixing on their way out of the sun, then the MSW solution is probably valid and we are in the realm discussed above.

It might also be noted that a simple application of the Gell-Mann–Ramond–Slansky see-saw model[102] for neutrino masses yields some interesting implications. If we assume that there is a mass hierarchy in the neutrinos with the electron neutrino having negligible mass, the μ the intermediate mass and the τ the heaviest, and we assume that the mixing of the ν_e in the sun goes to its nearest neighbor family, the ν_μ, then the ν_μ is carrying most of the mass of the MSW δ_m^2. The see-saw mechanism argues that

$$m_{\nu_i} \sim \frac{m_{f_i}^2}{M}$$

for a given family, or, in other words,

$$m_{\nu_\tau} \sim m_{\nu_\mu}(\frac{m_{f_3}}{m_{f_2}})^2.$$

If we use lepton masses for the fermion masses, this yields a ν_τ mass in the neighborhood of a few eV. However, if we use heavy quark masses, then, since the top quark mass is $\gtrsim$ 100 times that of the charm quark, this yields ν_τ masses in the neighborhood of 10 to 100 eV, making it perfect hot dark matter. It might also be noted that the see-saw mass scale, M, in this picture, ends up being the order of 10^9 to $10^{12} GeV$, which happens to be the only window allowed for the DFS-axion[103] scale. It might further be noted that if the non-baryonic dark matter is indeed the τ neutrino, then one is required to dismiss primordial gaussian fluctuations.

Note that even if the MSW mixing is $\nu_e - \nu_\tau$, the LTPT possibility is still there, but then all neutrinos would be light and could not serve as HDM. It is interesting that in this latter case the see-saw M is the GUT scale.

Structure from LTPT

LTPT can produce vacuum fluctuations of the random gaussian character just as could be generated at the end of inflation.[76] However, as emphasized in references 80 and 81, these structures will have a quantum scale that is the order of a galaxy size, and the bosons associated with the fluctations might even serve as the dark matter of the universe.

The other alternative for LTPT is to produce topological structures. Just as early universe phase transitions can produce strings and/or textures, LTPT can also produce such objects. Furthermore, LTPT can produce walls which are a problem for primordial phase transitions. However, there is a problem for some walls, depending on the nature of the interaction potential. LTPT that have a $\lambda\phi^4$

potential will end up with one wall dominating as was demonstrated in references 104, 105 and 106. However, this problem of one wall dominating can be surmounted in a variety of ways which have varying degrees of attractiveness, depending on the eyes of the beholder. For example, in the HSF phase transition, the walls are sine-Gordon rather than $\lambda\phi^4$. As Widrow has shown,[107] the sine-Gordon walls can yield "bags" of wall or "balls" of wall which survive several expansion times. These bags or balls can then serve as seeds in galaxy formation, and thus, it is their amplitude that becomes a deciding factor for $\frac{\Delta T}{T}$ limits as opposed to the energy scale of the infinite walls which can be made quite small. This latter point was emphasized by Hill, Schramm and Widrow.[82] Another way of avoiding single wall dominance is the decaying wall model of Kawano[108] where the walls serve as seeds and then decay away. It is also possible to escape one-wall domination with a large number of minima in the potential. Perhaps the most dramatic way of escaping one-wall domination, thus keeping a network of walls, as shown in Figure 8A, is if the walls have friction with the ambient medium, whether it be neutrinos or the remaining baryonic and/or non-baryonic matter in the universe.[109] Alessandro Massarotti has shown that friction can in many reasonable cases slow the walls down sufficiently that they do not evolve to the one-ball domination situation. In this case, one retains a complex network with L for the wall being much less than the horizon size.

It might be noted that long walls gravitationally repel rather than attract,[110,111] whereas balls of wall are attractive seeds. Thus, a combined network of balls and slowed-down long walls can yield a complex structure which may be even of a fractal character in agreement with the claims of Schramm and Szalay[92] from cluster correlations.

In addition to walls, LTPT can also produce textures[112] or non-topological solitons.[113] In these latter cases, or with the bags of wall dominating, one will have networks more closely ressembling Figure 8B and Figure 8A. It should be noted that the parameters L and δ and the nature of the structures generated are dependent on the model for the LTPT. It should also be noted that questions of the detailed physics of imbedding the LTPT into some larger GUT or TOE are dependent on the unification model. HSF have shown that a reasonable toy model can be constructed which can give a phase transition. These phase transitions in many ways are quite analogous to the axion-producing phase transition which has a coupling at a scale near to the order of $10^{11} GeV$, far above the QCD phase transition scale of the order of GeV. And like the axion, the particle involved in the LTPT of HSF has a pseudo-Nambu-Goldstone boson. However, instead of being related to the strong interaction and quarks, in the LTPT case it is related to the neutrinos and probably to family symmetry.

Generating seeds at an LTPT might be advantageous for producing the multiple walls of Broadhurst *et al.*[85,86] In particular, Icke and Weygaert,[114] and Coles[115] have independently demonstrated that the phenomenological Vornoi tessalations of the intersection of expanding rarefaction shells give a very good fit to large scale structure if the nodes of these tessalations are fit to the Abell clusters. In particular,

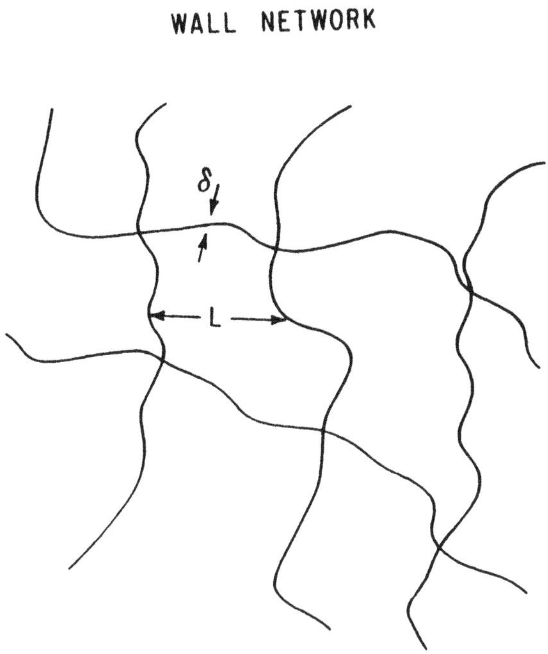

Figure 8a. A generic wall network defining the wall thickness δ and the characteristic spacing of structure L.

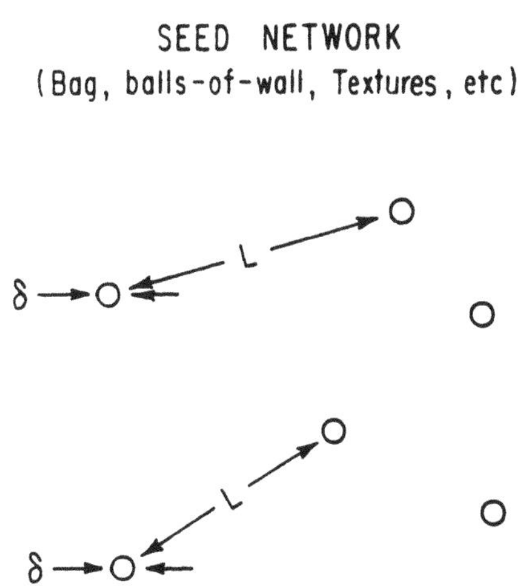

Figure 8b. A generic network for seed generation with seed size δ and seed separation L.

they note that one gets quasi-periodic walls at $\sim 130 Mpc$ with cluster correlation functions that are quite strong and follow the fractal behavior of Schramm and Szalay. However, the seed distribution required to give this tessalation causes a conflict with the microwave background radiation, if the seeds are generated prior to the decoupling. However, an LTPT could remedy that. Similarly, an LTPT can provide the seeds to enable hot dark matter to work as a galaxy formation model (see, for example, reference 116.) It might be noted that the typical bag of wall can easily yield a galaxy or a quasar-forming seed.

We can estimate the mass associated with a wall in the following way:

Let σ = energy density per unit area, that is:

$$\sigma \equiv \rho_w \delta \sim \frac{4 \times 10^{-5} s \rho_o \delta}{h_0^2} (1+z_c)^4$$

where

$$\delta \equiv \text{thickness}$$

$$\rho_w = \text{density of the wall}$$

$$s = \frac{\rho_w}{\rho_r} \text{ at } z_c$$

$$\rho_0 = 3 \times 10^{11} h_0^2 M_\odot / Mpc^3$$

then

$$M_w \sim \pi \sigma L^2 \sim 3 \times 10^7 (1+z_c)^4 (\frac{\delta}{Mpc})(\frac{L}{Mpc})^2 M_\odot$$

and for stable walls

$$\Omega_w(z) \sim \frac{3}{4} \frac{\sigma}{\rho_0 L (1+z)^3}.$$

Note that Ω_w at the present epoch, can be the order of unity. Wall domination can occur at present epoch if

$$z_c \gtrsim 11(\frac{L}{\delta})^{\frac{1}{4}}(\frac{ho^2}{s})^{\frac{1}{4}} - 1$$

for stable walls. It might be noted that if wall domination occurs at the present epoch, as long as there are multiple walls, rather than just one wall dominating, one has the interesting situation where the expansion of the universe is no longer following the normal matter-dominated relationship, and, in particular, one can achieve ages greater than $\frac{1}{h_o}$. Such a situation may be a solution to the age-Hubble constant problem if h_o is ever shown to be greater than 0.7.

It might also be noted for topological structure generated by LTPT that the structure is relatively independent of whether the non-baryonic dark matter is hot or cold.

Microwave Anisotropies

Since LTPTs provide no fluctations on the surface of last scattering, all fluctuations from the microwave background must be due to the differential redshift-blueshift non-cancellation due to a changing potential in the transparent medium or due to scattering of the microwave photons off of moving objects. One can estimate the potential change due to the ϕ field itself generated in the phase transition and by the dynamic motion of the structures and the Doppler shift thereby produced. One can also do the classical Rees-Sciama and Sachs-Wolf calculations for the $\frac{\Delta T}{T}$ generated by existing objects.[118,119] We can estimate its effects roughly in the following way: The static effects will dimensionally go as

$$\frac{\delta T}{T} \sim \Omega_w(\frac{L}{R_H})^2 \sim G\sigma L.$$

The time-changing effects can be estimated by multiplying the static effect by $\frac{V}{c}$. While different people remember different formulations of these things, one can show that because of the nature of walls and other topological systems, the effects can be reduced to the form $G\sigma L$ times $\frac{V}{c}$ or $\frac{V^2}{c^2}$. Since any walls or topological seeds we ever see must be moving with $V < c$, the dominant effect will in general go like $G\sigma L$, which can be shown to yield the result:

$$\frac{\delta T}{T} \sim 10^{-8}(\frac{1+z_c}{10})^4 \; s(\frac{\delta}{Mpc})(\frac{L}{Mpc})$$

$$\sim 10^{-6} \text{ for } L \sim 100Mpc,\ \delta \sim 1Mpc,\ z_c \sim 10.$$

Note that this yields $\frac{\delta T}{T} \sim 10^{-6}$ even for an L of $100Mpc$. The distribution, however, of these fluctuations depends very much on the detailed topological nature of the structures produced. In particular, Turner, Watkins and Widrow[119] have shown that balls of wall tend to produce spikes very similar in nature to the spikes that textures produce. A general formalism showing the wide range of structures in non-gaussian microwave background fluctuations has been developed by Goetz and Noetzold.[120]

In general, one can see that if structure of size L is generated by a late-time phase transition, and L is the maximum size of structure produced in that transition, then the late-time transition does give the minimum $\frac{\delta T}{T}$ for that structure. Of course the question is what is the characteristic size L of structure generated in a transition. For $\lambda\phi^4$ structures, L goes to the horizon size, in which case $\frac{\delta T}{T}$ gets larger than current observational limits. However, as mentioned above, many other possibilities can be generated in LTPT, with L at present being a somewhat freely adjustable parameter, depending on the model, the amount of friction, the decay of the walls, etc.

Conclusion

In these lectures we have seen that the basic big bang model is in excellent shape with the recent collider results helping to confirm it in the same way they've helped with $SU_3 \times SU_2 \times U_1$. We've also seen that the prediction that the bulk of the matter in the universe is in the form of some exotic non-baryonic species obviously remains to be confirmed. Furthermore, we've examined the current problems of generating structure in the universe, mentioning the traditional primordial scenarios as well as the new exotic idea of a late-time phase transition.

All in all, we've seen that cosmology is tremendously active with new data coming from both astronomical and particle physics techniques. We've also seen that some of the best current indications for particle physics beyond $SU_3 \times SU_2 \times U_1$ are coming from astrophysical arguments.

Acknowledgments

I would like to thank my recent collaborators, Xiaochun Luo, Chris Hill, John Ellis, Savas Dimopoulos, Gary Steigman, Keith Olive, Michael Turner, Rocky Kolb and Terry Walker, for many useful discussions. I would also like to acknowledge useful communications with Denys Wilkinson, Walter Mampe, Jim Byrne and Stuart Friedman on the neutron lifetime and with Bernard Pagel and Jay Gallagher on helium abundances and Alessandro Massarotti and Neil Turok on phase transitions. This work was supported in part by NSF grant AST 88-22595, by NASA grant NAGW 1321 and by DoE grant DE-FGO2-91ER40606 at the University of Chicago and by the DoE and NASA grant NAGW 1340 at the NASA/Fermilab Astrophysics Center.

Appendix I - The 17 keV "Thing"

A few years ago Simpson[122] reported the possible existence of a 17 keV mass state mixing at the 1% level with an electron neutrino during tritium decay. Although initially met with some skepticism, it could not be trivially dismissed. Recently, attention has been refocused on this object as a result of a series of other nuclear β decay experiments. In particular, Norman[123] and his collaborators reported the existence of a similar 17 keV component at the 1% mixing level in carbon-14 decay, and, most dramatically, Hime and Jelly[124] reported a similar 17 keV mass mixing at the 1% level in sulphur-35 decay. While this is all still somewhat preliminary, and there are worries that there could be some sort of non-obvious instrumental effect occurring (for example, it might be noted that the effect has been seen only using solid state detectors rather than magnetic spectrometers[125]), nonetheless, the possibility that an electron neutrino mixes at 1% level with something that has a 17 keV mass has raised much excitement.

There's a problem, however, in what this object can be. Numerous papers have been written discussing the problems and trying to come up with exotic models that might be able to fit it.[126] Basic problems are as follows:

1. It is well known that there are only three families of neutrinos from the LEP experiments. Thus, if the 17 keV object is another neutrino, it must be either the

μ or the τ. But limits on ν_e - ν_μ mixing show that it is much less than 1%, so the only neutrino possibility is ν_τ.

2. It is also known that neutrinoless double β decay has not been seen. This limits the majorana mass for the electron neutrino to be less than approximately 1 electron volt. Since the 17 keV mass times the 1% mixing yields a mass of 170eV, this particle cannot have a majorana mass in the normal interpretation. (The exotic option of having the majorana mass term be a combination of ν_e $(C\bar{\nu}_\mu + S\bar{\nu}_\tau)$ and thus maximally violate the CP cannot be easily excluded.[127])

3. If the 17 keV does not have a majorana mass, then it must have a Dirac mass if it is a neutrino.

4. If it has a Dirac mass, then big bang nucleosynthesis would count the spin-flip component as an extra neutrino state and, as we saw earlier, big bang nucleosynthesis does not allow more than a total of 3.4 neutrinos, thus excluding any of the normal neutrinos from having right-handed components that interact with normal neutrino-like interactions.

5. The remaining option would be that this right-handed component must interact much more weakly than normal left-handed neutrinos. Olive *et al.*[128] showed that the nucleosynthesis limits on exotic neutrinos can be surmounted if those exotic neutrinos interact much more weakly than normal neutrinos. Thus, the right-handed component would have to couple to a $\mu c Z'$ with a mass greater than about $\sim TeV$.

6. This leads to a dilemma in another astrophysical area. If the right-handed neutrino is so weakly coupled, then it would have freely escaped from SN 1987A. Gandhi and Burrows[129] have argued that right-handed neutrinos with masses greater than 14 keV are excluded (a recent numerical error in their calculation may push their limit up to 28 keV). While this limit at first appears only marginal, it should be noted that calculations treating neutrino processes in more detail (including neutrino bremstrahlung induced spin-flip which Turner[130] has shown will significantly enhance production) as well as the slightly higher temperatures encountered in the course of other supernova collapse calculations (see, for example, Mayle, Wilson, Schramm[131]) appear to strengthen this limit and seem to push it down significantly below $\sim 10keV$, thus severly constraining the existence of a right-handed Dirac neutrino. Furthermore, one can't have it both ways. If the neutrino is sufficiently weak to escape the cosmological bound, then it makes it easier for it to get out of the supernova and make the time-scale of the neturino burst in SN 1987A shorter than what was observed to be. The basic physics in the time-scale is simply that the ten second duration of the neutrino burst (see review by Truran and Schramm[132]) requires that neutrinos diffuse out rather than freely stream out. If any component of the neutrinos is able to stream out freely, then there would be a leakage of energy out of the core and the duration of the neutrino pulse would be much less. Similar arguments to this were used to set limits on axions and any other exotic particles that might have been produced in the collapsing core.

7. From the cosmological constraints on Ω, any 17 keV neutrino would have to be unstable, since a 17 keV stable neutrino would yield an Ω of approximately 200,

which would have led to the Big Crunch many eons ago.

8. If the neutrino were unstable, it would have to be sufficiently unstable that it did not escape the supernova core, and it could not decay to photons since no gamma rays were seen to accompany the neutrinos from SN 1987A. Thus, in the 10^5 year lifetime of transit from 1987A to the solar system, the neutrinos did not produce significant radiative products.

9. This further contrains any model since an invisible decay that did not produce photons would require some other new particles such as a majoron, which then should have been counted in big bang nucleosynthesis and, as we've already seen, that limit seems to be quite formidable.

All in all, this seems to mean that whatever the 17keV thing is, it is not a neutrino in any normal sense of the word.

References

1. J. Mather *et al.*, COBE preprint, Goddard Space Flight Center (1990) and *Astrophys. J.*, in press (1991).
2. D.N. Schramm, in *Proc. of the 1990 Rencontre de Physique at La Thuile, March 1990*, ed. M. Greco (in press, 1991);
   D.N. Schramm, in *Proc. 1990 Nobel Symposium, Gräftvållen, Sweden* (in press, 1991);
   D.N. Schramm, in *Proc. of the Moriond Astrophysics Meeting on Particle Astrophysics at Les Arcs, Savoie, France, March 1990*, ed. T. Piran (in press, 1991);
   D.N. Schramm, in *Z^0 Physics - Cargèse 1990*, eds. M. Levy *et al.* (Plenum, London, 1991).
3. R.A. Alpher, H. Bethe and G. Gamow, *Phys. Rev.* **73** (1948) 803.
4. R.A. Alpher, J.W. Follin and R.C. Herman, *Phys. Rev.* **92** (1953) 1347.
5. R. Taylor and F. Hoyle, F., *Nature* **203** (1964) 1108.
6. P.J.E. Peebles, *Phys. Rev. Lett.* **16** (1966) 410.
7. P. Wagoner, W.A. Fowler and F. Hoyle, F., *Astrophys. J.* **148** (1967) 3.
8. D.N. Schramm and R.V. Wagoner, *Ann. Rev. of Nuc. Sci.* **27** (1977) 37;
   K. Olive, D.N. Schramm, G. Steigman, M. Turner and J. Yang, *Astrophys. J.* **246** (1981) 557;
   A. Boesgaard and G. Steigman, *Ann. Rev. of Astron. and Astrophys.* **23** (1985) 319.
9. J. Yang, M. Turner, G. Steigman, D.N. Schramm and K. Olive, *Astrophys. J.* **281** (1984) 493.
10. K. Olive, D.N. Schramm, G. Steigman and T. Walker, *Phys. Lett B.* **236** (1990) 454.
11. T. Walker, G. Steigman, D.N. Schramm, H.-S. Kang and K. Olive, *Astrophys. J.* (submitted, 1991).
12. L. Kawano, D.N. Schramm and G. Steigman, *Astrophys. J.* **327** (1988) 750.

13. J.W. Truran, A.G. Cameron and A. Gilbert, *Canadian Journal of Physics* **44** (1966) 563.
14. L. Krauss and P. Romanelli, P., Yale University preprint (1990) and *Ap.J.*, in press (1991).
15. W. Fowler, J. Greenstein and F. Hoyle, *Geophys. J.R.A.S.* **6** (1962) 6.
16. C. Ryter, H. Reeves, E. Gradstajn and J. Audouze, *Astron. and Astrophys.* **8** (1970) 389.
17. J. Geiss, and H. Reeves, *Astron. and Astrophys.* **18** (1971) 126;
    B. Black, *Nature* **234** (1973) 148.
18. J. Rogerson and D. York, *Astrophys. J.* **186** (1973) L95.
19. H. Reeves, J. Audouze, W.A. Fowler and D.N. Schramm, *Astrophys. J.* **179** (1973) 909.
20. R. Epstein, J. Lattimer and D.N. Schramm, *Nature* **263** (1976) 198.
21. J.R. Gott, III, J. Gunn, D.N. Schramm B.M. Tinsley, *Astrophys. J.* **194** (1974) 543.
22. R.T. Rood, G. Steigman and B.M. Tinsley, *Astrophys. J.* **207** (1976) L57.
23. J. Yang, D.N. Schramm, G. Steigman and R.T. Rood, *Astrophys. J.* **227** (1979) 697.
24. T. Wilson, R.T. Rood and T. Bania, in *Proc. of the ESO Workshop on Primordial Healing*, eds. P. Shaver and D. Knuth (European Southern Observatory, Garching, 1983).
25. M.R. Hartoog, *Ap.J.* **231** (1979) 161;
    J. Ostriker and D.N. Schramm, Fermilab/Princeton preprint (1991).
26. J. Spite and F. Spite, *Astron. and Astrophys.* **115** (1982) 357;
    R. Rebolo, P. Molnaro and J. Beckman, *Astron. and Astrophys.* **192** (1988) 192.
    L. Hobbs and C. Pilachowski, *Astrophys. J.* **326** (1988) L23.
27. G. Steigman, D.N. Schramm and J. Gunn, *Phys. Lett.* **66B** (1977) 202.
28. V.F. Schvartzman, *JETP Letters* **9** (1969) 184;
    P.J.E. Peebles, *Physical Cosmology* (Princeton University Press, 1971).
29. D.N. Schramm and L. Kawano, *Nuc. Inst. and Methods* **A284** (1989) 84.
30. B. Pagel, *Proc. of 1989 Rencontres de Moriond* (1990).
31. R. Scherrer, J. Applegate and C. Hogan, *Phys. Rev. D* **35** (1987) 1151;
    C. Alcock, G. Fuller and G. Mathews, *Astrophys. J.* **320** (1987) 439;
    W.A. Fowler and R. Malaney, *Astrophys. J.* **333** (1988) 14.
32. H. Kurki-Suonio, R. Matzner, K. Olive and D.N. Schramm, *Astrophys. J.* **353** (1990) 406.
33. M. Peimbert and S. Torres-Peimbert, *Astrophys. J.* **193** (1974) 327.
34. B. Pagel, in *ESO/CERN Proc.* (1990).
35. G. Steigman, D.N. Schramm and J. Gallagher, *Comments on Astrophys.* **14** (1989) 97.
36. W. Mampe, P. Ageron, C. Bates, J.M. Pendlebury and A. Steyerl, *Phys.*

Rev. Lett. **63** (1989) 593.
37. J. Byrne, *et al., Phys. Lett. B*, submitted (1991).
38. H. Abele, M. Arnould, H.A. Borel, J. Dohner, D. Dubbers, S. Freedman, J. Last and I. Reichert, in *Proc. Grenoble Workshop on Slow Neutrons* (1989).
39. G. Steigman, K. Olive, D.N. Schramm and M. Turner, *Phys. Lett. B* **176** (1986) 33.
40. ALEPH, DELPHI, L3, OPAL and MARK II collaboration papers, in *Proc. of the Int. High Energy Conference, Singapore* (1990).
41. D.N. Schramm and G. Steigman, *Phys. Lett. B* **141** (1984) 337.
42. E. Kolb and R. Scherrer *Phys. Rev. D* **25** (1982) 1481;
E. Kolb, M.S. Turner, A. Chakravorty and D.N. Schramm, FERMILAB-PUB-91/28-A (1991) and *Phys. Rev. Lett*, in press (1991).
43. S. Dimopoulos, R. Esmailzadeh, L. Hall and G. Starkman, *Astrophys. J.* **330** (1988) 545.
44. L. Brown and D.N. Schramm, *Ap.J.* **329** (1988) L103.
45. D. Pilachowski, L. Hobbs and D. De Young, *Ap.J. Lett.* **345** (1989) L39.
46. H. Reeves, K. Sato and M. Tarasawa, U. of Tokyo preprint (1989);
Tarasawa, N. and Sato, K., U. of Tokyo preprint (1990).
47. C. Alcock, G. Fuller, G. Mathews and B. Meyer, Livermore preprint (1990).
48. F. Adams and K. Freese, MIT preprint (1990).
49. L. Kawano, W. Fowler and R. Malaney, Caltech-Kellogg preprint (1990);
J. Applegate, Columbia University preprint (1989).
50. K. Freese and D.N. Schramm, *Nucl. Phys.* **B233** (1984) 167.
51. M. Strauss, M. Davis and A. Yahil, U.C. Berkeley preprint (1989);
N. Kaiser and A. Stebbins, CITA preprint (1990);
E. Bertschinger, A. Dekel and A. Yahil, in *Proc. Blois Symposium on the Microwave Background* (1990);
M. Rowan-Robinson and A. Yahil in *Proc. Rencontres de Moriond* (1989).
52. A. Guth, *Phys. Rev. D* **23** (1981) 347;
see also K. Olive, *Physics Reports* **190** (1990) 309-403 or A. Lindei, *Particle Physics and Inflationary Cosmology*, (Harwood, New York, 1990).
53. D. Hegyi and K. Olive, K. *Astrophys. J.* **303** (1986) 56.
54. J. Gunn, talk at ITP Santa Barbara (1988);
D. York, talk at University of Chicago (1988).
55. G. Mathews and D.N. Schramm, *Astrophys. J.*, submitted (1991).
56. L. Cowie, in *Proc. IUPAP Conference on Primordial Nucleosynthesis and Evolution of Early Universe, Tokyo, September 1990*, in press (1991).
57. R. Bond and A. Szalay, *Proc. Texas Relativistic Astrophysical Symposium, Austin, Texas* (1982).

58. M. Turner, F. Wilczek and A. Zee, *Phys. Lett. B* **125** (1983) 35 and **125** (1983) 519.
59. M. Crawford and D.N. Schramm, *Nature* **298** (1982) 538.
E. Witten, *Phys. Rev. D* **30** (1984) 272;
see also C. Alcock and A. Olinto, *Ann. Rev. Nuc. Part. Phys.* **38** (1988) 161.
60. Ya. Zeldovich, *Adv. Astron. and Astrophys.* **3** (1965) 241.
61. B. Lee and S. Weinberg, *Phys. Rev. Lett.* **39** (1977) 165.
62. H.-Y. Chiu, *Phys. Rev. Lett.* **17** (1966) 712.
C.P. Hut, *Phys. Lett. B* **69** (1977) 85;
K. Sato and H. Koyayashi, *Prog. Theor. Phys.* **58** (1977) 1775.
63. S. Dimopoulos, R. Esmailzadeh, L. Hall and N. Tetradis, *Nucl. Phys. B*, submitted (1990).
64. D. Brahm and L. Hall, *Phys. Rev. D* **41** (1990) 1067.
65. K. Griest, M. Kamionkowski, and M. Turner, *Phys. Rev. D* **41** (1990) 3565.
K. Olive and M. Srednicki, M., *Phys. Lett. B* **230** (1989) 78.
66. D. Caldwell *et al.*, in *Proc. La Thuile*, ed. M. Greco;
see also Kamiokande II Collaboration, in *Proc. IUPAP Sumposium on Primodial Nucleosynthesis and Evolution of Early Universe, Tokyo, September 1990*, in press (1991).
67. L. Krauss, *Phys. Rev. Lett.*, in press (1990);
K. Griest and J. Silk, J., U.C. Berkeley preprint (1990).
68. J. Ellis, D. Nanopoulos, L. Roskowski and D.N. Schramm, *Phys. Lett. B* **245** (1990) 251-257.
69. D. N. Schramm, in *Proc. 1988 Berkeley Workshop on Particle Astrophysics*, ed. E. Norman (World Scientific, Singapore, 1989).
70. S.M. Faber and J.S. Gallagher, *Ann. Rev. Astron. and Astrophys.* **17** (979) 135.
71. E. Loh and E. Spillar, *Astrophys. J.* **329** (1988) 24.
72. C. Hill, D.N. Schramm and J. Fry, *Comments Nucl. Part. Phys.* **19** (1989) 25.
D.N. Schramm, in *Proc. IUPAP Symposium on Primordial Nucleosynthesis and Evolution of Early Universe, Tokyo, September 1990*, in press (1991).
73. SAGE collaboration, in *Proc. Int. High Energy Physics Meeting, Singapore, August 1990*, in press (1991).
74. G. Smoot, in *Proceedings of the Tokyo meeting on Primordial Nucleosynthesis, September 1990.* (1990).
75. D. Schneider, M. Schmidt and J. Gunn, *Astron. J.* **98** (1989) 1951-1958.
76. G. Efstathiou and M.J. Rees, *Mon. Not. Royal Astro. Soc.* **230** (1988) 5-11.
77. K. Olive, *Physics Reports* **190** (1990) 309-403.
78. E. Kolb and M. Turner, *The Early Universe,* (Addison Wesley, San

Francisco, 1989).

79 D.N. Schramm, in *Proc. of the 1990 Nobel Şymposium, Graäftåvalen, Sweden* (1990).

80. I. Wasserman, *Phys. Rev. Lett.* **57** (1986) 2234.

81. W. Press, B. Ryden and D. Spergel, *Phys. Rev. Lett.* **64** (1990) 1084.

82. C.T. Hill, D.N. Schramm, L.M. Widrow, in *Proc. of the XXVth Rencontres de Moriond at Les Arcs, Savoie, France, March 1990,* in press, and *Science*, submitted (1990).

83. J. Frieman, C. Hill and R. Watkins, FERMILAB-PUB, in press (1991).

84. M. Geller and J. Huchra, *Science* **246** (1989) 897.

85. T. Broadhurst, R. Ellis, D. Koo and A. Szalay, *Nature* **343** (1990) 726.

86. D.C. Koo and R.G. in *Towards Understanding Galaxies at Large Redshifts*, eds. F.G. Kron and A. Renzini (Kluwer Academic Publishers, Dordrecht, 1990) p. 209.

87. A. Szalay, in: *Proc. IUPAP Conf. Primordial Nucleosynthesis and Evolution of Early Universe, Tokyo, September 1990* (1990).

88. S. Faber and A. Dressler, in *Proc. Texas Symposium on Rel. Astrophysics, Dallas* (1988).

89. N. Bahcall and R. Soneira, *Astrophys. J.* **270** (1983) 20-38.

90. A. Klypin and A. Kopylov, *Soviet Astr. Lett.* **9** (1983) 41-46.

91. J. Primack and A. Dekel, U. of California at Santa Cruz preprint (1990).

92. A. Szalay and D.N. Schramm, *Nature* **314** (1985) 718-719.

93. J. Huchra, Harvard University preprint (1990).

94. S. Vandenburgh and J. West, Dominion Observatory preprint (1990).

95. G. Fuller and D.N. Schramm, University of California at San Diego preprint (1990).

96. C. Hill, and G. Ross, *Phys. Lett. B* **203** (1988) 125.

97. S. Dimopoulos, private communication (1990).

98. S. P. Mikheyev and A. Yu Smirnov, *Nuovo Cimento* **9C** (1986) 17.

99. L. Wolfenstein, *Phys. Rev. D* **17** (1978) 2369.

100. J. Bahcall, *Neutrino Astrophysics* (Cambridge University Press, Cambridge, England and New York, 1989).

101. SAGE colloboration, Int. High Energy Meeting, Singapore, 1990 (1990).

102. M. Gell-Mann, P. Ramond and R. Slansky, R., in *Supergravity*, eds. P. Van Nieuwenhuizen and D. Freedman (North Holland Pub. Co., Dordrecht, 1979),

103. M. Dine, W. Fischler and M. Srednicki, *Phys. Lett. B* **104** (1981) 199.

104. W. Press, B. Ryden and D. Spergel, *Astrophys. J* **397** (1989) 590.

105. H. Hodges, *Phys. Rev. D* **37** (1988) 3052-5.

106. A.J. Stebbins and M.S. Turner, *Astrophys. J.* **339** (1989) L13.

107. L. Widrow, CfA/Harvard preprint (1990).

108. L. Kawano, *Phys. Rev. D* **41** (1990) 1013.

109. A. Massarotti, FERMILAB-PUB-90/77-A (1990) and *Phys. Rev. D*, in press (1991).

110. J. Ipser and P. Sikivie, *Phys. .Rev.* *D* **30** (1984) 712.
111. G. Götz, FERMILAB-PUB-90/35-A (1990).
112. N.Turok, *Phys. Rev. Lett.* **63** (1989) 2625.
113. T.D. Lee, *Phys. Rev.* *D* **35** (1987) 3637.
114. V. Icke and R. van de Weygaert, *Quarterly Journal of the Royal Astronomical Society*, in press (1990).
115. P. Coles, *Mon. Notices Royal Astronomical Society*, submitted (1990).
116. E. Bertschinger, R. Scherrer, J. Vilenson, and A. Van Dalen, Ohio State preprint (1990).
117. M. Rees and D. Sciama, *Nature* **217** (1968) 511.
118. R.K. Sachs and A.M. Wolf, *Astrophys. J.* **147** (1967) 74.
119. M. Turner, R. Watkins and L. Widrow, L., *Astrophys. J. Lett.*, submitted.(1990).
120. G. Götz and D. Nötzold, FERMILAB-PUB-89/235-A (1989).
G. Götz and D. Nötzold, FERMILAB-PUB-89/236-A (1989).
121. K. Olive, D. Schramm, D. Thomas and T. Walker, U. of Minn. preprint (1991).
122. J.J. Simpson, *Phys. Lett.* *B* **174** (1986) 113;
J.J. Simpson and A. Hime, *Phys. Rev.* **D39** (1989) 1825.
123. E. B. Norman, in *Proc. 14th Europhys. Conf. on Nucl. Phys., Bratislava, Czecko-Slovakia* (1990);
B. Sur *et al.*, Lawrence Berkeley Laboratory preprint 30109 (1990).
124. A. Hime and N.A. Jelly, Oxford University preprint OUNP-91-01 (1990).
125. See, for example, discussion by F. Boehm, California Institute of Technology preprint (1991).
126. E.W. Kolb and M.S. Turner, FERMILAB-PUB-91/50-A (1991);
L. Krauss, Yale preprint YCTP-P13-91 (1991);
D. Caldwell and P. Langacker, UCSB-HEP 91-02 (1991);
S.L. Glashow, Harvard University preprint HUTP-90/A075 (1990);
S.K. Babu and R.N. Mohapatra, University of Maryland preprint UMO-PP-91-186 (1991);
L. Dickson and Y. Nir, Stanford preprint SLAC-Pub-5496 (1991);
G. Ross, Oxford preprint (1991).
R. Barbieri and L. Hall, Lawrence Berkeley Laboratory preprint 30465 (1991);
T. Walker, Ohio State preprint (1991).
127. S. Dimopoulos, private communication (1991).
128. K. Olive, D.N. Schramm and G. Steigman, *Nucl. Phys.* **180** (1981) 497-515.
129. R. Gandhi and A. Burrows, *Phys. Lett.* *B* **242** (1990) 77.
130. M.S. Turner, private communication (1991).
131. R. Mayle, J. Wilson and D.N. Schramm, *Astrophys. J.* **318** (1987) 288.
132. D.N. Schramm and J. W. Truran, *Phys. Rep.* **189** (1990) 89.
133. For a summary of basic astrophysical constraints on neutrions see E.W.

Kolb, D.N. Schramm and M.S. Turner, in *Neutrino Physics*, ed. Klaus Winter, in press (Cambridge University Press, 1991).

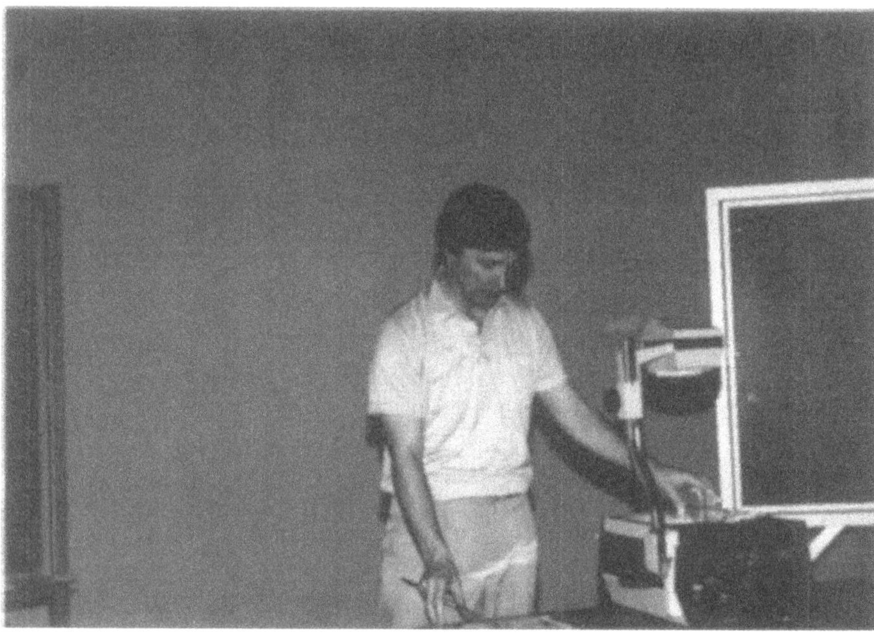

JOHN BARROW

BARYOSYNTHESIS, GRAVITATION AND HOT BIG-BANG COSMOLOGY

John D Barrow
Astronomy Centre
University of Sussex
Brighton BN1 9QH
UK

1. Introduction

This first lecture deals with mathematical and formal aspects of Newtonian and general relativistic gravitation as well as some higher-order lagrangian theories of gravity. It introduces the Friedman metrics and places them in the context of all solutions to Einstein's equations. The emphasis is upon basic principles and ideas that do not lean too heavily upon specific models. Subsequent lectures will discuss the basic physics of the early universe and various schemes for the generation of a primordial baryon asymmetry.

2. Newtonian Gravitation

2.1 Newtonian versus General Relativistic Gravity

Most of intuition regarding the behaviour of self-gravitating systems, like universes, derives from our experience with Newtonian gravitation and so it is important to appreciate the differences between Newtonian and general relativistic theories of gravity. These are summarized in the table below. Most of the contrasts are self-explanatory but some, like the ill-posedness of Newtonian cosmology, require a little more explanation and this will be given below. We stress that if a cosmology were based upon special relativity then it would be defined by a metric (Minkowski space-time, η) and a 4-manifold (R^4) whereas in general relativity the pair (η, R^4) is generalized to (g, M) where g is any Riemannian metric and M any 4-manifold (although additional properties might be required - eg orientability, differentiability etc).

General relativity	Newtonian Gravity
* 10 field equations	1 field equation
* 10 potentials	1 potential
* field equations non-linear (sum of 2 solutions not necessarily a solution)	Field equations linear (sum of solutions is a solution)
* intrinsically geometrical	absolute space (R^3) and time (R)
* permits infinite space	requires finite space
* all energy sources gravitate	only density gravitates
* cosmological problem well-posed	cosmological problem ill-posed
* consistent with special relativity	instantaneous signal propagation
* singularities of space-time	singularities in space-time
* horizons and black holes	no horizons or black holes
* gravitational waves	no gravitational waves

We remark that general relativity is non-linear in an additional sense to that listed above. It is self-interacting in that the carrier of the gravitational force (the graviton) also carries the gravitational "charge" and feels the gravitational force. Gauge theories of this sort are non-Abelian. For example, QCD has this property because gluons carry the colour charge, whereas U(1) electromagnetism does not because the photon does not carry electric charge. As a consequence there could not be a gauge theory of photons alone but there could be a gauge theory of gluons only, or a vacuum theory of general relativity. Another subtlety in the tabulated properties is the veto on black holes in Newtonian theory. The Newtonian "black hole" of Mitchell and Laplace (see Barrow & Tipler 1986a for original references and discussion) is often employed for explanatory purposes but it should be recognised, aside from the illegal extrapolation of the theory to deal with potentials of order c^2, that if a Newtonian object has a radius $R = 2GM/c^2$ then this merely prevents light reaching infinity. There is no veto upon photons crossing the radius $2GM/c^2$, whereas in the general relativistic Schwarzschild solution the barrier created by the event horizon at $2GM/c^2$ is absolute. Finally, the intrinsic coupling of the geometry of space-time to its matter content ensures that infinities in physical quantities create singularities in the fabric of space-time in general relativity and not merely infinities on some indestructible absolute space as is the case in Newtonian theory. Space and time can come to an end in general relativity; they cannot in Newtonian theory.

2.2 The Newtonian Potential and the Cosmological Constant

Most textbooks regard the cosmological constant as an embarrassing creation of general relativity that should be excluded by a demand that the Newtonian limit of general relativity reduce to Poisson's equation. However, it is instructive to see that the cosmological constant also arises inevitably in the right approach to Newtonian theory. Suppose, like Newton, we were concerned to be able to treat spherical masses as though they were point masses located at their centres; then we could ask, what is the most general form of the gravitational potential $\Phi(r)$ such that the external gravitational field of a spherical shell is the same as that of an equal point mass located at its centre? If we consider a shell of radius a, mass M, and surface density σ, and equate the potential due to the shell at some point $r > a$ from its centre to that of a point mass $M = 4\pi a^2\sigma$ at $r = 0$ then we have an integral equation for the potential $\Phi(r)$.

$$M\Phi(r) + 2\pi\sigma a\lambda(a) = \frac{2\pi\sigma a}{r}\int_{r-a}^{r+a} x\Phi(x)\,dx \qquad (2.1)$$

where λ is a constant which can depend upon the (fixed) value of a. Differentiating (2.1), the solution is given by $M = 4\pi\sigma a^2$, $\lambda = 2Ba^2$, B constant, and

$$\Phi = \frac{-GM}{r} + \frac{1}{6}\Lambda r^2 \qquad (2.2)$$

where Λ is an arbitrary constant that we have labelled and prefixed by a numerical factor in order to make it exactly correspond to the weak-field limit of general relativity with a cosmological constant Λ. The potential (2.2) derives from the generalization of Poisson's equation to

$$\nabla^2\Phi + \Lambda = 4\pi G\rho \qquad (2.3)$$

On some occasions (*eg* when interpreting the Newtonian limit of static, straight cosmic strings, (Vilenkin 1985)) it is useful to have a generalization of (2.3) that includes the contributions of the three principal pressures, p_i. In this case (2.3) becomes

$$\nabla^2\Phi + \Lambda = 4\pi G\left[\rho + \sum_{i=1}^{3}\frac{p_i}{c^2}\right] \qquad (2.4)$$

In the standard derivations of general relativity (Weinberg 1972), a linear combination of second-rank tensors involving g_{ab}, Rg_{ab} and R_{ab} is set proportional to the energy momentum tensor T_{ab} and the constant of proportionality $(8\pi G/c^4)$ is derived from the Newtonian limit after associating g_{00} with Φ/c^2. The 'miracle' of general relativity is that a purely mathematical assembly of second-rank tensors should have anything to do with Newtonian gravity in any limit. In fact, this limit could have failed to exist. The 'Newtonian limit' of Einstein's equations in 2 + 1 dimensional space-time does not lead to the analogue of (2.3) in two space-dimensions (Barrow, Burd & Lancaster 1985), but to $\nabla^2\Phi + \Lambda = 0$. However, in N + 1 dimensional space-time with N > 2 the correct Newtonian limit always arises.

2.3 Newtonian Cosmology - fluid formulation

Newtonian cosmology is often employed for ease of explanation in cosmological expositions and calculations. (Detailed treatments can be found in Heckmann & Schücking 1955, 1959, Raychaudhuri 1955, Zeldovich 1965, Ellis 1971, Shikin 1971, 1972 Narlikar & Kembhavi 1980, Barrow & Götz 1989.) For this reason it is important to appreciate when and why it can be believed.

First, let us formulate the cosmological problem as a problem in non-relativistic fluid mechanics in which there exists an external gravitational field $\Phi(\underset{\sim}{x})$. We consider a pressureless fluid with density ρ, velocity vector $\underset{\sim}{v} = \{v^\alpha\}$ and so the three equations defining the theory are:

$$\dot{\rho} + \underset{\sim}{\nabla}.(\rho\underset{\sim}{v}) = 0 \qquad (2.4)$$

$$\dot{\underset{\sim}{v}} + (\underset{\sim}{v}.\underset{\sim}{\nabla})\underset{\sim}{v} = -\underset{\sim}{\nabla}\Phi \qquad (2.5)$$

$$\nabla^2\Phi + \Lambda = 4\pi G\rho \qquad (2.6)$$

The Poisson equation (2.6) is elliptic and Newtonian gravity acts instantaneously. As a result, boundary conditions must be prescribed for Φ at infinity in order to determine the solutions uniquely for given values of $\rho(\underset{\sim}{x})$ and $\underset{\sim}{v}(\underset{\sim}{x})$ at $t = 0$. Unfortunately, there is no mathematical or physical reason *a priori* to dictate which boundary conditions should be chosen. This is in essence the so-called 'gravitational paradox'. We can understand it more fully if we analyse (2.4)-(2.6) further in a homogeneous Euclidean space with Cartesian coordinates $\underset{\sim}{x} = \{x^\alpha\}$ and absolute time, t. This ensures that there exists a simply transitive group of motions for the Euclidean metric space $\delta_{\alpha\beta}dx^\alpha dx^\beta$, and hence that the velocity field obeys a generalized 'Hubble law'.

$$v^\alpha = h^\alpha{}_\beta\, x^\beta \qquad (2.7)$$

The matrix $h^\alpha{}_\beta$ can be decomposed into symmetric ($\theta_{\alpha\beta} = \theta_{\beta\alpha}$) and antisymmetric parts ($\omega_{\alpha\beta} = -\omega_{\beta\alpha}$):

$$h^\alpha{}_\beta = \theta^\alpha{}_\beta + \omega^\alpha{}_\beta \qquad (2.8)$$

and then $\theta^\alpha{}_\beta$ can be split into its trace free (shear) and trace (volume dilation) parts

$$\theta^{\alpha}{}_{\beta} = \sigma^{\alpha}{}_{\beta} + \delta^{\alpha}{}_{\beta}\frac{\dot{a}}{a} \quad : \quad \sigma^{\alpha}{}_{\alpha} = 0 \tag{2.9}$$

where the volume is taken as a^3 for some characteristic length-scale factor $a(x^{\alpha},t)$. If $\omega_{\alpha\beta} \neq 0$ then the flow has non-zero vorticity; if $\sigma_{\alpha\beta} \neq 0$ then it has non-zero shear (ie distortion at constant volume). If $\sigma_{\alpha\beta} = \omega_{\alpha\beta} = 0$ then it will either expand or contract isotropically if $\dot{a} \neq 0$. Using these kinematic quantities we can rewrite the continuity, momentum equations (2.4)–(2.5), as

$$M = \frac{4\pi}{3}\rho a^3 = \text{constant} \tag{2.10}$$

$$3\ddot{a} = -4\pi Ga(\rho + \Lambda - \sigma_{\alpha\beta}\sigma^{\alpha\beta} + \omega_{\alpha\beta}\omega^{\alpha\beta}) \tag{2.11}$$

In addition, the integrability condition for Φ,

$$\frac{\partial^2\Phi}{\partial x^{\alpha}\partial x^{\beta}} = \frac{\partial^2\Phi}{\partial x^{\beta}\partial x^{\alpha}}, \tag{2.12}$$

leads to a vorticity equation whose first integral prescribes the conservation of angular momentum

$$\dot{\omega}_{\alpha\beta} + \frac{2\dot{a}}{a}\omega_{\alpha\beta} + \omega_{\alpha\beta}\sigma^{\gamma}{}_{\beta} + \omega_{\gamma\beta}\sigma_{\alpha}{}^{\gamma} = 0 \tag{2.13}$$

from which we see that if our fluid is shear-free ($\sigma_{\alpha\beta} = 0$) then $\omega \propto a^{-2}$ and the angular momentum $I\omega \propto Ma^2\omega$ is constant (note that such a state is impossible in general relativity, Ellis 1971).

The gravitational potential is a sum of four pieces

$$\Phi = \frac{2\pi G}{3}\rho r^2 + \frac{\Lambda r^2}{6} + \Phi_1 + \Phi_2 \tag{2.14}$$

where $r^2 = \delta_{\alpha\beta}\, x^{\alpha}\, x^{\beta}$. The first term on the right of (2.14) is just the potential for a finite homogeneous sphere of constant density ρ and radius r. The second term, also isotropic, is the contribution of the cosmological constant (see (2.2)). The third term, Φ_1, satisfies Laplace's equation, $\nabla^2\Phi_1 = 0$, and is determined by $\sigma_{\alpha\beta}$ and $\omega_{\alpha\beta}$ as

$$\Phi_1 = -\tfrac{1}{2}(\dot{\sigma}_{\alpha\beta} + \frac{2\dot{a}}{a}\sigma_{\alpha\beta} + \sigma_{\alpha\gamma}\sigma_{\beta}{}^{\gamma} - \frac{1}{3}\delta_{\alpha\beta}\sigma_{\gamma\epsilon}\sigma^{\gamma\epsilon} \qquad (2.15)$$

$$+ \delta^{\kappa\lambda}\omega_{\alpha\kappa}\omega_{\lambda\beta} - \frac{1}{3}\delta_{\alpha\beta}\omega_{\kappa\lambda}\omega^{\kappa\lambda})x^{\alpha}x^{\beta}$$

It is not associated with presence of matter and resembles the (quadratic) potential within a homogeneous ellipsoid. The function Φ_2 is an arbitrary function of time and is a pure integration constant of the Poisson equation.

The equations (2.10)–(2.13) display the 'Newtonian paradox' explicitly. There are only five equations for ten unknowns and the five components of $\sigma_{\alpha\beta}$ remain completely undetermined by the differential equations. We see from (2.15) that this indeterminancy is associated with Φ_1, which accordingly contains five undetermined functions. If homogeneous and isotropic solutions ($\sigma_{\alpha\beta} = \omega_{\alpha\beta} = 0$) solutions are sought (see Milne & McCrea (1934) or Bondi (1960)) then the indeterminancy is removed and Newtonian cosmology is well–posed. However, when the universe is anisotropic ($\sigma_{\alpha\beta} \neq 0$ and, or $\omega_{\alpha\beta} \neq 0$) the cosmological problem is ill–posed. A good deal of the effort expended in studying Newtonian cosmology has focussed upon ways of closing this general ill–posedness; for example, by appending to the theory the shear propagation equations which arise in general relativity, or by adding some other condition which ensures that solutions are the weak–field limit of some general relativistic solution.

Because of the 'gravitational paradox' we have just described there exist Newtonian solutions with no general relativistic analogue (for example there are solutions of (2.10)–(2.13) with $\sigma_{\alpha\beta} = 0$ and $\omega_{\alpha\beta} \neq 0$). Of course, there are also many general relativistic solutions which have no Newtonian analogue. Thus we might schematize the solution spaces of general relativity and Newtonian gravity as sets of which have an intersection but there are Newtonian solutions with no relativistic analogue (which we would regard as unphysical) as well as relativistic solutions with no Newtonian analogue (for example those which possess a non–zero "electric' part of the Weyl curvature tensor associated with anisotropic spatial curvature).

2.4 The Newtonian N–body Problem

The indeterminacy of Newtonian cosmology is not a problem peculiar to the fluid description. It is possible to draw some interesting conclusions about it by studying the Newtonian N–body problem in general. No symmetry assmptions are made here. We assume simply that N is finite. Let m_k, $\mathbf{r}_k$ and $\mathbf{v}_k$ denote the mass, position vector and velocity vector of the k + h particle and let $r_{ij} = |\mathbf{r}_i - \mathbf{r}_j|$. The total kinetic energy T is defined by

$$T = \tfrac{1}{2}\Sigma m_k v_k{}^2 \qquad (2.16)$$

where Σ denotes a sum over k from 1 to N. The total energy, h, is the sum of T and the potential energy, −U, so

$$T = U + h \tag{2.17}$$

We define the moment of inertia of the system by

$$I = \Sigma m_k r_k^2 \tag{2.18}$$

and the angular momentum by

$$\underline{J} = \Sigma m_k(\underline{r}_k \times \underline{v}_k) \tag{2.19}$$

The virial theorem of Lagrange and Jacobi is

$$\ddot{I} = 4T + 2\Sigma \underline{r}_k . \frac{\partial U}{\partial \underline{r}_k} \tag{2.20}$$

If we define $M = \Sigma m_k$ to be the total mass then,from (2.18), we see that

$$IM = \Sigma^* m_j m_k r_{jk}^2 \tag{2.21}$$

where Σ^* denotes the restricted double sum of j and k over $1 \leq j < k \leq N$

Using these definitions we can prove an interesting general singularity theorem. First, take the modulus of equation (2.19), so

$$J \leq \Sigma m_k |\underline{r} \times \underline{v}| \leq \Sigma m_k r_k v_k = (\Sigma m_k^{\frac{1}{2}} r_k)(\Sigma m_k^{\frac{1}{2}} v_k) \tag{2.22}$$

now use the Cauchy-Schwarz inequality so

$$J \leq (\Sigma m_k r_k^2)(\Sigma m_k v_k^2) = 4IT \tag{2.23}$$

This is Sundman's inequality (Pollard 1966).

We will say a <u>strong singularity</u> occurs if all $r_{jk} \to 0$. (This is the analogue of a strong curvature or crushing singularity in general relativity, see Tipler, Ellis & Clarke 1980). From (2.21) we see that it requires $I \to 0$. This cannot occur at $t = \infty$ because if so then since for Newtonian gravity (with $\Lambda = 0$)

$$U = -\Sigma^* \frac{G m_j m_k}{r_{jk}} \tag{2.24}$$

$r_{jk} \to 0$ implies $U \to \infty$ hence, by (2.20), that $\ddot{I} \to \infty$ and so $I \to \infty$ as $t \to \infty$ in contradiction to $I \to 0$. So the strong singularity must occur at some $t = t_* < \infty$. Now, for some interval $t_1 < t < t_*$ we must have $\dot{I} \leq 0$ so we can multiply (2.23) by $-\dot{I}/I \geq 0$ and integrate to obtain

$$\tfrac{1}{4}J^2\log(I^{-1}) \leq EI + k \tag{2.25}$$

where E and k are constants. Now, since $I \to 0$ as $t \to t_*$, the inequality (2.25) ensures that we must have $J^2 \to 0$, but since J is a constant of the motion we must have $J = 0$. Thus an all-encompassing singularity requires zero angular momentum in Newtonian gravity. No such restriction on the rotation is required in general relativity (see Tipler *et al* 1980).

It is possible to ascertain various general results about the asymptotic behaviour of the N-body problem in Newtonian gravity, both with $\Lambda = 0$ and $\Lambda \neq 0$, (see Barrow & Götz 1989). These results all amount to a determination of the asymptotic behaviour of the moment of inertia I, defined by (2.18), as a function of time. The fact that no information is available about the detailed shape of the system, only its coarse-grained volume evolution, is the discrete manifestation of the 'gravitational paradox' and incompleteness of the problem we found in the fluid picture.

3. General Relativity

In this section we shall introduce natural units (Barrow 1983) such that the fundamental constants G, c, $\hbar$ and k_B have values

$$8\pi G = c = \hbar = k_B = 1 \tag{3.1}$$

There are a variety of ways to derive Einstein's equations. The most familiar textbook derivation (*eg* Weinberg 1972), following Einstein, is to equate the most general linear combination of second-rank tensors involving the metric g_{ab} and Ricci curvature tensor R_{ab}, to the energy momentum tensor T_{ab}. Use of the contracted Bianchi identities and the fact that T_{ab} is covariantly constant then gives (in N-dimensional space-time)

$$R_{ab} - \tfrac{1}{2}Rg_{ab} = T_{ab} \tag{3.2}$$

where we have suppressed the cosmological constant because it can be reinterpreted as an effective part of T_{ab} and we have used the units (3.1) after obtaining correspondence with the Newtonian limit to set the Einstein constant $8\pi G/c^4 = 1$.

It is important to note that (3.2) gives a linear relation between the curvature and T_{ab}. Because of this it is possible to contract (3.2) and obtain an equivalent form of the Einstein equations.

$$R_{ab} = T_{ab} - \frac{1}{(N-2)} g_{ab}T \qquad (3.3)$$

This elementary step has profound consequences. All the singularity theorems (see Hawking & Ellis 1973) make no use of the Einstein equations. They use the geodesic equations to establish that if $R_{ab}V^aV^b \geq 0$ for a timelike vector V^a then a singularity is inevitable (given some other technical conditions). In general relativity - and only general relativity - the linearity of (3.2) permits the re-expression (3.3) and hence the Einstein equations can be used to translate the sufficient condition geometrical $R_{ab}V^aV^b \geq 0$ into a statement of physics:

$$\left[T_{ab} - \frac{g_{ab}T}{N-2}\right] V^aV^b \geq 0 \qquad (3.4)$$

which is known as the *strong energy condition* in the N = 4 case. In more complicated theories of gravity, for example those derived from quadratic lagrangians, this nice retranslation of the timelike convergence condition $R_{ab}V^aV^b \geq 0$ into a condition on T_{ab} is not possible.

A second point to make about the equations (3.2), or equivalently (3.3), is that they are quasi-linear. All the familiar equations of physics have this property. It means that they have a generic form.

$$f(g_{ab}, \partial g_{ab})\partial^2 g_{ab} + F(g_{ab}, \partial g_{ab}) = 0 \qquad (3.5)$$

Where ∂ denotes derivatives with respect to any of the space-time coordinates and f and F are some continuous functions.

The essential feature of (3.5) is that the highest (*ie* second) derivatives of g_{ab} occur linearly. Quasi-linear equations have a variety of 'nice' mathematical properties and well-behaved initial value problems.

The second, and perhaps most fundamental, route to Einstein's equations is via an action principle. In the presence of a cosmological constant the gravitational lagrangian is taken as the linear function of the curvature

$$L_g = -2\Lambda + R \qquad (3.6)$$

The matter lagrangian, L_m, is related to the energy-momentum tensor by

$$T^{ab} = \frac{2}{\sqrt{-g}} \frac{\partial}{\partial g_{ab}} (\sqrt{-g}L_m) \qquad (3.7)$$

The standard Hilbert action employed to derive the Einstein equations, with $\Lambda \neq 0$, is

$$S_H = -\tfrac{1}{2}\int\sqrt{-g}\, d^N x (L_g + L_m) \qquad (3.8)$$

However, it has only been shown surprisingly recently (York 1972, Madsen & Barrow 1989) that the variation of S_H leaves an unwanted and uncancelled surface contribution equal to $2\int K d\Sigma$, where K is the extrinsic scalar curvature of the boundary surface Σ. The Einstein equations are therefore obtained from the stationary variation, not of S_H, but of

$$\bar{S}_H = S_H - 2\int K d\Sigma \qquad (3.9)$$

provided only that the variations of the metric, δg_{ab}, vanish on the boundary Σ. The surface element in (3.9) could be rewritten as $dx^{N-1}\sqrt{h}$ where h is the determinant of the spatial metric in (N-1) dimensions. The presence of the "boundary term" in the action (3.9) is an important consideration in attempts to quantize general relativity.

4. Higher-order lagrangians

The derivation of general relativity from an action principle is suggestive of a natural generalization of Einstein's theory to a (possibly quantum) regime of large space-time curvature. We can replace the linear Hilbert lagrangian (3.6) by a higher-order polynomial in one or more curvature scalar invariants. For example, to quadratic aorder we could consider the lagrangian (see Barrow & Ottewill 1983)

$$L_g = -2\Lambda + R + \alpha R^2 + \beta R_{ab}R^{ab} + \gamma R_{abcd}R^{abcd} \qquad (4.1)$$

Where α, β and γ are arbitrary constants. However, in four space-time dimensions we have a variational identity

$$\frac{\delta}{\delta g_{ab}}\int d^4x \sqrt{-g}\,(R^{abcd}R_{abcd} - 4R_{ab}R^{ab} + R^2) = 0 \qquad (4.2)$$

which means that we can set either α, β or γ equal to zero in (4.1) without loss of generality because its contribution to the variation differs from that of the other two quadratic pieces by only a pure divergence. In fact, if our space-time is also homogeneous and isotropic (for example, a Friedman or de Sitter universe) then we have also (in 4-dimensional space-time) that

$$\frac{\delta}{\delta g_{ab}}\int d^4x \sqrt{-g}\,(3R^{ab}R_{ab} - R^2) = 0 \qquad (4.3)$$

and so (4.2) and (4.3) allow us to set β and γ zero without loss of generality.

If we allow the lagrangian L_g to be an arbitrary analytic function of R alone, then the vacuum field equations that result are

$$0 = f'(R)R_{ab} - \tfrac{1}{2}f(R)g_{ab} - \nabla_a\nabla_b f'(R) + g_{ab}\Box f'(R) \qquad (4.4)$$

where ∇_a is the covariant derivative with respect to x^a and $\Box = g_{ab}\nabla^a\nabla^b$. It has been shown by Barrow & Cotsakis (1988 that under a conformal transformation of the metric g_{ab} the field equations (4.4) become those of general relativity with a scalar field matter source. The scalar field has a potential $V(\varphi)$ which is a function of f and f'. This result is of considerable interest for studies of inflation since it reveals which higher-order lagrangian corrections to general relativity produce inflationary behaviour without the presence of explicit scalar matter field sources.

The structure of the boundary term in these lagrangian generalizations of general relativity is extremely complicated. Recently, it has been shown by Madsen and Barrow (1989) that the boundary term can only be expressed in very particular cases. One of them is the case of general relativity that we have just examined (equ (3.9)) but if we have a higher-order lagrangian theory then the boundary term can only be evaluated for particular space-times with high symmetry. For example, if we take $L_g = f(R)$ again, then the boundary term can be calculated if the space-time is maximally symmetric. This result suggests that quantum gravity theories may only be well-posed in very particular cases (like general relativity) or when there is a high level of symmetry.

Solutions possessing maximal symmetry (like de Sitter) are characterized by the property that all the curvature invariants can be expressed in terms of the Ricci scalar:

$$R_{klmn} = \frac{1}{N(N-1)} R(g_{kl}g_{mn} - g_{kn}g_{ml}) \qquad (4.5a)$$

$$R_{mn} = \frac{1}{N} Rg_{mn} \qquad (4.5b)$$

If the gravitational lagrangian is an arbitrary function F(X,Y,Z) where $X = R$, $Y = R_{ab}R^{ab}$ and $Z = R_{abcd}R^{abcd}$ then the condition for the vacuum gravity theory derived from this lagrangian to possess a maximally symmetric (de Sitter) solution (or solutions) is that,

$$\frac{1}{2}F - \frac{1}{N}\frac{\partial F}{\partial X} - \frac{z}{N^2}R^2\frac{\partial F}{\partial Y} - \frac{4R^2}{N^2(N-1)}\frac{\partial F}{\partial Z} = 0 \qquad (4.6)$$

All the higher-order lagrangian theories we have been discussing are fourth-order theories as can be seen from the field equations (4.4) of the f(R) theory. It is of some interest to ask the form of the most general theory that gives rise to second-order field equations. This question has been answered by Lovelock (1971). The Lovelock lagrangian in N-dimensional space-time, is

$$L_* = \sum_{0\leq r<\frac{1}{2}N} \alpha_r L_{(r)} \tag{4.7}$$

where the α_r are arbitrary constants with dimensions of $[\text{length}]^{-2r-2}$ and

$$L_{(r)} = \frac{1}{2^r}\delta^{i_1\ldots i_{2r}}_{j_1\ldots j_{2r}} R^{j_1 j_2}{}_{i_1 i_2} \cdots R^{j_{2r-1} j_{2r}}{}_{i_{2r-1} i_{2r}} \tag{4.8}$$

Here $0 \leq i,\ j \leq N-1$ and $\delta^{i_1\ldots i_{2r}}_{j_1\ldots j_{2r}}$ is the generalized Kronecker symbol. Given the dimension N, we can see from (4.7) that L_* is always a finite sum. In particular $L_{(0)} = 1$, $L_{(1)} = R$, $L_{(2)} = R_{abcd}R^{abcd} - 4R_{ab}R^{ab} + R^2$. When $N = 4$ the term $L_{(2)}$ is a total divergence (cf (4.2)) and $L_{(r)}$ vanishes for $r > 2$. The form of $L_{(r)}$ for $R \geq 3$ in $N \geq 6$ dimensions is prohibitively complicated. The vacuum field equations arising from the Lovelock lagrangian are not quasi-linear, but they do have some special properties that make the initial value problem comparitively well-behaved; they have the form

$$0 = \sum_{0\leq r<\frac{1}{2}N} \alpha_r \, {}^{(r)}G^m{}_n \tag{4.9}$$

where

$${}^{(r)}G^m{}_n = -\frac{1}{2^{r+1}}\delta^{m i_1\ldots i_{2r}}_{n j_1\ldots j_{2r}} R^{j_1 j_2}{}_{i_1 i_2} \cdots R^{j_{2r-1} j_{2r}}{}_{i_{2r-1} i_{2r}} \tag{4.10}$$

So for example,

$${}^{(1)}G^m{}_n = R^m_n - \tfrac{1}{2}\delta^m_n R \tag{4.11}$$

The Lovelock lagrangian has recently attracted considerable interest because of the form of the field theory limit of superstring theories. It appears that the first two terms in the resulting gravitational theory have the form of $L_{(1)}$ and $L_{(2)}$, however it does not appear that the full gravitational theory is given by the Lovelock lagrangian. Various investigations of the solutions of the Lovelock theory up to $L_{(2)}$ order have been made. Cosmological solutions have been found, as well as black hole solutions. One of the novelties of the latter appears to be the possibility of the black hole specific heat becoming positive so that the Hawking evaporation does not evaporate the black hole completely. It ends after a finite time leaving a stable massive relic (Myers & Simon

1988). Finally, we should add that the Lovelock lagrangian is one for which the boundary term in the associated gravitational action can be calculated.

5. The Initial Value Problem

We have introduced both general relativity and several classes of higher-order lagrangian theory with far more complex field equations. In order to appreciate the depth of these theories as well as the likely generality of any particular cosmological solution of them that we may be using it is useful to have a characterization of the general initial data that may be prescribed in each theory. Let us consider a metric in synchronous coordinates

$$ds^2 = dt^2 - g_{\alpha\beta}dx^{\alpha}dx^{\beta}; \quad 1 \leq \alpha, \beta \leq 3 \qquad (5.1)$$

then on any space-like hypersurface of constant t-time we require the values of the six independent components of $g_{\alpha\beta}$ and six independent components of $\dot{g}_{\alpha\beta}$. If we assume that the matter source is a perfect fluid with density ρ, pressure $P(\rho)$ and normalized 4-velocity U^a, then the Cauchy problem also requires that we specify three components of U^{α} and $p(\rho)$. Thus we require 16 spatial functions of 3 variables in 4-dimensional space-time. However, we may eliminate eight of these by using the four Einstein constant equations ($G^0{}_a$ equations) and the four coordinate covariances of the theory and this leaves the general solution specified by eight functions of three spatial variables (four in vacuum). In higher-order lagrangian theories the generic situation (excluding particular forms like the Lovelock one) yields fourth-order field equations and so the initial value problem requires the specification of $g_{\alpha\beta}$, $\dot{g}_{\alpha\beta}$, $\ddot{g}_{\alpha\beta}$ and $\dddot{g}_{\alpha\beta}$ on a t = constant slice. The remaining constraints remain the same and so a four-dimensional perfect fluid model is prescribed by 20 spatial functions of 3 variables. If there are D dimensions of space then the number of arbitrary functions of D variables that must be specified on a spacelike slice of constant time are given in the following table:

Gravitation Theory	Vacuum	Perfect Fluid
general relativity	$(D-2)(D+1)$	D^2-1
higher-order lagrangian	$2(D^2-1)$	$(D+1)(2D-1)$

For a further discussion, including exceptional degenerate cases like the pure R^2 lagrangian see Starobinsky & Schmidt (1987) and Barrow & Sirousse-Zia (1989).

6. **The Friedman Metric**

Most cosmological interest focusses upon the isotropic and homogeneous universes first investigated by Friedman and Robertson and Walker (see Weinberg 1972). They have the metric

$$ds^2 = dt^2 - a^2(t)\left[\frac{dr^2}{1-kr^2} + r^2(d\theta^2 + \sin^2\theta d\varphi^2)\right] \qquad (6.1)$$

where the constant k can be set equal to 0 or ±1 without loss of generality (by using the coordinate transformations $r \to |k|^{\frac{1}{2}}r$ and $a(t) \to |k|^{-\frac{1}{2}}a(t)$ if $|k| \neq 1$). When $k = 0$ the space-like slices of constant t-time have Euclidean geometry. If the spatial topology is R^3 then the volume is infinite. However, with 3-torus topology (identify (x, y, z) with (x + A, y + B, z + C) for some constants A, B, C) the volume is finite. When $k = -1$ the coordinate transformation $r = \sinh\chi$ yields

$$ds^2 = dt^2 - a^2(t)\{d\chi^2 + sh^2\chi\ (d\theta^2 + \sin^2\theta d\varphi^2)\} \qquad (6.2)$$

The t = constant slices are spaces of constant negative curvature. The spatial volume is infinite if the natural R^3 spatial topology is chosen but would be finite for the 3-torus topology. When $k = +1$ the transformation $r = \sin\chi$ produces the form

$$ds^2 = dt^2 - a^2(t)\ \{d\chi^2 + \sin^2\chi(d\theta^2 + \sin^2\theta d\varphi^2)\} \qquad (6.3)$$

The geometry of the t = constant slices is that of a space of constant positive curvature. The natural topology is S^3 and the volume of space is equal to $2\pi^2a^3(t)$.

In all of these space-times the 3-space may be considered as a curved 3-dimensional surface relative to an 'artificial' flat 4-dimensional space. The scale factor, a(t), is the curvature radius of this surface in the fourth dimension. The time derivative $\dot{a}$ describes the expansion speed of the universe in this 'extra' dimension. This expansion does not take place "into" any external 3-space. The scale factor a(t) is not an observable.

Einstein's equations for the Friedman metric when the matter source is a perfect fluid with energy momentum tensor

$$T_{ab} = (\rho + p)U_aU_b - pg_{ab}; \quad U_aU^a = 1 \qquad (6.4)$$

and $\Lambda = 0$, reduce to the pair ($8\pi G = 1 = c$)

$$\ddot{a} = -\frac{a}{6}(\rho + 3p) \qquad (6.5)$$

and

$$\frac{\dot{a}^2}{a^2} = \frac{\rho}{3} - \frac{k}{a^2} \qquad (6.6)$$

The vanishing covariant divergence of T_{ab} gives the fluid conservation equation

$$\dot{\rho} + \frac{3\dot{a}}{a}(\rho + p) = 0 \qquad (6.7)$$

However, this is not independent of (6.5)–(6.6) and any two of (6.5)–(6.7) imply the third. We recognise (6.6) as the conservation of energy for a Newtonian cosmology, but there the constant k has no geometrical interpretation. It merely determines the sign of the total energy of the Newtonian system. Equation (6.5) is a generalization of Newton's second law of motion in which the role of the Newtonian mass $4\pi\rho a^3/3$ is played by the combination $4\pi(\rho + 3p)a^3/3$ because of the special relativistic requirement that all sources of energy (including pressure) have an associated mass (via $E = Mc^2$) and hence gravitate.

If the universe contains many non-interacting fluids, p_i, ρ_i, then ρ and p in (6.5)–(6.6) are replaced by $\Sigma_i\rho_i$ and $\Sigma_i p_i$ and each pair ρ_i, p_i satisfies an equation of the form (6.7).

The coordinates in the Friedman metric are called 'comoving coordinates'. Each point retains its (r, θ, φ) label as the universe expands or contracts with changing a(t). This is equivalent to the choice of 4-velocity $U_a = \delta_a{}^0$, (cf eqn. (6.4)).

In order to solve the system (6.5)–(6.7) we need to specify an equation of state linking p and ρ. If we do not do this then the quations have no content because <u>any</u> function a(t) will solve (6.5)–(6.7) for some p and $\rho > 0$. If we specify a perfect fluid

$$p = (\gamma - 1)\rho: \quad \gamma \text{ constant} \qquad (6.8)$$

then $\rho \propto a^{-3\gamma}$ and there exists a collection of exact solutions involving elementary functions

<u>k = 0</u>: $a(t) \propto t^{2/3\gamma}$

$$\rho = \frac{4}{3\gamma^2 t^2} \qquad \gamma \neq 0 \qquad (6.9)$$

$$a(t) = \exp(H_0 t): \qquad \gamma = 0 \qquad (6.10)$$

$$\rho = 3H_0^2 = \text{constant}$$

$\underline{k \neq 0}$: $\quad a(t) \propto (t - kt^2)^{\frac{1}{2}}$: $\qquad \gamma = 4/3 \qquad$ (6.11)

$$a(t) \propto t \qquad \gamma = 2/3 \qquad (6.12)$$

The solutions (6.11)–(6.12) are the only values of $2 \geq \gamma > 0$ where solutions for all k can be found for a(t). When $\gamma = 1$ the solutions with $k \neq 0$ can be found in terms of elliptic functions or, more transparently, in parametric form

$\underline{k = +1}$:

$$a(\eta) \propto (1 - \cos\eta) \qquad (6.13)$$

$$t(\eta) \propto (\eta - \sin\eta) \qquad (6.14)$$

$\underline{k = -1}$:

$$a(\eta) \propto (\cosh\eta - 1) \qquad (6.15)$$

$$t(\eta) \propto (\sinh\eta - \eta) \qquad (6.16)$$

where the parameters η and t have been chosen so that $\eta = 0$ and $t = 0$ when $a = 0$. Equations (6.13)–(6.14) are the parametric equations of a cycloid in the $t - a$ plane.

It is also worth noting that the only vacuum ($\rho = p = 0$) solution of the Friedman equations with $a \neq$ constant occurs when $k = -1$

$$a(t) \propto t \qquad (6.17)$$

This is called the Milne universe. All the $k = -1$ models with $\rho + 3p > 0$, $\rho > 0$ approach (6.17) as $t \to \infty$ (eg (6.15)–(6.16) as η or $t \to \infty$).

With the equation of state (6.8) and $\gamma > 2/3$, $\rho > 0$ the qualitative behaviour of the Friedman models is dictated by the sign of k. When $k \leq 0$ the models begin at an initial singularity and expand forever. When $k > 0$ they reach an expansion maximum and recollapse. It is conventional to call the $k < 0$, $k > 0$ and $k = 0$ Friedman models 'open', 'closed' and 'flat' respectively. However, one must beware of being brainwashed by this terminology. The $k = +1$ Friedman models do not recollapse if $0 \leq \gamma \leq 2/3$ (one can see from (6.5) that this ensures $\ddot{a} > 0$ and hence a(t) can have no expansion maximum when $\rho + 3p < 0$). Technically, a closed universe is one with compact spatial slices of constant time and in general it need not have either positive curvature or an expansion maximum (Barrow & Tipler 1985, Barrow, Galloway & Tipler 1986, Barrow 1988). It so happens that for the isotropic Friedman models the curvature is always of one sign (determined by ka^{-2} only) and if $\rho + 3p > 0$ the sign of k determines whether there is an expansion maximum or not. In anisotropic cosmologies this correspondence does not necessarily hold.

In general, when there is no equation of state we can show that if $\rho > 0$, $p + 3p > 0$ and $|p| \leq C|\rho|$, C positive constant (or alternatively, if $dp/d\rho$ continuous) then $k > 0$ Friedman models recollapse to a second singularity (Barrow, Galloway & Tipler 1986). The last condition on p and ρ is necessary to prevent a pressure singularity occurring before recollapse is achieved. Under these conditions there is always an initial singularity with $p = \infty$. When $p > 0$ and $p + 3p < o$ there will never be an expansion maximum for any sign of k, and there may or may not be an initial singularity.

6.1 Conformal Time

The most transparent representation of the Friedman universe containing perfect fluid (6.4), (6.8) is obtained by introducing the conformal time, τ, where $dt = a d\tau$, so (Barrow & Tipler 1986a, Barrow 1987),

$$\tau = \int \frac{dt}{a(t)}; \quad ' \equiv d/d\tau \tag{6.18}$$

and defining

$$y = a^{(3\gamma-2)/2} \tag{6.19}$$

Then the Friedman equations reduce to a single equation, the simple harmonic oscillator, for $y(\tau)$

$$y'' = -k\left[\frac{3\gamma-2}{2}\right]^2 y \tag{6.20}$$

For example, the $k = +1$ models with $a(0) = 0$ have $y(0) = 0$, hence

$$a(\tau) = a_{max}\left\{\sin\left[\left[\frac{3\gamma-2}{2}\right]\tau\right]\right\}^{\frac{2}{3\gamma-2}} \tag{6.21}$$

$$t(\tau) = a_{max}\int_0^\tau \left[\sin\left[\frac{3\gamma-2}{2}\right]x\right]^{\frac{2}{3\gamma-2}} dx \tag{6.22}$$

When $3\gamma - 2 > 0$, the $k = +1$ models all have the same total life-time in τ-time whereas they do not have this property in t-time. In τ-time they are distinguished only by their size at the expansion maximum. This general reduction to the simple harmonic oscillator is a useful property to exploit in quantum problems.

6.2 Observable Parameters

For astronomical purposes it is expedient to work with functions of $a(t)$ which are observable. The scale factor $a(t)$ is not an observable and its value may be set equal to unity at any time t by a redefinition of coordinates. We define the *Hubble parameter*, H,

$$H(t) = \frac{\dot{a}}{a} \tag{6.23}$$

and *deceleration parameter*, q

$$q(t) = -\frac{\ddot{a}a}{\dot{a}^2} \tag{6.24}$$

and the *redshift*, z,

$$1 + z = \frac{a_o}{a} \tag{6.25}$$

where the subscript 'o' denotes the present day value of a quantity. If we evaluate the equations (6.5)–(6.6) at the present time, t_o, with $p = 0$ then they become the algebraic relations

$$H_0^2(2q_0 - 1) = ka_0^2 \tag{6.26}$$

$$2q_0 = \frac{8\pi G}{3H_0^2}\rho_0 \tag{6.27}$$

where we have not set $8\pi G = 1$ explicitly in (6.27). The *critical density*, ρ_c, is defined by

$$\rho_c = \frac{3H_0^2}{8\pi G} \tag{6.28}$$

and the *density parameter*, Ω_o, by

$$\Omega_o = \frac{\rho_o}{\rho_c} \tag{6.29}$$

so

$$2q_o = \Omega_o \tag{6.30}$$

we see from (6.26) that the sign of k is determined by the sign of $(q_o - \frac{1}{2})$ or, equivalently, by the sign of $(\Omega_o - 1)$.

When $\Lambda \neq 0$ the Friedman equations (6.5)–(6.6) generalize to

$$\frac{\ddot{a}}{a} = -\frac{1}{6}(\rho + 3p) + \frac{\Lambda}{3} \tag{6.31}$$

$$\frac{\dot{a}^2}{a^2} = \frac{\rho}{3} - \frac{k}{a^2} + \frac{\Lambda}{3} \tag{6.32}$$

and so, for p = 0, the algebraic relations (6.26)–(6.27) generalize to

$$\frac{\Lambda}{3} = \frac{4\pi G}{3}\rho_0 - q_0 H_0{}^2 \tag{6.33}$$

$$\frac{k}{a_0{}^2} = H_0{}^2(2q_0 - 1) + \Lambda = \frac{H_0{}^2}{2}(3\Omega_0 - 2q_0 - 2) \tag{6.34}$$

so the sign of k is fixed by that of $(3\Omega_0 - 2q_0 - 2)$ and $\Omega_0 \neq 2q_0$ when $\Lambda \neq 0$.

7. The Material Content of the Universe

7.1 The Friedman Metric and the Cosmological Principle

The Friedman metric (6.1) is taken as a good first-order approximation to the state of the expanding universe today. Sometimes this step is called the implementation of the *Cosmological Principle* — that the universe be spatially homogeneous and isotropic. Many text-book descriptions of the observational evidence for the Cosmological Principle are rather confused and appeal to the homogeneous distribution of matter in the Universe as the principal justification. As a result, in recent years, when significant non-uniformities have been detected in the distribution of luminous galaxies, the status of the Cosmological Principle may appear to have been undermined. However, the distribution of matter is not the primary evidence for the Cosmological Principle. Although a spatially homogenous distribution of matter is a sufficient condition for the use of the Friedman metric (6.1) to high accuracy it is not a necessary one.

If the over-density above the mean, ρ, on a length scale L is $\delta\rho$, then the most important quantity is the gravitational potential perturbation $\delta\Phi/\Phi$ associated with a density non-uniformity $\delta\rho/\rho$, which is given by

$$\frac{\delta\Phi}{\Phi} \sim \frac{\delta\rho}{\rho}\left[\frac{L}{ct}\right]^2 \tag{7.1}$$

To see this, recall the dimensions of Poisson's equation $\nabla^2\delta\Phi = 4\pi G\delta\rho$ and in a p = 0, k = 0 Friedman background $6\pi G\rho t^2 \sim 1$ and $\Phi \sim c^2$. Equation (7.1) tells us the gravitational potential perturbation (i.e., in general relativity, the metric perturbation) associated with a density inhomogeneity on length scale L where $ct \sim cH_0{}^{-1} \sim 10^{27}$ cm is the horizon size today. We see that we can have very large non-uniformities (e.g. $\delta\rho/\rho \sim 10^6$ on the scale of galaxies and below) but their perturbations to the Friedman metric are completely negligible. Thus it is possible for there to exist significant density inhomogeneities on very large scales without undermining the assumption of the Friedman metric. However, the isotropy of the microwave background temperature around the sky,

$$\frac{\Delta T(\theta,\varphi)}{T} \lesssim 10^{-3} \tag{7.2}$$

couples directly to the gravitational potential perturbation to the space-structure with

$$\frac{\delta\Phi}{\Phi} \sim \frac{\Delta T}{T} \tag{7.3}$$

Hence, we see that the microwave background isotropy, (7.2), over a wide range of angular scales provides the primary evidence for the use of the Friedman metric – i.e. the Cosmological Principle.

7.2 The Age of the Universe

By integrating the Friedman equation (6.32), in the case when $\Lambda = 0$, we can determine the expansion age of the Universe, t_u, since the moment $t = 0$ which has been defined by the condition $a(t = 0) = 0$, in terms of the two observables H_0 and Ω_0 (or equivalently, when $\Lambda = 0$, in terms of H_0 and q_0)

$$t_u = H_0^{-1} f(\Omega_0) \approx 10^{10} h_0^{-1} \text{ yrs} \tag{7.4}$$

where

$$f(\Omega_0) = \frac{2}{3} \qquad \text{when } \Omega_0 = 1 \tag{7.5}$$

$$f(\Omega_0) = \frac{\Omega_0}{2(\Omega_0 - 1)^{3/2}} \cos^{-1}\left[\frac{2}{\Omega_0} - 1\right] - \frac{2}{\Omega_0}(\Omega_0 - 1)^{1/2}; \ \Omega_0 > 1 \tag{7.6}$$

$$f(\Omega_0) = (1 - \Omega_0)^{-1} - \frac{1}{2}\Omega_0(1 - \Omega_0)^{-3/2} \cosh^{-1}(2\Omega_0^{-1} - 1); \ \Omega_0 < 1 \tag{7.7}$$

so we see that $f(\Omega_0)$ is a continuous monotonic function ($f'(\Omega_0) < 0$) with $f(2) = 0.57$, $f(0) = 1$ and $f \lesssim \frac{1}{2}\pi\Omega_0^{-\frac{1}{2}}$. Observations of H_0 and Ω_0 imply that

$$13 \times 10^9 \text{ yr} < t_u < 20 \times 10^9 \text{ yr} \tag{7.8}$$

and $\Omega_0(H_0/100\text{kms}^{-1}\text{Mpc}^{-1})^2 \lesssim 2.5(10^{10}\text{yr}/t_u)^2$. Other data is available: terrestrial rocks tell us $t_u > 3.9 \times 10^9$yr; meteoritic material and lunar rocks give $t_u > 4.6 \times 10^9$ yr, whilst more uncertain age estimates of globular clusters in the Galaxy give values like $t_u \sim 8\text{–}15 \times 10^9$ yr.

7.3 Matter Content

Up to the scale of star clusters the characteristic sizes and masses of material aggregates in the Universe can be understood as equilibrium states between competing forces of gravitational and electromagnetic origin. The characteristic masses, for example, of stars, planetary bodies and planet-based life-forms are determined approximately by the fine structure constant, $\alpha = e^2/\hbar c \sim 1/137$, the gravitational structure constant, $\alpha_G \equiv (Gm_N^2)/\hbar c \sim 10^{-39}$ and the proton mass $m_N \sim 10^{-24}$ gm:

$$m(\text{star}) \sim \alpha_G^{-3/2} m_N;\ m(\text{planet}) \sim \alpha^{3/2} \alpha_G^{-3/2} m_N;\ m(\text{life form}) \sim \alpha^{3/4} \alpha_G^{-3/4} m_N. \tag{7.9}$$

On larger scales the characteristic objects in the universe (e.g. star clusters, galaxies, galaxy clusters) are not solid bodies and we do not yet know if they are stable equilibria determined essentially by constants of physics or whether they owe their sizes to cosmological initial conditions or to random symmetry breakings in the early universe. The principal scales of astronomical structure are given in the table below.

Object	$\delta\rho/\rho$	L	$M/M_\odot$
Globular Star Cluster	10^{10}	0.1 kpc	10^5
Galaxies	10^6	30 kpc	10^{11}
Galaxy Clusters	10^3	1 Mpc	10^{13}
Superclusters	~2-3	30 Mpc	10^{17}

where $1\ M_\odot \equiv 2 \times 10^{33}$ gm and 1 pc $= 3.09 \times 10^{13}$cm.

The matter density of the Universe can be bounded below by direct observations of luminous matter and by indirect deductions of the total amount of gravitating material by the observed motions in systems assumed to be in dynamical equilibrium. Local observations in the disk of the Milky Way yield Ω_0(solar nhd) ~ 0.004; observations of bright spirals give Ω_0(galaxies) $\simeq 0.006$–0.02; studies of the dynamics of binaries and groups give Ω_0(groups) $\simeq 0.04$–0.13; studies of galaxy clusters give Ω_0(clusters) $\simeq 0.1$–0.2. Typically, the density of luminous material is less than the total by a factor ~10 over the scales spanning galaxies to clusters. The identity of the 'missing' dark matter, and whether or not it contributes $\Omega_0 \sim 1$ are major problems that remain unresolved. In what follows we shall regard the total density of the Universe ($\rho_0 = \Omega_0\rho_{0c} = 2 \times 10^{-29}\ \Omega h_0^2$ gmcm^{-3}) as an uncertain parameter lying the the range (Kolb & Turner 1990)

$$\Omega_0 \approx 10^{-2} - 1 \tag{7.10}$$

The oft-repeated story of primoridal nucleosynthesis places important limits on the possible baryonic contribution to the present total density

$$0.01 < \Omega_{BO} < 0.1 \tag{7.11}$$

We shall find it useful to use the baryon number density to describe the density. Assuming the baryons to be protons we may represent the present baryon number density as

$$n_{B0} = 1.1 \times 10^{-5}\,\Omega_0 h_0^2 \text{ cm}^{-3} \tag{7.12}$$

8. Radiation Content

The dominant form of radiation in the Universe has in the microwave band. The recent COBE satellite measurements confirm that this microwave background radiation has a thermal spectrum to high precision with temperature

$$T_{\gamma 0} = 2.735 \pm 0.06 \text{ K} \tag{8.1}$$

This corresponds to a photon number density

$$n_{\gamma 0} = 20T^3 \text{ cm}^{-3}\text{ K}^{-3} \sim 488 \text{ cm}^{-3} \tag{8.2}$$

We see that (7.12) and (8.2) give us a dimensionless number characterizing the Universe

$$\frac{n_{\gamma 0}}{n_{B0}} \sim 10^9 - 10^{10} \tag{8.3}$$

This photon to baryon ratio is closely related to the (conserved) photon entropy per baryon $S \equiv s_\gamma / n_B$ via

$$\frac{n_\gamma}{n_B} = 0.28\ S \tag{8.4}$$

The entropy per baryon S can be regarded as a fundamental cosmological parameter whose numerical value requires an explanation. It describes the balance of matter and radiation in the Universe and hence the statification of the history of the Universe into matter and radiation dominated eras. The radiation density evolves as $\rho_\gamma \propto a^{-4}$ whilst the matter density scales as $\rho_B \propto a^{-3}$. Hence, using the temperature evolution $T_\gamma \propto a^{-1}$ we can compute the age and temperature of the Universe at the epoch, $t = t_{eq}$, when $\rho_B = \rho_\gamma$. We find in terms of S and fundamental constants

$$t_{eq} \simeq \left[\frac{Gm_N^2}{\hbar c}\right]^{-\frac{1}{2}} S^2 \frac{\hbar}{m_N c^2} \simeq 10^{-5}\, S^2 \text{ s} \simeq 10^{10}\,\Omega_0^{-2} \text{ s} \tag{8.5}$$

where the physical constants have been displayed explicitly and m_N is the proton mass. This time corresponds to a universal radiation temperature $T = T_{eq}$, with

$$T_{eq} \simeq 10^5\,\Omega_0 \text{ K} \tag{8.6}$$

Another fundamental epoch which must be identified is that of 'recombination' when the temperature exceeds the ionization temperature of the hydrogen atom, $\Gamma_{rec} \sim 4000$K. This occurs at a cosmic time determined by fundamental constants and S given by

$$t_{rec} \sim \left[\frac{Gm_N^2}{\hbar c}\right]^{-\frac{1}{2}} \alpha^{-3} \left[\frac{\hbar}{m_e c^2}\right] \left[\frac{m_N}{m_e}\right]^{\frac{1}{2}} \simeq 10^8\ S^{\frac{1}{2}} s \simeq 3 \times 10^{12} \Omega_0^{-\frac{1}{2}} s \qquad (8.7)$$

where m_e is the electron mass.

We note the interesting coincidence that, roughly, $t_{eq} \simeq t_{rec}$. This is a consequence of the apparent coincidence that

$$S \sim \alpha^{-2} \left[\frac{m_N}{m_e}\right] \sim 10^9 \qquad (8.8)$$

It remains to be seen if a deduction of S from baryon non-concerning interactions in the very early Universe might replace this coincidental relationship by a consequence of deeper relations between fundamental constants. For example, some recent investigations claim that baryon non-conserving interactions at the electro-weak unification scale lead naturally to an entropy per baryon $S \sim \alpha^{-4}$.

At $t < t_{rec}$ the Universe is an ionized plasma of photons, electrons, nucleons and light nuclei together with various collisionless populations of weakly interacting neutrinos and possibly other unknown elementary particles and gravitons. At time $t < t_{eq}$ the total density is dominated by the radiation content and we say the Universe is radiation dominated. During the radiation era the universe behaves as a good quasi-neutral plasma. This can be confirmed by computing the Debeye number, $N_D \propto n\lambda_D^3$ which gives the number of electrons in a sphere of radius equal to the Debeye length, λ_D, defined by

$$\lambda_D = \frac{k_B T}{4\pi n e^2} \qquad (8.9)$$

where k_B is Boltzmann's constant. We find the Debeye number to be a time-independent quantity and (Barrow 1983a)

$$N_D \simeq 10^4\ S^{\frac{1}{2}} \gg 1 \qquad (8.10)$$

so is required for good quasi-neutral behaviour.

During the radiation era the total density $\rho = \rho_B + \rho_\gamma \simeq \rho_\gamma$ and the curvature term is negligible in the Friedman equation. Hence the dynamics are described by the $k = 0$ Friedman model with blackbody radiation content ($p = {}^1/_3\ \rho$)

$$a(t) \propto t^{\frac{1}{2}} \qquad (8.11)$$

$$\rho = \frac{3}{32\pi G t^2} \qquad (8.12)$$

In general the Universe will contain a radiation sea composed of many relativistic or massless species which may or may not be interacting. If the total effective number of relativistic degrees of freedom at temperature T is g then

$$g = \sum g_b + \frac{7}{8}\sum g_f \tag{8.13}$$

where Σg_b and Σg_f are the totals of bosonic and fermionic species. The total density is

$$\rho = \frac{g}{2}\rho_\gamma = \frac{g\pi^2}{30}T^4 \tag{8.14}$$

(in units where $\hbar = k_B = c = 1$) if all species have temperature T. If we equate (8.12) and (8.14) we obtain the temperature-time relation of the early universe

$$\frac{t}{1s} = 2.4\, g^{-\frac{1}{2}}\left[\frac{1MeV}{T}\right]^2 \sim \frac{m_p g^{-\frac{1}{2}}}{T^2} \tag{8.15}$$

where $m_p = G^{-\frac{1}{2}} \sim 10^{19}$GeV $\sim 10^{32}$K is the Planck mass. Our analysis has assumed the existence of a discrete sum of spin states in (8.13). If the spectrum were continuous then the standard black body radiation theory changes, as do relations (8.11)–(8.12). For example, if $g(m) \propto m^a$, $a > 0$, then the equation of state and density temperature relations become (Barrow & Hawthorne 1990)

$$\rho \propto T^{a+5} \;; \qquad p = \rho/(a+4) \tag{8.16}$$

This temperature-time relation is all that is required to investigate a whole range of elementary particle processes in the early Universe. The occurrence of interactions will depend upon the expansion rate ($H = \dot{a}/a = 1/2t$) of the Universe at temperature T. To orient ourselves concerning the early Universe environment let us discover if it behaves as an ideal gas. At high energies the generic behaviour of elementary particle interactions is that determined by asymptotic freedom so the interaction cross-section is of order

$$\sigma \sim \alpha^2 T^{-2} \tag{8.17}$$

and the number density of equilibrium species is of order

$$n \sim gT^3 \tag{8.18}$$

Hence the mean free path of relativistic species is

$$\lambda \sim (\sigma n)^{-1} \sim (\alpha^2 gT)^{-1} \tag{8.19}$$

By contrast, the mean interparticle spacing is

$$L \sim n^{-1/3} \sim g^{-1/2}\, T^{-1} \tag{8.20}$$

and so

$$\frac{\lambda}{L} \simeq \alpha^{-2} g^{-2/3} \tag{8.21}$$

For ideal gas behaviour we require $\lambda/L \gg 1$. Since typically $\alpha \sim 10^{-2}$ in unified gauge theories these conditions can be satisfied unless the particle populations increase enormously ($g > 10^6$) except for very brief intervals of time when phase transitions occur.

The cross-section (8.17) allows us to determine when such interactions are in equilibrium during the early universe. Equilibrium requires the interaction time $t = (\sigma n)^{-1}$ to be less than the expansion time (8.15). This condition yields

$$T < g^{1/2}\alpha^2 m_p \tag{8.22}$$

For typical value provided by grand unified gauge theories at very high energies ($g \sim 10^2$, $\alpha \sim 10^{-2}$) this reduces to $T < 10^{16}$GeV. Hence, we see that asymptotically free interactions will not in general be in equilibrium at very early times and in order to be confident of applying concepts like "temperature" we must assume that other processes, say, gravitational interactions at $t_p \sim 10^{-43}$s can create equilibrium at $T \sim T_p \sim 10^{19}$GeV. However, as $T \to T_p$ we must beware of pushing conventional consquences of statistical mechanics too far. The number of particles in the causal horizon is just

$$N_\gamma = n_\gamma t^3 \sim g^{-\frac{1}{2}} \left[\frac{m_p}{T}\right]^3 \tag{8.23}$$

and thus becomes very small for $T \to m_p$. A fully quantum treatment of the gravitational field is required as $T \to m_p$. The total mass inside the causal horizon is

$$M_H(t) = \frac{4\pi}{3}\rho t^3 \sim \frac{t}{8\pi G} \sim 10^{38}\left[\frac{t}{1s}\right] \text{gm} \tag{8.24}$$

and we see that the Compton wavelength of this horizon mass $\sim M_H^{-1}$ exceeds its size $\sim t$ when $t < G^{\frac{1}{2}} \equiv t_p \sim 10^{-43}$s.

9. The Baryon Asymmetry of the Universe

9.1 Antimatter

One aspect of the cosmic mass distribution which is of relevance to elementary particle processes is the absence of antimatter in space. Most astronomical observations are made by collecting photons, and so because the photon is a Majorana particle (*ie* its own antiparticle) we cannot tell whether a distant object is made of matter or antimatter just by "looking" at the electromagnetic radiation emitted by it. Instead, we must rely on indirect limits associated with the gamma-ray emission expected from matter-antimatter annihilation and the material content of cosmic rays [Steigman (1976), Stecker (1978), Barrow (1981, 1983a)]. A variety of strong limits exist: none of the planets encountered by the Apollo, Pioneer, Venera and Voyager space probes are made of antimatter and the limits on γ-ray emission by the outer planets show them to be made of matter - the spaceprobes survived their encounters with the inner planets and if the outer planets were made of antimatter the annihilation of solar wind ions on their surface would make them more than a million times brighter in γ-rays than observed. The gamma-ray background also implies that the ratio of antistars to stars in the Milky Way and M31 galaxies is less than 10^{-3}–10^{-4}, and less than $\sim 10^{-10}$ of the interstellar medium is in the form of antimatter. The X-ray clusters, Perseus and Virgo have less than 10^{-5} of their mass in the form of antimatter. If the inter-galactic gas is hot ($T \sim 10^6$K) then less than 10^{-5} can be antimatter, but if it is cool and annihilation of matter and antimatter will be so weak that no limit is obtained. Finally, no antinuclei heavier than antihydrogen ($\bar{p}$) have ever been found in cosmic rays. At energies ~5–12 GeV, $\bar{p}/p \sim 5.10^{-4}$ as would be expected (Stephens, 1981; Szabelski et al., 1980) from secondary interactions of primary cosmic ray protons ($p+p \to p+\bar{p}+p+p$). Past claims by Buffington *et al* to have detected antiprotons with energies too low to make them secondaries have not survived scrutiny of the experimental procedure.

9.2 The baryon number of the Universe

The implication of the observed specific entropy of the University, $S \sim 10^9$, is that during the first microsecond of its expansion history, when $T > m_{pr} \sim 1$ GeV the Universe possessed a non-zero baryon number and an excess of quarks over antiquarks roughly equal to

$$\frac{n_q - n_{\bar{q}}}{n_q + n_{\bar{q}}} \sim 10^{-9} - 10^{-10} \tag{9.1}$$

Confronted with this observation there are three sorts of conclusion that and have been drawn by theorists:

(a) Baryon number is absolutely conserved and since it is proportional to $(n_q - n_{\bar{q}})$, both (9.1) and the matter-antimatter asymmetry of the Universe are consequences of initial conditions at the big bang singularity. This has been the prevailing opinion until the late 1970s, but clearly it does not 'explain' anything.

(b) If baryon number is conserved it might be more "natural" to believe that the baryon number of the Universe if zero rather than a small finite number like $\sim 10^{-9}$. In order to reconcile this with the observed asymmetry we require the baryon number to be spatially-varying, $\nabla_x(n_q - n_{\bar{q}}) \neq 0$. However, this initial state results in prolific annihilation of matter into radiation of symmetry still exists when $T \sim 1$ GeV at $t \sim 10^{-4}$s and predicts a present value of $n_B/n_\gamma \sim 10^{-18}$, too small by $\sim 10^8$. Omnes (1972), tried to argue that a phase transition, occurring when $T \sim 1$ GeV, could produce a natural separation of matter from antimatter in the early Universe before annihilation occurs and consequently annihilation would be far less efficient. Unfortunately this interesting idea did not work. Firstly, the reality of the phase transition was ruled out by Bogdanova and Shapiro (1974); but, aside from that, the "horizon problem" completely undermines Omnes' programme. If the phase transition is a causal process then it can only separate matter and antimatter within regions smaller than the light travel distance at the moment of the transition. The horizon size after $\sim 10^{-3}$s is only about 100 kilometres. The expansion of the Universe expands this to only $\sim 10^{15}$ km by the present, yet matter and antimatter are observed to be separated over gigaparsecs of space. The simple Omnes picture clearly fails. Stecker (1982) has argued that in cosmological models possessing a period of de Sitter expansion ("inflation") early on, the size over which correlation of matter and antimatter domains can be correlated is enormously increased.

(c) The failure of possibilities (a) and (b) leads us naturally to examine the possibility that baryon number is not conserved. In this case the asymmetry (9.1) might arise in a calculable way from some other, or perhaps any, initial choice of the baryon number. However, as we shall see, baryon non-conservation alone is not sufficient to generate a non-zero baryon asymmetry like (9.1); for every interaction generating a net baryon number there might be another destroying it at the same rate.

9.3 Criteria for a baryon asymmetry

If we assume unitarity (all probabilities of interactions add up to 1) and CPT is a good symmetry, then a nonzero baryon asymmetry can be generated if the following conditions hold, (Sakharov, 1967; Kuzmin, 1970).

i) baryon number is not conserved
ii) C and CP are not conserved
iii) there is departure from thermal equilibrium

Condition (i) is necessary if we are to pass from a state with zero baryon number (B), to one with $B \neq 0$; however, it is not a sufficient condition. The C operator changes $n_q - n_{\bar{q}}$ so if C is conserved we must have $n_q = n_{\bar{q}}$ in the system and so $B = 0$. Since the P operator leaves both n_q and $n_{\bar{q}}$ unchanged the CP operator also requires $n_q = n_{\bar{q}}$ and hence $B = 0$. Hence the condition (ii) is necessary. Finally, if thermal equilibrium holds then T is a good symmetry and so CPT symmetry would imply CP symmetry and $B = 0$ by (ii): therefore, we require condition (iii). [For more detail see especially Kolb and Wolfram (1980) and Kolb & Turner (1990)].

An interesting example of a system in which the gravitational interaction meets requirement (i) but neither (ii) nor (iii) is black hole formation and evaporation – the Hawking effect. If we begin with a cloud composed solely of matter ($B > 0$) and allow it to form a black hole then an external observer will be unable to learn anything concerning the baryon number of the matter inside the black hole event horizon (the 'no-hair' theorem). Only the mass, charge and angular momentum can be measured by external observers (this does not mean that the matter inside the black hole ceases to have any other properties, only that they are inaccessible to outside observers). Now, subsequently the black hole can evaporate via the Hawking effect (Hawking, 1974). It does so symmetrically to create a $B = 0$ final state since equal numbers of particles and antiparticles are emitted thermally by the black hole. The process of black hole formation and evaporation has changed the baryon number of a closed system from $B > 0$ to $B = 0$. The gravitational interaction does not conserve baryon number. However, this process alone can never generate a final state in which $B \neq 0$ because CP is conserved in the evaporation process and the evaporation obeys the laws of equilibrium thermodynamics. If the black hole evaporates particles that subsequently undergo B and CP non-conserving interactions it is possible for the decays of these particles to create a net baryon asymmetry (Barrow, 1980; Barrow and Ross, 1981).

9.4 Baryosynthesis in the Early Universe

The general picture for the generation of a non-zero baryon number in the early Universe goes back to the work of Sakharov (1967), who proposed a model in which hypothetical heavy ('lepto-quark') bosons mediated interactions between quarks and leptons. He also pin-pointed the three necessary and sufficient conditions for baryon number generation given above. Later, in 1970, Kuzmin proposed an ad hoc model which added an arbitrary baryon non-conserving part to the lagrangian for CP-violating K^0 $\bar{K}^0$ decay.

With the advent of GUT's it is now possible to produce a more complete and compelling picutre of baryon-number generation because particular theories, like SU(5), predict interactions that violate C, CP and B and which occur out of equilibrium during the early stages of the Universe (Yoshimura, 1978; Ignatiev *et al* 1978; Dimopoulos and Susskind, 1978; Toussaint *et al* 1979; Ellis *et al* 1978; Weinberg, 1979; Dolgov, 1980).

Let us suppose that there exists a population of B, CP and C non-conserving superheavy scalar or vector bosons labelled X and $\bar{X}$ when $T \sim m_X \sim 10^{14}$GeV. Equal numbers of X and $\bar{X}$ bosons should exist (let us assume for the sake of argument that this is the case), and the Universe will be baryon symmetric. After a characteristic time, $\sim 10^{-35}$s, these heavy bosons will decay asymmetrically (due to C and CP violation) into unequal numbers of quarks and antiquarks. These decays will be out of equilibrium if the X-bosons are heavy enough. For simplicity, suppose the X bosons decay into quarks (q) and leptons (ℓ) through two channels with baryon number B_1 and $B_2 \neq B_1$ with branching ratios r and (1−r) respectively. The $\bar{X}$ bosons will decay into similar channels with branching ratios $\bar{r}$ and $(1-\bar{r})$ say; because C and CP are not conserved we will have $r \neq \bar{r}$, (Weinberg, 1979)

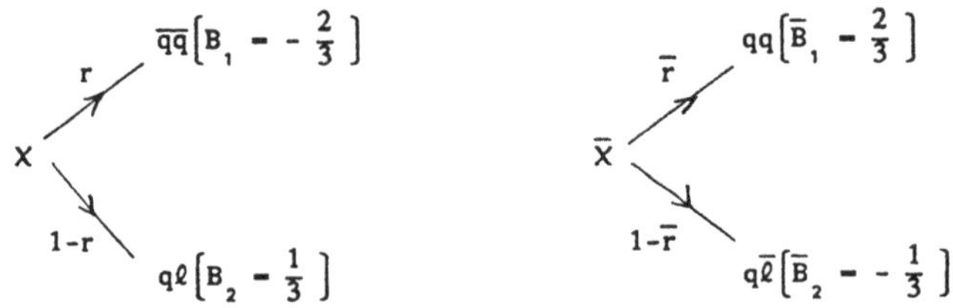

(9.2)

The mean net baryon number created by each X and $\bar{X}$ decays is respectively

$$B_X = rB_1 + (1-r)B_2$$
$$B_{\bar{X}} = -\bar{r}B_1 - (1-\bar{r})B_2 \qquad (9.3)$$

So, the decay of an $X\bar{X}$ pair produces, on average, a baryon number ϵ

$$\epsilon \equiv B_X + B_{\bar{X}} = (r-\bar{r})(B_1-B_2) \qquad (9.4)$$

If C and CP are not conserved then $r \neq \bar{r}$; (note that baryon minus lepton number is conserved in these decays which is a feature of some GUTs, Langacker, 1981).

The X and $\bar{X}$ bosons decay at a rate

$$\Gamma_d \sim g_d a_X m_X^2 (T^2+m_X^2)^{-\frac{1}{2}} \qquad (9.5)$$

where g_d is the number of decay channels (=2 x no. of generations = 6 in the minimal model). The $X\bar{X}$ decay proceeds efficiently until the cosmological expansion rate

$$H = \frac{1}{2t} = 1.66 g_*^{\frac{1}{2}} T^2 m_p^{-1} \qquad (9.6)$$

equals Γ_d. This occurs at a temperature, T_D, where

$$T_D \sim (a_X g_d^{\frac{1}{2}} m_X m_p)^{\frac{1}{2}} \qquad (9.7)$$

and a time,

$$t_D \sim (a_X g_d m_X)^{-1} \sim 10^{-35} s \qquad (9.8)$$

To generate the baryon number (9.4) these decays must occur out of equilibrium. Originally Weinberg (1979) gave the criterion for non-equilibrium decay as

$$m_X > T_D \qquad (9.9)$$

which leads to, (Barrow, 1980a),

$$m_X > 2.10^{-2} g_d \alpha g_*^{-\frac{1}{2}} m_p \qquad (9.10)$$

that is to (and this is significantly smaller than Weinberg's original estimate.)

$$m_X > 0.3\alpha m_p \equiv m_* \qquad (9.11)$$

For gauge vector bosons (X,Y), $\alpha \equiv \alpha_X \sim 1/45$ so $m_X \gtrsim 10^{17}$ GeV; whereas for scalar (Higgs) bosons, $\alpha \equiv \alpha_H \sim \alpha_X(m_f/m_w)^2 \sim 10^{-4}$, (where m_f and m_w are light fermion and W boson masses), we have a more acceptable, $m_H \gtrsim 10^{12}$–10^{14} GeV.

The number density of heavy bosons before decay is thermal

$$n_X \sim \frac{\zeta(3)}{\pi^2} g_X T_D^3 \tag{9.12}$$

and the total entropy

$$s \sim \frac{4\pi^2}{45} g_*(T_D) T_D^3 \tag{9.13}$$

Therefore, if ϵ is the mean net baryon number produced in an $X\overline{X}$ decay, we end up generating as specific entropy, (Weinberg, 1979)

$$S^{-1} = \frac{n_B}{s} \sim \frac{45\zeta(3)}{4\pi^4} g_* \sim 10^{-2} \epsilon \tag{9.14}$$

A CP violation of order $\epsilon \sim 10^{-7}$ is necessary to explain the observed specific entropy $S \sim 10^{-9}$. In the standard three generation SU(5) model, which we use merely for illustration, as it is known to be in conflict with other observations. Calculations show $\epsilon \angle 10^{-16}$ and extra fermionic families need to be added to yield $\epsilon \sim 10^{-8}$ (other complications also exist which might help, Kolb & Turner, (1990). In general one knows only that $|\epsilon| \angle \alpha$ and in fact the sign of ϵ is also unknown in the sense taht a "definition" of matter in $X\overline{X}$ decay may not necessarily agree with the convention adopted in other CP-violating systems, for example neutral kaon decay.

Since the original paper of Weinberg (1979a) it has been realized that condition (9.9), although *sufficient* for non-equilibrium decay, is by no means necessary. Accordingly, the result (9.14) give the *maximum* value of n_B/s that can be generated in GUT's. When $m_X \angle T_D$ there is still only partial equilibrium because the equilibrium temperature is constantly falling as the Universe expands; the Boltzmann exponential cut-off is moderated into a power-law by the expansion and we can gauge the effectiveness of decays by a parameter, (Fry et al., 1980),

$$K = \left[\frac{\Gamma_X}{H}\right]_{T=m_X} = \frac{m_*}{m_X} \tag{9.15}$$

$$\sim \frac{3.10^{17}\alpha}{m_X} \text{ GeV} \tag{9.16}$$

K gives the effectiveness of the $X\overline{X}$ (or $Y\overline{Y}$ or $H\overline{H}$) decays. When $K<1$ the non-equilibrium is strong, condition (9.9) holds, and the maximal baryon asymmetry (9.14) is generated. When $1 < K < K_c$, where K_c is a critical value $\sim 10^2$–10^4, then the non-equilibrium is partial, and numerical studies reveal a simple prediction that

$$\frac{n_B}{s} \sim \frac{g_X}{g_*} \epsilon(1+K)^{-1.3} \tag{9.17}$$

When $K>K_c$ the baryon number is driven to zero very rapidly by baryon non-conserving

2–2 scatterings of quarks and leptons, their rate is

$$\Gamma_{22} \sim g_d\, \alpha^2 T^5 (T^2 + m_X^2)^{-2} \qquad (9.18)$$

and thus becomes equal to the cosmological expansion rate, H, when

$$K = K_c \sim \frac{g_d}{g_*}\, \alpha^{-1}\ . \qquad (9.19)$$

When $K \gtrsim K_c$ the baryon number is damped exponentially (Barrow, 1980b; Fry et al, 1980) and

$$\frac{n_B}{s} \sim \epsilon\, \frac{g_X}{g_*}\, K^{5/6} \exp(-\alpha^{1/3} K^{1/3}) \qquad (9.20)$$

Various features of the evolution of the baryon asymmetry with temperature are worth noting : first, there is a steady build-up of n_B/s in the $K < 1$ cases, leading to a peak value that subsequently falls by a factor of 10–10^2 because of the 2–2 scatterings occuring in equilibrium. Since they are B violating, they drive $B \to 0$ but are eventually cut-off because they proceed too slowly to compete with cosmological expansion. Thereafter the baryon number is steady. The smaller the value of K, the bigger the final asymmetry until the baryon number saturates and it is insensitive to the numerical value of the coupling α except when K is very large.

The calculations that have been performed in supersymmetric theories to determine the value of m_X from the energy at which all interactions have the same effective strength, yield a value of $m_X \sim 10^{16}$ GeV, which corresponds to a K value for X decays of,

$$K_X \sim 5 \qquad (9.21)$$

9.5 Baryosynthesis in vacuum bubble collisions

Recent studies of the possible occurence of phase transitions in the early universe have led to the proposal of new mechanisms for the generation of a baryon asymmetry. The most effective appears to be the occurrence of a first order phase transition that is completed by bubble collisions. This type of process was originally proposed by Guth (1981) as a means of effecting ('old') inflation. Unfortunately, in his original inflationary universe model the phase transition could not proceed to completion and the bubbles of true vacuum could not completely percolate to change the state of the Universe from false to true vacuum everywhere. Recently, it has proved possible to rescue this idea in the context of inflationary models which expand with power-law ($a \sim t^n$, $n > 1$) rather than exponential ($a \sim \exp(H_0 t)$, H_0 constant) evolution of the scale factor (La & Steinhardt 1989, Barrow & Maeda 1990). These models (or at least those in which the power-law expansion arises because of a Brans–Dicke scalar field) have become known as 'extended' inflationary universes. Here, we shall be interested in the possibility of generating baryon non-conserving super-massive scalar Higgs bosons in the bubble wall collisions which complete the phase transition. Their out-of-equilibrium decay generates a natural baryon asymmetry and cosmic entropy per baryon ratio. The following account is based upon the investigation by Barrow, Copeland, Kolb and Liddle (1991).

The thermalization process of bubble wall collisions at the end of extended inflation provides a natural arena for baryogenesis in the early Universe, as it automatically creates conditions far from thermal equilibrium, exactly as required for B, C, and CP violating GUT processesto create a asymmetry.

In order to keep our discussion as general as possible, we will not specify any particular inflation model, but rather describe the salient features of the potentials in terms of a few parameters that can be easily identified with any scalar potential that undergoes spontaneous symmetry breaking. We denote the inflaton field throughout to be σ, which has a potential of the general form suitable to provide a first-order phase transition necessary for extended inflation. The parameters of the potential are asssumed to be :

1. σ_0, the energy scale for SSB, i.e., the VEV of the scalar field.
2. λ, a dimensionless coupling constant of the inflation potential. We will assume that the potential is proportional to λ
3. ξ, a dimensionless number that measures the difference between the false and the true vacuum energy density via $\rho_V = \xi\lambda\sigma_0^4$.

The parameter ξ must be less than unity for sufficient inflation to occur; this is also precisely the condition that allows the thin-wall approximation (to be discussed below) to be made.

From these few parameters we can find all the information we require about the bubbles formed in the phase transition (Callan & Coleman 1977). For instance, an important parameter is the size of bubbles nucleated in the tunnelling to the true vacuum. In the thin-wall approximation, the size of the nucleated bubble is given by

$$R_C \sim (\xi\lambda^{\frac{1}{2}}\sigma_0)^{-1} \tag{9.22}$$

Bubbles smaller than this critical size will not grow, and it is exponentially unlikely that we shall nucleate bubbles larger than this critical size. We will assume that all the true-vacuum bubbles are initially created with size $R = R_C$.

Another crucial parameter is the thickness of the bubble wall separating the true-vacuum region inside from the false-vacuum region outside the bubble. For the potential described above, the bubble wall thickness is

$$\Delta \sim (\lambda^{\frac{1}{2}}\sigma_0)^{-1} \tag{9.23}$$

Note that the ratio of the bubble thickness to its size is $\Delta/R_c \sim \xi$; so, if $\xi \ll 1$, the thin-wall approximation is valid.

Finally, the energy per unit area of the bubble wall is

$$\eta \sim \lambda^{\frac{1}{2}}\sigma_0^3. \qquad (9.24)$$

At the end of extended inflation all of the energy is in these bubble walls.

We must have some idea of the size of bubbles at the end of inflation, when bubbles of true vacuum percolate, collide, and release the energy density tied up in the bubble walls, so creating the entropy of the Universe. The bubbles of true vacuum are nucleated with size $R = R_c$. After nucleation the bubble will grow until it collides with other bubbles. We can show that the size of the bubble at the end of extended inflation is still approximately R_c.

The physical size of the bubble nucleated at time t_{NUC} is related to its coordinate size by $R(t_{NUC}) = r(t_{NUC})a(t_{NUC}) = R_c$. If there is negligible growth in the coordinate size of the bubble between the t_{NUC} and the end of inflation t_{END}, then at the end of inflation the bubble will have a physical size

$$R(t_{END}) \equiv R = r(t_{NUC})a(t_{END}) = R_c[a(t_{END})/a(t_{NUC})] \qquad (9.25)$$

With the expectation that successful extended inflation will produce a burst of bubble nucleation at the end of inflation, and that there is not much growth in the size of a typical bubble between the time of nucleation and when it collides with other bubbles at the end of inflation, we will assume that at the end of extended inflation all bubbles have the same size, $R = \alpha R_c$, where $\alpha \equiv a(t_{END})/a(t_{NUC}) \sim 0(1)$.

At the end of extended inflation the Universe is percolated by bubbles of true vacuum of size $R = \alpha R_c$, with all the energy density residing in the bubble walls. We have spoken of the "end of inflation" as if it was a well-defined time, but in fact it is not. We simply define the end of inflation to be when the Universe is first percolated by bubbles of true vacuum. The "filling fraction" of the Universe at percolation is some fraction f_V. Based upon Monte-Carlo simulations, (Hawking, Moss & Stewart 1982), it is expected that $f_V \sim 0.3$. The final step is the release of this energy into radiation via bubble-wall collisions. Let us concentrate on a single bubble of radius R. The total mass of the bubble is

$$M = 4\pi\eta R^2 \sim 4\pi\lambda^{\frac{1}{2}}\sigma_0 R^2 \qquad (9.26)$$

The collisions of the bubble walls produce some spectrum of particles, which are subsequently thermalized. The mean energy of a particle produced in the collision is of order of the thickness of the wall, $\langle E\rangle \sim \Delta^{-1}$, and hence the mean number of particles produced in the collisions is

$$\langle N \rangle \simeq M/\langle E \rangle \sim 4\pi\Delta\lambda^{\frac{1}{2}}\sigma_0^3 R^2. \qquad (9.27)$$

In general, the bubble collisions will produce all species of particles, or at least all species with masses not too large compared to $\langle E \rangle$. In what follows we will assume that this is the case for the baryon-number violating Higgs particles. If the Higgs mass exceeds Δ^{-1} by a significant amount, we can expect some suppression, presumably exponential, in the number of Higgs formed. This possibility will be discussed at the end of this section. For now, we simply parameterize the fraction of the primary annihilation products that are supermassive Higgs by a fraction f_H. The typical number of Higgs particles produced per bubble is

$$\langle N_H \rangle \sim f_H \langle N \rangle \sim 4\pi f_H \Delta\lambda^{\frac{1}{2}} \sigma_0^3 R^3. \qquad (9.28)$$

We will now assume that the only source of supermassive Higgs is from the primary particles produced in the bubble-wall collision. This will be true if the reheat temperature, T_{RH}, is below the Higgs mass. The validity of this approximation will also be discussed later.

The Higgs particles produced in the wall collisions decay, producing a net baryon asummetry ϵ per decay. hence, the excess the baryons over antibaryons produced from a single bubble,

$$N_B = \epsilon \langle N_H \rangle \sim 4\pi\epsilon f_H \sigma_0^3 R^2, \qquad (9.29)$$

where we have substituted in for the bubble thickness from Eq. (9.23). This results in a baryon number density of

$$n_B = f_V N_B/(4\pi R^3/3) = 3\epsilon f_V f_H \sigma_0^2 R^{-1}. \qquad (9.30)$$

We must now calculate the entropy generated in bubble-wall collisions. As stated above, the mass of a bubble is $M = 4\pi\sigma_0^3\lambda^{\frac{1}{2}}R^2$. The thermalization of the mass in the bubble walls will redistribute this energy throughout the bubble, resulting in a radiation energy density (recall that at percolation, a fraction f_V of the Universe is taken up by bubbles of radius R) of order

$$\rho_\gamma \sim f_V M/(4\pi R^3/3) \sim 3 f_V \lambda^{\frac{1}{2}} \sigma_0^3/R. \qquad (9.31)$$

We may now express the reheat temperature to the radiation energy density via

$$\rho_\gamma = \frac{g_* \pi^2}{30} T_{RH}^4 , \tag{9.32}$$

where g_* is the effective number of degrees of freedom in all the species of particles which may be formed in the thermalization process. From this we obtain the entropy density s, produced by the thermalization of the debris from bubble-wall collisions:

$$s = \frac{2\pi^2}{45} g_* T_{RH}^3 \sim 2.3 g_*^{\frac{1}{4}} f_V^{\frac{3}{4}} \lambda^{\frac{3}{8}} \sigma_0^{9/4} R^{-\frac{3}{4}}. \tag{9.33}$$

From Eqs (3.5) and (3.8) we can calculate the baryon asymmetry B as

$$B = n_B/s \sim \epsilon f_V^{\frac{1}{4}} f_H g_*^{-\frac{1}{4}} \lambda^{-\frac{3}{8}} (\sigma_0 R)^{-\frac{1}{4}}. \tag{9.34}$$

As stated above, a successful extended inflation model is expected to have the feature that a typical bubble does not grow significantly between nucleation and percolation, so we can assume that the bubble size at percolation is $R = \alpha R_c = \alpha(\xi \lambda^{\frac{1}{2}} \sigma_0)^{-1}$. The baryon asymmetry B is then given by

$$B = \epsilon f_V^{\frac{1}{4}} f_H \alpha^{-\frac{1}{4}} g_*^{-\frac{1}{4}} \lambda^{-\frac{1}{4}} \xi^{\frac{1}{4}}. \tag{9.35}$$

Provided the mass of the Higgs is less than T_{RH}, the fraction f_H is just g_H/g, where g_H is the number of Higgs degrees of freedom. Substituting this gives the final result

$$B = \epsilon f_V^{\frac{1}{4}} g_H \alpha^{-\frac{1}{4}} g_*^{-5/4} \lambda^{-\frac{1}{4}} \xi^{\frac{1}{4}}. \tag{9.36}$$

This allows us to make numerical estimates of B based on sample values of these parameters. Notice that the dependence on both λ and ξ, which are the two parameters on which the inflaton's potential depends, is very weak, as is that on the filling factors f_V and α. The important contributions are ϵ, the degree of asymmetry in CP-violating Higgs decays, and the number of particle species available for production in the wall collisions.

Now we examine some typical numbers for the baryon asymmetry which may be obtained from Eq. (9.36), in the light of the experimental limits discussed in the introductory section setting B at around 10^{-10}. The two filling factors f_V and α both appear to the quarter power, and hence as they are both of order one, they can be neglected to a good approximation. The number of Higgs degrees of freedom, g_H, is expected also to be of order one, with simply one degree of freedom for each polarization in the case of a single Higgs and further degrees in the case of a doublet or more of Higgs particles. The total number of degrees of freedom g_* is epected to be of order 100–800 in a grand unified theory. This implies

$$B \sim 10^{-2} \, \epsilon \left[\frac{\xi}{\lambda} \right]^{\frac{1}{4}} . \tag{9.37}$$

The remaining parameters ϵ, λ and ξ are less certain, with some dependence on the particlar unified paticle theory under examination, though it is reassuring that both λ and ξ also enter only to the quarter power and hence the dependence on these quantities is weak. This does however have the further implication that ϵ should be very small, as we shall shortly see. That a suitable baryon asymmetry can be produced with such a small ϵ indicates that the bubble wall collisions are very efficient in producing a baryon asymmetry. A reasonable estimate of ξ is that it may be of order 10^{-2} (recalling $\xi < 1$ is the condition both for sufficient inflation and for the thin-wall approximation to be valid). The parameter λ should probably be less than of order 1, though nothing in principle prevents it from being much smaller; note that a small λ increases the baryon asymmetry as it leads to a less efficient production of entropy, though λ must also be sufficiently large that Higgs particles can be produced in the wall collisions. From these arguments, it seems likely that the ratio ξ/λ will be within a few orders of magnitude of unity, implying that if this mechanism is to generate the appropriate baryon asymmetry ϵ must be of order 10^{-3}. This argument will be made tighter below.

There are constraints that must be satisfied in order for this scenario to work. As mentioned at the end of proceeding section, the typical energy of particles produced in wall collisions, Δ^{-1}, should exceed the Higgs mass. (If this does not hold, then f_H will have an extra suppression factor. While this may allow a large ϵ it will most likely require some fine tuning of the amount of suppression). A further constraint is that the reheat temperature be less than the Higgs mass in order to prevent the Higgs particles produced in the wall collisions reaching a state of thermal equilibrium. These two constraints translate into an upper and lower bound for λ_H (neglecting the volume factors and substituting for g_* as before)

$$\lambda^{\frac{1}{2}} > \lambda_H^{\frac{1}{2}} > \frac{1}{\sqrt{10}} \left[\frac{\xi}{\lambda} \right]^{\frac{1}{4}} \lambda^{1/3} . \tag{9.38}$$

Clearly suitable values of λ_H are only possible provided

$$\left[\frac{\xi}{\lambda} \right]^{\frac{1}{4}} < \sqrt{10}, \tag{9.39}$$

but to allow a range of λ_H this bound should be stronger. Although this provides the interesting constraint on ξ and λ that $\xi < 100\lambda$, it is not a particularly strong one and it leads to a weak lower band on ϵ of around 10^{-8}. Hence this baryogenesis scenario appears satisfactory for a large set of possible model parameters.

We should discuss more general requirements of the model in order for this type of entended inflation scenario to be considered as a sensible candidate for barygenesis. One important requirement is that the symmetry breakings do not lead to an unacceptable density of relic monopoles (monopoles being the inevitable outcome of any symmetry breaking from a semi-simple group to the standard model). In several theories this can be arranged by creating the monopoles in a pre-inflationary breaking. The monopoles are then subsequently diluted during the inflationary era and present no further problems. For example, in a specific model proposed by Olive and Turok (1982) and Turok (1984), SO(10) is broken in a two step process

$$\begin{aligned} SO(10) &\to SU(3) \otimes SU(2)_L \otimes SU(2)_R \otimes U(1)_{B-L}, \\ &\to SU(3) \otimes SU(2)_L \otimes U(1)_V \otimes Z_2, \end{aligned} \qquad (9.40)$$

where the first breaking is through the 45-dimensional representation of SO(10), and the second breaking is though the 126-dimensional representation (Harvey et al 1982). In the symmetry breaking scheme of Eq.(9.40), monopoles are produced at the first transition, but not at the second (existing monopoles are converted rather than new ones formed, (Turok 1984)). Hence, in such a picture at least two symmetry breakings are needed to reach the standard model, the latter causing both inflation and the out of equilibrium conditions required for baryogenesis. It is even possible for this later transition to produce cosmic strings as defects in the inflaton field (Copeland et al 1990).

We should also comment on the possible role of the baryon-number violating anomalous currents in the non-perturbative sector of the electroweak theory. (A more detailed study can be found in the lectures by Turok in this volume). It has been conjectured by Kuzmin et al (1985) that this anomalous current may cause a "wash-out" of any pre-existing baryon asymmetry at temperatures above the electroweak phase transition as the barrier height between sectors of different baryon number may only be around 10 TeV, and baryon number non-conserving interactions could then occur in thermal equilibrium. If these calculations prove correct, then this may destroy any baryon asymmetry generated in the wall collisions (this would wash out *any* primordial baryon asymmetry of course, and is not a problem specific to the scenario we are proposing here). It has also been proposed that the electroweak phase transition may actually create a suitable baryon asymmetry but so far these models have exhibited only limited success. The effect of sphalerons may be mitigated if a non-zero value of B − L is generated, as is possible with the breaking scheme of Eq.(9.40).

In conclusion then, we see that the result of the first-order phase transition bringing extended inflation to an end is an environment well out of thermal equilibrium. In such conditions baryogenesis via the decay of baryon number violating Higgs particles can procede, and for a wide range of model parameters produces a baryon asymmetry of the required magnitude.

9.6 Baryosynthesis by Primordial Black Holes

One can also derive a baryon asymmetry from violent primordial events like bubble wall collisions by the medium of mini black hole formation and Hawking evaporation (Barrow, Copeland, Kolb, and Liddle 1991a). One of the implications of the "no-hair" theorems of black hole physics is that baryon number is not conserved in black hole evaporation. In 1975 Hawking suggested that evaporating primordial black holes might radiate an excess of baryons over antibaryons (see Carr. 1976) This idea was taken up again in the context of GUTs by many people (Barrow & Tipler 1978, Toussaint et al 1979).

The violence of bubble wall collisions may well produce a significant number of black holes as well as relativistic particles, because of the gravitational potential fluctuations generated by the collisions. When such black holes decay by the emission of Hawking radiation, they may emit baryon number violating Higgs particles whose decays generate the baryon asymmetry.

The mechanism of baryogenesis by evaporation of primordial black holes divides into two sub-categories (Barrow 1980, Barrow et al 1991a). In the first version, evaporation occurs while radiation dominates the energy density of the Universe, with the black holes providing the baryon asymmetry but with the entropy arising from the background radiation. Since radiation energy density falls off faster than that of matter, the contribution of the black hole energy density becomes more important as time goes by. If the time until radiation domination is less than the black hole lifetime, we get a second scenario in which the black hole density dominates at the time of evaporation. Here the black holes provide both radiation entropy and the baryon number.

Once formed, the black holes evaporate at a rate given by

$$\dot{M}_{BH} = -\frac{g_*}{3}\frac{m_p^4}{M_{BH}^2}, \tag{9.41}$$

which leads to a time dependence of black hole mass of

$$M_{BH}^3(t) = M_0^3 - g_* m_p^4\,(t - t_{RH}). \tag{9.42}$$

It is convenient to define a black hole lifetime,

$$\tau \equiv M_0^3/g_* m_p^4, \tag{9.43}$$

and the expression for the mass as a function of time becomes $M(t) = M_0[1 - (t - t_{RH})/\tau]^{1/3}$. The evaporation ends at time $t_{BH} = t_{RH} + \tau$.

Black holes radiate as black bodies with temperature $T_{BH} = m_p{}^2/8\pi M_{BH}$. This allows us to calculate what is, for our purposes, the most important quantity - the number of particles emitted during the course of the evaporation. Let us first calculate the number of particles emitted while the black hole is between the temperatures T and T + δT. The change in mass of the black hole, δM, which is the amount of energy radiated as particles, is given by

$$\delta M = \frac{m_p^2}{8\pi} \left[\frac{1}{T} - \frac{1}{T + \delta T} \right] . \tag{9.44}$$

Each emitted particle has energy 3T (the mean energy of a particle in a Maxwell–Boltzmann distribution at temperature T), so the number of emitted between those temperatures is just

$$\delta N = \frac{m^3{}_P}{24\pi T} \left[\frac{1}{T} - \frac{1}{T + T\delta} \right] = \frac{m^3{}_P}{24\pi T^3} \delta T. \tag{9.45}$$

Integrating this, we find that the number of particles emitted as the black hole tmperature increases from its initial temperature to T_0 to ∞ is $N = 4\pi M_0{}^2/3m_p{}^2$.

There are two possible sources for the formation of small primordial black holes. Firstly, holes may form via the grativitational instability of inhomogeneities created during the thermalization phase, particularly during the wall collisions themselves where we can expect high local densities to prevail. Secondly, there is the possibility of trapped regions of false vacuum (within their Schwarzschild radii) caught between bubbles of true vacuum.[18] In the context of our model, this latter possibility seems unlikely for the following reason. As we know the false vacuum energy density, we can calculate the radius r_s which a region would require in order to be within the Schwarzschild radius. As a ratio to the critical bubble size, R_C, this radius is $r_s/R_C = \xi^{1/3} m_p/\sigma_0$, which is much greater than one (perhaps 100 for the typical model parameters we shall consider below). In our picture bubbles have little time to grow before the rapid nucleation burst brings inflation to an end; clearly it is extremely unlikely for these bubble to nucleate so as to surround a false vacuum region large enough to form a black hole.

Black hole evaporation affects the evolution of both components of the total mass density. Since the hole mass is decreased by evaporation, the evolution of the black hole density, which in the absence of evaporation would be that of nonrelativistic matter ($\rho_{bh} \propto a^{-3}$, where a is the scale factor), is altered. The production of radiation from the hole evaporation also modifies the evolution of radiation energy density, which normally scales as a^{-4}. Of course, the departure of the energy densities from the normal evolution is most pronounced around the time $t \sim t_{RH} + \tau$. An exact treatment of this effect is given in Barrow et al (1991a), (1991b) where a network of equations is derived describing the evolution of the different components of the energy density and also the evolution of the baryon asymmetry. In order to understand the general results, let us for the moment ignore the complication resulting from the decrease of the black hole mass.

Two different situations arise, depending on whether black holes or radiation dominate the energy density of the Universe at the time the holes evaporate. We assume that the fraction of the total density of the Universe forming black holes at the end of inflation is given by β. If $\beta < \frac{1}{2}$, then the evolution of the scale factor is that appropriate to a radiation-dominated Universe, ie., $a(t) \sim t^{\frac{1}{2}}$, amd the energy density of black holes goes as $a^{-3} \propto t^{3/2}$, while that of radiation goes as $a^{-4} \propto t^{-2}$. Therefore, provided their lifetime is sufficiently long, black holes will to dominate the Universe at a time $t_* = t_{END}\,(1 - \beta)^2/\beta^2$, and hence if $\tau > t_* - t_{END}$, they will come to dominate before their evaporation. If $\beta > \frac{1}{2}$, black holes dominate even initially. If the black holes dominate before evaporation, then their evaporation produces not only the baryons, but also the entropy. For full details see Ref.(45). Here we shall simply summarize the results.

First consider the case where black hole evaporation occurs before domination. This corresponds to small β and initially light black holes. Since the black holes never dominate, the Universe expands like a radiation-dominated Universe, with $a \propto t^{\frac{1}{2}}$. If the black holes evaporate before domination, their radiation will not significantly change the background entropy density.

In this case the final baryon asymmetry is

$$B_A = \frac{n_B}{s} = \frac{1}{2}\,\epsilon f_H \left[\frac{45\pi}{g_*}\right]^{\frac{1}{4}} \left[\frac{M_0}{m_P}\right]^{\frac{1}{2}} \left[\frac{M_0}{M_{HOR}}\right]^{\frac{1}{2}} \frac{\beta}{(1-\beta)^{\frac{1}{4}}}, \qquad (9.46)$$

Note that the penultimate factor gives the initial black hole mass as a fraction of the horizon mass at the time the black holes form.

Now consider the second possibility, that holes evaporate after they dominate the energy density. This divides into two further sub-cases; in the former, black holes come to dominate at time t_* as defined earlier, while in the latter black holes dominate immediately after formation.

In the first of these sub-cases, once $t > t_*$ the scale factor evolves as appropriate for a matter-dominated Universe, $a(t) \sim t^{2/3}$, and so $\rho_{BH}(t) = \rho_{BH}(t_*)(t_*/t)^2$ and $\rho_\gamma(t) = \rho_\gamma(t_*)(t_*/t)^{8/3}$, where the energy densities of black holes and radiation become equal at t_*.

As before, the evaporation of a single black hole gives a baryon number $n_B = \epsilon f_H \langle N \rangle n_{BH}(t_{BH})$. This time, though, the entropy is also determined by the other black hole evaporation products, as they provide the dominant contribution. The result for the baryon number is

$$B_1 = \frac{1}{2}\,\epsilon f_H \left[\frac{45\pi}{g_*}\right]^{\frac{1}{4}} \left[\frac{M_0}{m_{pl}}\right]^{\frac{1}{2}} \left[\frac{M_0}{M_{HOR}}\right]^{\frac{1}{2}} (1-\beta)^{\frac{1}{4}} \left[1+\frac{\tau}{t_{RH}}\right]^{-\frac{1}{2}}. \qquad (9.47)$$

This expression is very similar to that obtained in the first "evaporation before domination" scenario; in particular the black hole mass appears in the same function form, and the prefactors are all the same with the exception of the β term, which naturally has changed as we move to a different physical situation. The last factor, demonstrates how long a black hole lifetime dilutes the baryon asymmetry obtained; if τ is very small this factor is just equal to one, while for $\tau > t_{RH}$ we get a reduction in the baryon asymmetry by a factor of about $M^3{}_0/(M_{HOR} m_p{}^2 g_*)^{\frac{1}{2}}$. Clearly, this factor can be important for long-lived (initially massive) black holes. These are also exactly the type of holes that one might expect to survive long enough to come to dominate even if β is originally substantially less the $\frac{1}{2}$.

We now examine the second sub-case of black hole domination - that in which they black holes dominate even initially. The black hole energy density is now given by $\rho_{BH}(t) = \rho_{BH}(t_{RH})(t_{RH}/t)^2$, and

$$B_2 = \frac{1}{2}\,\epsilon f_H \left[\frac{45\pi}{g_*}\right]^{\frac{1}{4}} \left[\frac{M_0}{m_{pl}}\right]^{\frac{1}{2}} \left[\frac{M_0}{M_{HOR}}\right]^{\frac{1}{2}} \beta^{\frac{1}{4}} \left[1+\frac{\tau}{t_{RH}}\right]^{-\frac{1}{2}}. \qquad (9.48)$$

which is just Eq.(9.47) multiplied by $(\beta/(1 - \beta))^{\frac{1}{4}}$. This factor represents the dilution of the black hole energy density up to domination. As expected, Eqs. (9.47) and (9.48) match in the case of marginal domination where $\beta = \frac{1}{2}$. The β dependence in Eq.(9.48) simply areflects the fraction of the horizon mass contributed by black holes. It differs from Eq. (9.47) because here there is no evolution in the initial radiation-dominated phase, hence no era of dilution before domination. In the case of Eq. (9.47) an extra multiplier of $[(1 - \beta)/\beta]^{\frac{1}{4}}$ is needed to account for the evolution in the initial radiation-dominated phase.

This completes the set of results for the different regions of domination more details are to be found in the paper, Barrow, Copeland, Kolb and Liddle (1991a).

Acknowledgements

I would like to thank the organisers for their invitation to participate in the Winter School and for their heroic organisational efforts in the face of many unforseen difficulties.

References

Barrow, J.D. (1980) Mon. Not. Roy. astron. Soc. 192, 427.

Barrow, J.D. (1980a) Mon. Not. Roy. astron. Soc. 192, 19p.

Barrow, J.D. (1981) Surveys in High Energy Physics 1, 183

Barrow, J.D. (1983) Quart. Jl. Roy. astron. Soc. 24, 24.

Barrow, J.D. (1983a) Fund. Cosmic Phys. 8, 83

Barrow, J.D. (1988) Nucl. Phys. B 296, 697.

Barrow, J.D. & Tipler, F.J., (1978) nature 276, 453.

Barrow, J.D. & Ottewill, A. (1983) J. Phys. A 16, 2757.

Barrow, J.D., Burd, A. & Lancaster, D. (1985) Class. Q. Grav. 3, 551.

Barrow, J.D. & Ross, G., (1981) Nucl. Phys. 181, 461.

Barrow, J.D. & Tipler, F.J. (1985) Mon. Not. Roy. astron. Soc. 216, 395.

Barrow, J.D. & Tipler, F.J. (1986a) *The Anthropic Cosmological Principle*, OUP: Oxford.

Barrow, J.D., Galloway, G. & Tipler, F.J. (1986) Mon. Not. Roy. astron. Soc. 223, 835.

Barrow, J.D. (1987) in *Gravitation in Astrophysics*, eds B.Carter & J.B. Hartle, Reidel: Dordrecht.

Barrow, J.D. & Cotsakis, S. (1988) Phys. Lett. B 214, 515.

Barrow, J.D. & Götz, G. (1989) Class. Q. Grav. 6, 1253.

Barrow, J.D. & Sirousse-Zia, H. (1989) Phys. Rev. 39, 2187.

Barrow, J.D. & Maeda, K-I., (1990) Nucl. Phys. B341, 294.

Barrow, J.D. & Hawthorne, S.W. (1990) Mon. Not. Roy. astron. Soc.

Barrow, J.D., Copeland, E.J., Kolb, E.W., & Liddle, A.R., (1991) Phys. Rev. D 43, 977.

Barrow, J.D., Copeland, E.J., Kolb, E.W., & Liddle, A.R., (1991a) Phys. Rev. D 43, 984.

Barrow, J.D., Copeland, E.J., & Liddle, A.R., (1991b) Sussex Preprint: The Evolution of Black Holes in an Expanding Universe.

Bogdanova, L.N. & Shapiro, I.S., (1974) Sov. Phys. JETP Lett. 20, 217.

Bondi, H. (1960) *Cosmology*, CUP: Cambridge.

Buffington, A., & Schindler, S.M. (1981) Astrophys. J. Lett. 247, L105.

Callan, C., & Coleman, S., (1977) Phys. Rev. D 16, 1762.

Carr, B.J. (1976) Astrophys. J. 206, 8.

Copeland, E.J., Kolb, E.W., & Liddle, A.R., (1990) Phys. Rev. D 42, 2911.

Dimopoulos, S., & Susskind, L., (1978) Phys. Rev. D 18, 4500.

Dolgov, A. (1980) Sov. J. Nucl. Phys. 32, 831.

Ellis, G.F.R. (1971) in *General Relativity and Cosmology*, ed R.Sachs, Academic; NY.

Ellis, J., Gaillard, M.K., & Nanopoulos, D.V., (1978) Phys. Lett. B 80, 360 & 82, 464 (erratum).

Fry, J., Olive, K., & Turner, M.S., (1980) Phys. Rev. D 22, 2953.

Guth, A., (1981) Phys. Rev. D 23, 347.

Harvey, J., Reiss, D.B., & Ramond, P., (1982) Nucl. Phys. B199, 223.
Hawking, S.W., (1974) Nature 248, 30.
Hawking, S.W. & Ellis, G.F.R. (1973) *The large scale structure of space-time*, CUP: Cambridge.
Hawking, S.W., Moss, I., & Stewart, J., (1982) Phys. Rev D 26, 2681.
Heckmann, O. & Schücking, E. (1955) Zeit. Astrophys. 38, 95.
Heckmann, O. & Schücking, E. (1959) in *Handbuch der Physik* 53, 489, Springer: Berlin.
Ignatiev, A.Y., Krasinokov, N., Kuzmin, V., & Tavkhelkize, A., (1978) Phys. Lett. 76B, 436.
Kolb, E. & Turner, M.S. (1990) *The Early Universe*, Addison Wesley: Redwood City.
Kolb, W., & Wolfram, S., (1980) Nucl. Phys. B 172, 224.
Kuzmin, V.A. (1970) Sov. Phys. JETP Lett. 12, 228.
Kuzmin, V., Rubakov, V.A., & Shaposhnikov, M.E., (1985) Phys. Lett. B 155, 36.
La, D., & Steinhardt, P.J., (1989) Phys. Rev. Lett. 62, 376.
Langacker, P., (1981) Phys. Rep. 72C, 185.
Lovelock, D., (1971) J. Math Phys. 12, 498.
Madsen, M. & Barrow, J.D. (1989) Nucl. Phys. B 323, 242.
Milne, E. & McCrea, W.H. (1934) Quart. J. Math. 5, 73.
Myers, R.C. & Simon, J.Z. (1988) "Black Hole Thermodynamics in Lovelock Gravity", ITP preprint.
Narlikar, J.V. & Kembhavi, A.K. (1980) Fund. Cosmic Phys. 6, 1.
Olive, D., & Turok, N., (1982) Phys. lett. B 117, 193.
Omnes, R., (1972) Phys. Reports 36, 1.
Pollard, H. (1966) *A Mathematical Introduction to Celestial Mechanics*, Prentice Hall: NJ.
Raychaudhuri, A.K. (1955) Phys. Rev. 98, 1123.
Sakharov, A.D., (1967) Sov. Phys. JETP Lett. 5, 24.
Shikin, I.S. (1971) Sov. Phys. JETP 32, 101.
Shikin, I.S. (1972) Sov. Phys. JETP 34, 236.
Starobinskii, A.A. & Schmidt, H.-J. (1987) Class. Q. Grav. 4, 695.
Stecker, F. (1978) Nature 273, 493.
Stecker, F. (1980) Phys. Rev. Lett. 45, 1460.
Stecker, F. (1982) in *Cosmology; an Oxford Symposium*, ed Wolfendale, A., Reidel: Dordrecht.
Steigman, G. (1976) Ann. Rev. Astron. Astrophys. 14, 339
Stephens, S.A,, (1981) Astrophys. Sp. Sci. 76, 87.
Szabelski, J., Wdowczyk, J., & Wolfendale, A., (1980) Nature 285, 386.
Tipler, F.J., Ellis, G.F.R. & Clarke, C.J.S. (1980) in *General Relativity, One Hundred Years After the Birth of Albert Einstein*, Vol 2, Plenum: NY.
Toussaint, D., Treimann, S.B., Wilczek, F., & Zee, A., (1979) Phys. Rev. D 19, 1036.

Turok, N., (1984) Nucl. Phys. B242, 520.

Vilenkin, A. (1985) Physics Reports 121, 263.

Weinberg, S. (1972) *Gravitation and Cosmology*, Wiley: NY.

Weinberg, S. (1979) Phys. Rev. Lett. 42, 850.

York, J. (1972) Phys. Rev. Lett. 28, 1082.

Yoshimura, M., (1978) Phys. Rev. Lett. 41, 281.

Zeldovich, Y.B. (1965) Sov. Phys. JETP 21, 656.

$\Omega_B = 1$ UNIVERSE AND PRIMORDIAL NUCLEOSYNTHESIS

HO-SHIK KANG

Department of Physics
The Ohio State University
Columbus, OH 43210-1106, USA

ABSTRACT

It is explicitly shown that the primordial nucleosynthesis yields of deuterium, helium-3, helium-4, and lithium-7 are totally consistent with those observationally inferred abundances *with* a large baryon density, even $\Omega_B = 1$, if neutrinos are degenerate enough. Futhermore, the consistency obtained from the agreement of the predicted abundances of *all* the light elements with those observationally inferred abundances is not problematic in this degenerate nucleosynthesis model even in the case that $Y_p^{OBS} \leq 0.235$ in contrast to the case for standard nucleosynthesis.

1. Introduction

It is well known[1] that standard Big Bang Nucleosynthesis (BBN) is the only theory which is consistent with the observed abundances of deuterium, helium-3, helium-4, and lithium-7, whose abundances span about nine orders of magnitude. Along with the cosmic microwave radiation background observed first by Penzias and Wilson[2] long after the theoretical estimates[3] for the cosmic microwave radiation background in the 1940's, the theory of nucleosynthesis is one of two consequences of the standard Hot Big Bang Cosmological Model which can be used as tests of the model. The remarkable success of standard BBN in obtaining the agreement of the abundances of the light elements inferred from observational data and those primordial abundances predicted from the theory demonstrates not only that standard BBN is a unique window on the early Universe, but also that standard BBN can probe the validity of extensions to the standard model of elementary particle physics. Standard BBN, for example, can be used to study the number of neutrino families[1] (N_ν), the properties of hypothetical particles[4], the time dependences of fundamental constants[5], *etc.*

The primordial abundances of the light elements in standard BBN are calculated as a function of only one free parameter, the ratio of baryons (ordinary matter around us) to photons ($\eta \equiv n_B/n_\gamma$). Besides the synthesis of the light elements, standard BBN predicts a narrow allowed range of the free parameter which constrains the baryon density parameter $\Omega_B(\equiv \varrho_B/\varrho_c$, where ϱ_c is the present critical density) using the relation that[6]:

$$\Omega_B h^2 = 3.73 \times 10^{-3} \vartheta_o^3 \eta_{10}, \quad (1)$$

where $\eta_{10} = 10^{10}\eta$, $\vartheta_o = T_{\gamma o}/2.75K$, and $h = H_o/100kms^{-1}Mpc^{-1}$.

After updating with the latest nuclear reaction rates, T. Walker *et al* (hereafter WSSOK) calculated the primordial yields of the light elements and compared them with the observational inferred abundances, leading to bounds to η (and consequently Ω_B):

$$2.8 \leq \eta_{10} \leq 4.0. \quad (2)$$

On the other hand, WSSOK found that for $N_\nu = 3.0$ and $882 \leq \tau_n \leq 896$ sec. the predicted primordial abundance of helium-4 is bounded:

$$0.236 \leq Y_p \leq 0.243, \quad (3)$$

where Y_p is the helium-4 mass fraction to hydrogen and τ_n the neutron mean-life.

Although the synthesis of the light elements is consistent, these bounds, Eqs.2 - 3, can cause two *potential* problems which make the standard BBN falsifiable. First, the baryonic density bounds, Eq.2, indicate that, when we compare them with observational masses of luminous parts of galaxies, the bulk of the baryons in the Universe is dark matter. From the nucleosynthesis point of view, therefore, the existence of baryonic dark matter can not be denied. The dynamical (dark plus bright, as long as it interacts gravitationally) masses of galaxies are well fit to the baryonic density bounds. However, for clusters of galaxies the upper bound of the baryonic density is not large enough to fit well. For larger scale structure than clusters of galaxies, the dynamical masses can not be fit at all to the predicted baryonic density. Hence, standard BBN will be indeed in trouble if the bulk of dynamical masses for large scale structure turn out to be baryons (this should be confirmed by observations). Second, WSSOK showed that the observations were consistent with $Y_p^{OBS} = 0.23 \pm 0.01$. If the upper bound of Y_p^{OBS} turns out to be less than 0.235, i.e.

$$Y_p^{OBS} \leq 0.235, \quad (4)$$

then the predicted helium-4 primordial abundance in standard BBN can not be consistent with observations, see Eq.3.

The goal of this paper is to resolve these two *potential* problems by proposing a modified BBN without destroying any consistency obtained in standard BBN with the observationally inferred abundances of *all* the light elements. Furthermore, in the light of the philosophy that there is only one kind of dark matter, we particularly emphasize the hope that our alternative model can provide enough baryons to agree with recent observations of dynamical masses for large scale structure, and even to satisfy the prediction of inflation theories; $\Omega_{TOT} = \Omega_B = 1$. In order to estimate the range of η_{10} for $\Omega_B = 1$, we take for the present photon temperature, $T_{\gamma o} = 2.76 \pm 0.05\,K$, and for the Hubble parameter $0.40 \leq h \leq 0.65$. Here the upper limit of h is obtained by the requirement that $t_o \geq 10$ Gyr in an $\Omega_{TOT} = 1$ Universe. Hence, if $\Omega_B = 1$, η_{10} should be in the following range:

$$40 \lesssim \eta_{10} \lesssim 120. \quad (5)$$

In the next section we comment on standard BBN and its deviations for $\Omega_B = 1$. Our model is explored in section 3, and section 4 is devoted to conclusions.

2. Standard BBN and Its Deviations for $\Omega_B = 1$

Since the properties of neutrinos are less known than those of other particles in the cosmic plasma of photons, electrons and positrons, neutrinos and antineutrinos, and nucleons, we can categorize deviations from standard model into two types; neutrino-related and neutrino-unrelated modifications. It is worthwhile reviewing standard BBN with an emphasis on its assumptions and thermal history in order to find what assumptions have been changed and what have not, and what parts of the thermal history have to be changed in a specific model of nonstandard BBN. There are numerous neutrino-unrelated modifications such as inhomogeneous BBN[7,8], BBN with late-decaying massive particles[9] and BBN with photon cooling[10]. However, non of such deviations is successful as much as standard BBN to satisfy *all* the constraints on the abundances of the light elements *with* η_{10} values corresponding to $\Omega_B = 1$ although some deviations require more investigation to rule out completely. The reader is referred to, e.g., a recent review of non-standard primordial nucleosynthesis by R. Malaney and G. Mathews[11]. Hence, in this section we will briefly review standard BBN, and in the next section a neutrino-related modification.

Standard BBN: Standard Big Bang Nucleosynthesis, a cornerstone of the modern architecture of the Hot Big Bang Cosmology, consists largely of three parts; macroscopic physics, microscopic physics, and observational physics.

The macrophysics includes the Einstein theory of general relativity and the Cosmological Principle. The cosmological principle is the hypothesis (based on observations) that the Universe is spatially homogeneous and isotropic, i.e., every point in space is physically identical. Besides the correctness of general relativity, the cosmological constant is assumed to be zero.

On the other hand, the microphysics contains the standard theory of the electroweak interaction and nuclear physics with experimental data for nuclear reaction rates. This microphysics explains the interactions of the particles in the early Universe, and the production of the light elements. Three flavors of light, nondegenerate neutrinos are assumed, and the neutron mean-life is taken from experimental data.

We need to extract the *primordial* abundances of the light elements from observational data in order to compare them to the predicted abundances. In addition, we need to measure the cosmic microwave radiation background.

Under the preceding assumptions, we can review the evolution of the early Universe. Beginning at a temperature $T \sim 10$ MeV, the particles of the cosmic plasma are in thermal equilibrium (an assumption). Until the temperature drops to the neutrino decoupling temperature[6], $T_D \simeq 2.0,\ 3.3$ MeV, for electron neutrino and non-electron neutrinos respectively, neutrinos are kept in thermal equilibrium with the background radiation via the weak interactions,

$$\nu_\ell + \bar{\nu}_\ell \leftrightarrow e^- + e^+ \quad , \quad \nu_\ell + e^\pm \leftrightarrow \nu_\ell + e^\pm \quad , \quad \bar{\nu}_\ell + e^\pm \leftrightarrow \bar{\nu}_\ell + e^\pm . \tag{6}$$

The subscript ℓ denotes all possible flavors of leptons ($\ell = e, \mu, \tau$). Once neutrinos decouple from the thermal bath, their entropy is conserved separately. On the other hand, the neutrons and protons maintain equilibrium through the charged current weak interactions, i.e.,

$$p + \bar{\nu}_e \leftrightarrow n + e^+ \quad , \quad p + e^- \leftrightarrow n + \nu_e \quad , \quad n \leftrightarrow p + e^- + \bar{\nu}_e , \tag{7}$$

until the temperature drops to the freeze-out value, $T_f \sim 0.7$ MeV, at which the neutron-to-proton ratio freezes out about the equilibrium value, $n/p \sim 1/6$. Meanwhile, electron-positron annihilation, which occurs at $T \lesssim 0.5$ MeV, heats the photons, but not the already decoupled neutrinos. As a result, the neutrino temperature decreases relative to the photon temperature; i.e., $(T_\nu/T_\gamma)^3 = 4/11$. As the Universe expands and cools, at $T \lesssim 0.1$ MeV, the abundances of the light elements increase dramatically, which is triggered by the sufficient increase of the deuterium abundance through

$$p + n \leftrightarrow D + \gamma , \tag{8}$$

where the binding energy of the deuterium is 2.225 MeV. This occurs at a temperature much lower than the deutrium binding energy when the number of highly energetic photons which can destroy newly synthesized nuclei is reduced sufficiently (as the Universe cools) so that the photodestruction of the nuclei becomes insignificant. After the nucleosynthesis epoch the calculated primordial abundances of the light elements remain constant which should be tested with the observationally inferred abundances.

3. Degenerate BBN

The punch line of degenerate BBN is as follows: there are three kinds of quantum numbers[12] at the epoch we are interested in, at about 1 second after Big Bang; Charge, Baryon, and Lepton quantum numbers. Charge quantum number has been observed to be zero (i.e., the Universe is neutral in electric charge), Baryon quantum number (initial number of protons and neutrons, and the only free parameter in standard BBN) also has been observed to be very small ($\sim 10^{-10}$), but Lepton quantum numbers have been *assumed* to be zero since Lepton numbers are not directly observable. However, Harvey and Kolb[13] showed that large Lepton numbers can survive within some extended unified theories even when Baryon number is very small. Therefore, we disregard the assumption of null Lepton numbers so that both Baryon number and Lepton number are treated as free parameters in this alternative in contrast to standard BBN where only Baryon number is a free parameter. For a recent full calculation of this model, although this scenarino has been proposed by many authors[14], the reader is referred to the work by Kang and Steigman[6], particularly their discussion of the cosmological effects of degenerate neutrinos on the thermal history of the early Univese.

Lepton quantum numbers, L_ℓ, can be expressed in terms of the degeneracy parameters ξ (defined by the ratio of chemical potential of neutrinos to $k_B T$). Then, the total lepton asymmetry is obtained from the equilibrium Fermi-Dirac distribution function.

$$L \equiv \sum_\ell L_\ell = \frac{\pi^2}{12\zeta(3)} \sum_{\nu_\ell} (\frac{T_{\nu_\ell}}{T_\gamma})^3 [\xi_{\nu_\ell} + \frac{1}{\pi^2} \xi^3_{\nu_\ell}] + (\nu_\ell \rightarrow \ell), \tag{9}$$

where ζ is the Riemann zeta function and ξ_{ν_ℓ} (ξ_ℓ) is the degeneracy parameter of the neutral (charged) leptons.

The contribution of the charged leptons to BBN is negligible since μ and τ decay at the time when $T < m_{\mu,\tau}$, so $\xi_\mu = \xi_\tau = 0$. Further, due to the charge neutrality of the Universe, $\xi_e \sim \eta$ which is negligible.

However, the neutral leptons (neutrinos) have significant effects on BBN. For the electron-neutrino degeneracy, there is a crucial effect on the neutron-to-proton ratio. In thermal equilibrium,

$$\frac{n(n)}{n(p)} = exp(-\frac{\Delta m}{T} - \xi_{\nu_e} + \xi_e) \simeq exp(-\frac{\Delta m}{T} - \xi_{\nu_e}), \tag{10}$$

where $\Delta m = 1.29$MeV is the neutron-proton mass difference. It is found[6] that positive ξ_{ν_e} decreases the neutron abundance compared to the nondegenerate case, therefore, the ^4He abundance decreases; negative ξ_{ν_e} has somewhat more complex effects on the abundances of the light elements (see the ref. 6 for the detail).

On the other hand, the nonelectron-neutrino degeneracies increase the total energy density and hence the Universal expansion rate, which have cosmologically same effect with, e.g., the extra neutrino flavors. This can be expressed through an equivalent speed-up factor, S, defined by:

$$S \equiv \frac{H'}{H} = (\frac{\varrho'_{tot}}{\varrho_{tot}})^{1/2} = (\frac{g'_{eff}}{g_{eff}})^{1/2}, \tag{11}$$

where the primes are for the case of degenerate neutrinos and the unprimed quantities are for the standard case of 3 species of light, nondegenerate neutrinos. g_{eff} is the total effective number of degrees of freedom defined by $\varrho_{tot}/\varrho_\gamma \equiv g_{eff}/2$. It then follows that

$$g'_{eff}(T) = \sum_B g_B (\frac{T_B}{T_\gamma})^4 + \frac{7}{8} \sum_F g_F (\frac{T_F}{T_\gamma})^4 [1 + \frac{30}{7}(\frac{\xi_{\nu_\ell}}{\pi})^2 + \frac{15}{7}(\frac{\xi_{\nu_\ell}}{\pi})^4], \tag{12}$$

where B(F) indicates bosons(fermions), $g_{B(F)}$ is the number of boson(fermion) degrees of freedom, and $T_{B(F)}$ is the temperature.

Notice that the sign of ξ_{ν_ℓ} cannot be distinguished for its effect on S because the total density depends on even powers of ξ_{ν_ℓ}. Therefore, for ν_μ and ν_τ only the absolute value of ξ_{ν_ℓ} is relevant. However, the sign of ξ_{ν_e} does matter for BBN, because ξ_{ν_e} affects not only the total density but also the neutron-to-proton ratio; see Eq.10. Further, since the permitted values of ξ_{ν_e} are very small ($\xi_{\nu_e} \lesssim 1.6$) as we shall see in figs.1 - 5, the effect of ξ_{ν_e} on the spped-up factor is also very small. Therefore, we shall separate ξ_{ν_e} from S.

Numerical Results for $\Omega_B = 1$: Here we adopt the following set of the observationally inferred abundances of the light elements to compare with those calculated abundances[1]:

$$D/H \ \geq \ 1.8 \times 10^{-5} \tag{13a}$$

$$(D +^3 He)/H \ \leq \ 1.0 \times 10^{-4} \tag{13b}$$

$$0.22 \ \leq \ Y_p \ \leq \ 0.24 \tag{13c}$$

$$1.0 \times 10^{-10} \ \leq \ ^7Li/H \ \leq \ (1.4 \ - \ 2.3) \times 10^{-10}, \tag{13d}$$

where the abundances except for ^4He are the number ratios to hydrogen. Since the cosmological effects of electron-neutrinos are different from those of nonelectron-neutrinos (muon- and tau-neutrinos are comologically identical in this model), we have two additional free parameters, ξ_{ν_e} and S. Therefore, we optimize the parameter space of ξ_{ν_e} and S to find any allowed range satisfying all the constraints of the abundances, Eqs.13, *with* large η values, see Eq.5, to close the Universe with baryons alone.

We plot explicitly isoyield contours of the abundances, Eqs.13, of the light elements in the (ξ_{ν_e}, S) plane for η_{10} values relevant for $\Omega_B = 1$; $\eta_{10} = 40,\ 60,\ 80,\ 100,\ 120$ in figs.1 - 5 respectively. We plot $Y_p = 0.235$ with the dots, showing that $Y_p^{OBS} \leq 0.235$, Eq.4, can not be a *potential* problem in this model. Also, $^7Li/H \leq 1.4 \times 10^{-10}$ which fits the observed lithium in the hotter Pop II dwarfs is drawn with the dotted line. Notice that the constraint of the deterium abundance, Eq.13a, does not contribute to restrict the allowed ranges of the parameter space. The allowed ranges are shaded.

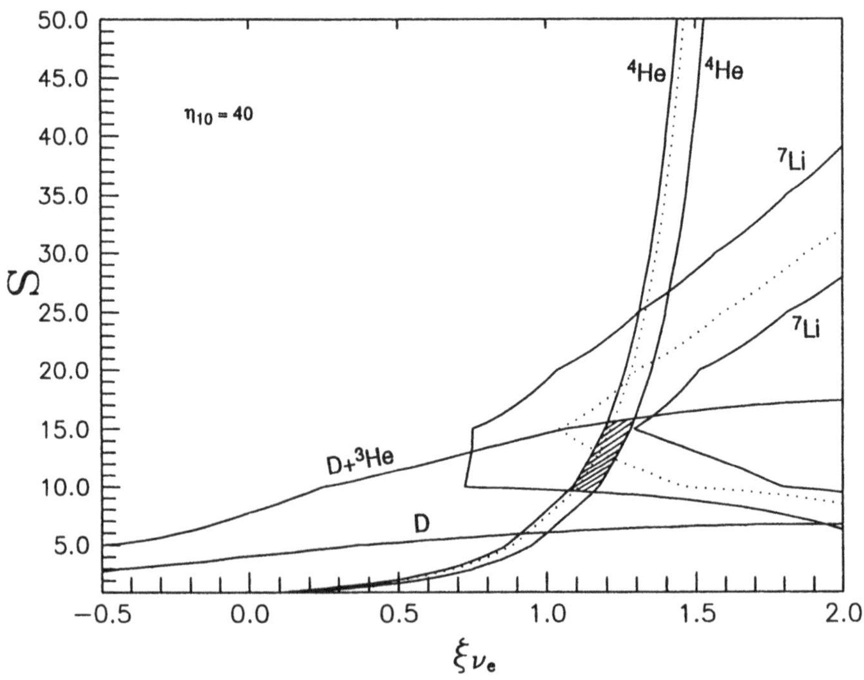

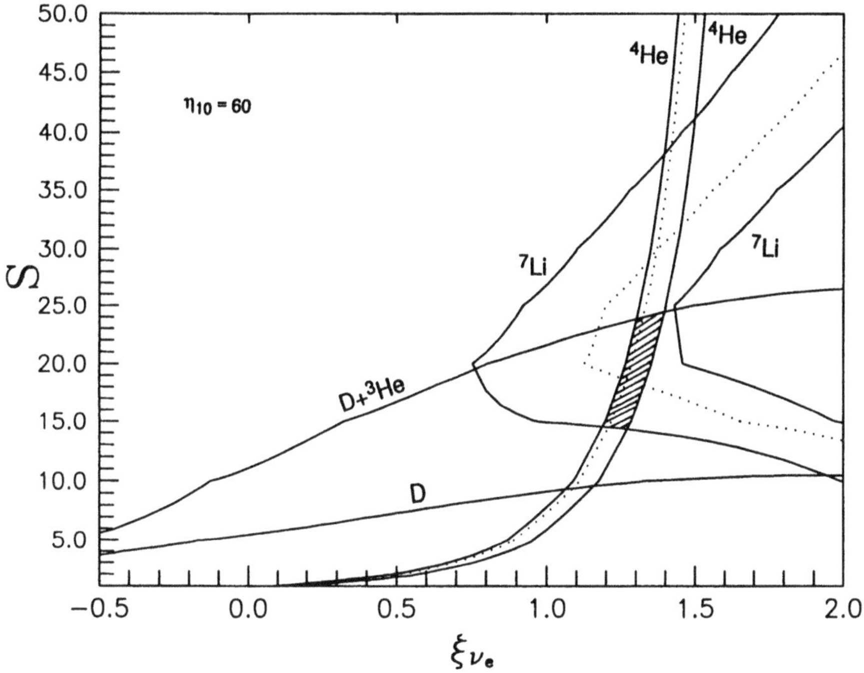
50.0
45.0
40.0
35.0
30.0
25.0
20.0
15.0
10.0
5.0
S
η10 = 60
4He
4He
7Li
7Li
D+3He
D
-0.5
0.0
0.5
1.0
1.5
2.0
ξνe

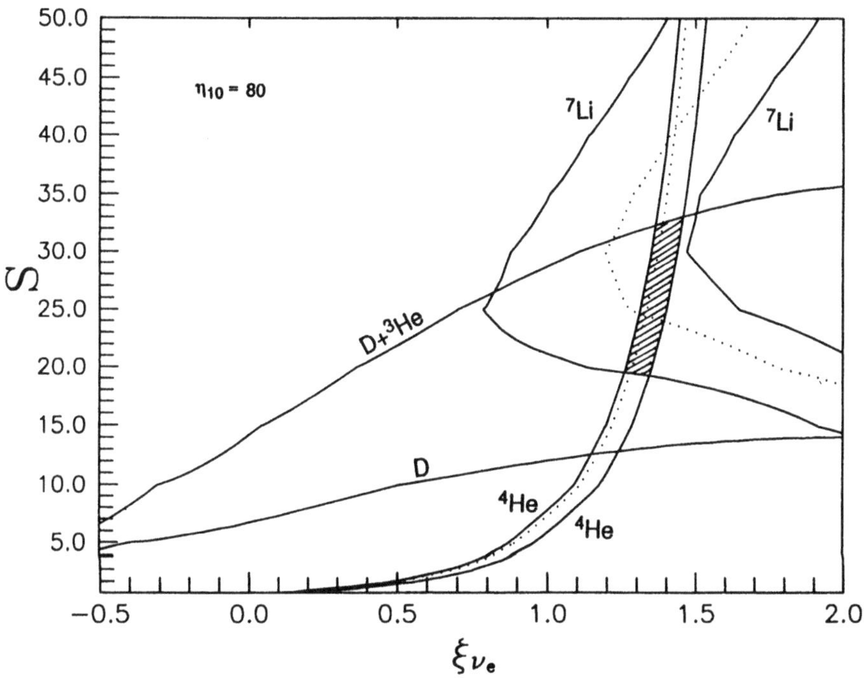
50.0
45.0
40.0
35.0
30.0
25.0
20.0
15.0
10.0
5.0
S
η10 = 80
7Li
7Li
D+3He
D
4He
4He
-0.5
0.0
0.5
1.0
1.5
2.0
ξνe

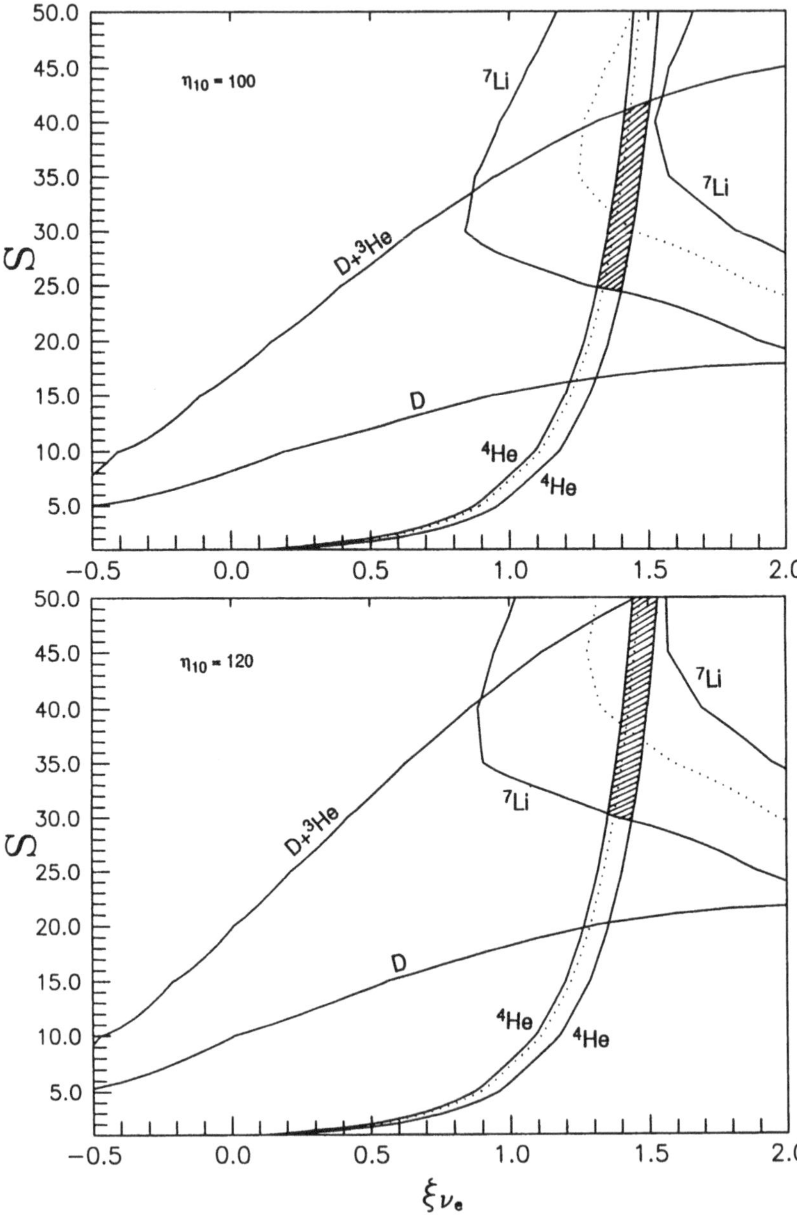

Figure 1 - 5. Iso-yield curves of the light elements abundances as a function of (ξ_{ν_e}, S) for $\eta_{10} =$ 40, 60, 80, 100, 120 respectively. The solid curves are for $D/H = 1.8\times10^{-5}$, $(D+^3He)/H = 1.0\times10^{-4}$, $Y_p = 0.22,\ 0.24$ and $^7Li/H = 2.3\times10^{-10}$. The dotted lines are for $Y_p = 0.235$ and $^7Li/H = 1.4\times10^{-10}$. The allowed ranges for $\Omega_B = 1$ are shaded.

4. Conclusions

It is shown that primordial nucleosynthesis is totally consistent with all the constraints on the light elements abundances with η_{10} values corresponding to $\Omega_B = 1$, if neutrinos are degenerate enough. Furthermore, these allowed ranges with large η_{10} values in the (ξ_{ν_e}, S) plane will not be in trouble even if $Y_p^{OBS} \leq 0.235$ in contrast to the case for standard BBN. Since ξ_{ν_e} and S in the allowed ranges increase *continuously* as η_{10} increases from the value corresponding to the case for standard BBN, i.e. Eq.2, to the value of a baryonic closure Universe, i.e. Eq.5, the rest of the parameters can be strictly restricted in this model if one of the parameters is restricted.

5. Acknowledgements

I wish to thank the Second Winter School of Physics for hospitality during my stay in the school and support, and G. Steigman for giving this problem and support. I have benefited from the series of inspired lectures of D. N. Schramm in the school. It is a great pleasure to thank R. J. Scherrer for valuable discussions and reading the paper. Parts of numerical calculations are carried out with the Cray X-MP and Y-MP at the Ohio Supercomputer Center. This work was supported in part by the Department of Energy.

6. References

1. T. P. Walker, G. Steigman, D. N. Schramm, K. A. Olive and H.-S. Kang, *Astrophys. J.* (1991) in press (OSU-TA-10/90), and references therein.

2. A. A. Penzias and R. W. Wilson, *Astrophys. J.* **142** (1965) 419.

3. G. Gamow, *Phys. Rev.* **70** (1946) 571; R. A. Alpher and R. C. Herman, *Rev. Mod. Phys.* **22** (1950) 153.

4. D. A. Dicus, E. W. Kolb, V. L. Teplitz and R. V. Wagoner, *Phys. Rev.* **D17** (1978) 1529.

5. J. Yang, D. N. Schramm, G. Steigman, and R. T. Rood, *Astrophys. J.* **227** (1979) 697.

6. H.-S. Kang and G. Steigman, *Nucl. Phys.* (1991) submitted (OSU-TA-7/91), and references therein.

7. J. H. Applegate, C. J. Hogan and R. J. Scherrer, *Phys. Rev.* **D35** (1987) 1151; C. R. Alcock, G. M. Fuller and G. J. Mathews, *Astrophys. J.* **320** (1987) 439.

8. R. A. Malaney and W. A. Fowler, *Astrophys. J.* **333** (1988) 14.

9. S. Dimopoulos, R. Esmailzadeh, L. J. Hall and G. D. Starkman, *Astrophys. J.* **330** (1988) 545.

10. J. G. Bartlett and L. J. Hall, *Phys. Rev. Lett.* **66** (1991) 541.

11. R. A. Malaney and G. J. Mathews, *Phys. Reps.* (1990) submitted (UCRL-JC-106097).

12. S. Weinberg, *Gravitation and Cosmology* (Wiley, New York, 1972) p. 531, and references therein.

13. J. A. Harvey, and E. W. Kolb, *Phys. Rev.* **D24** (1981) 2090.

14. A. Yahil and G. Beaudet, *Astrophys. J.* **206** (1976) 26; G. Beaudet and A. Yahil, *ibid.* **218** (1977) 253; Y. David and H. Reeves, *Phil. Trans. Roy. Soc.* **A296** (1980) 415.

BURT OVRUT

Instantons in Antisymmetric Tensor Theories, Cosine Potentials and Cosmology

Burt A. Ovrut[1,2]
Department of Physics, University of Pennsylvania
Philadelphia, Pennsylvania 19104-6396

ABSTRACT

An infinite class of theories involving both a scalar field and an antisymmetric tensor field is presented that contains topologically stable instanton defects with finite action. These theories are shown to be equivalent to a class of non-linear sigma models of a single complex scalar field. It is further shown that these instantons naturally induce a cosine potential in the effective Lagrangian of the associated scalar mean field theory and that this cosine potential energy provides a framework for understanding both inflationary cosmology and large scale structure. A new scenario for producing large scale structure, based on the late time effect of the Hawking temperature during inflation, is also discussed.

Introduction

It has long been known that topologically stable vortex defects exist in two Euclidian dimensions and induce a periodic cosine potential in the effective Lagrangian of the associated scalar mean field [1]. Polyakov has shown [2]that in three Euclidean dimensions topologically stable point defects exist in the theory of compact QED and, again, induce a cosine potential in the effective Lagrangian of the associated scalar mean field. The existence of topologically stable point defects associated with a massless antisymmetric tensor field in four Euclidian dimensions has been demonstrated by Rohm and Witten [3] within the context of superstring theory. Antisymmetric tensor point defects, when coupled to Euclidian four dimensional gravity, were shown to lead to so-called wormhole solutions [4]. In these solutions, the radial function $a(r)$ in the line element $ds^2 = dr^2 + a(r)^2 d\Omega^2$ acts as a regulator for the point defect core and renders the action finite. More recently, a new class of stable Euclidian four dimensional string instantons involving the graviton, dilaton and axion field has been constructed [5] and their role in string theory discussed.

In these talks we show that, in four Euclidian dimensions, the theory of a free antisymmetric tensor field necessarily contains point defects which are topologically

[1]Work supported in part by the Department of Energy under Contract No. DOE-ACO2-76-ERO-3071 and NATO Grant Number 860684.

[2]Lectures presented at the Second Winter School of Physics on Cosmology and Elementary Particles, University of Puerto Rico, Rio Piedras, Puerto Rico, March 27-April 5, 1991.

stable as long as the defects are elements of the cohomology group $H^3(\mathcal{R}^4 - \{0\}; \mathcal{Z})$. Since these defects are point-like we will refer to them as H-instantons. However, unlike non-Abelian gauge instantons, they are singular at their core. This singularity is sufficiently strong to cause their action to diverge. It follows that the contribution of these instantons to the partition function or any physical process will be suppressed. It is important, therefore, to try to modify the theory in such a way as to render the H-instanton action finite. We present a special infinite class of theories in four Euclidian dimensions that also contain topologically stable instanton defects [6]. As in wormhole solutions, these involve H-instantons and another scalar field whose role is to regulate the H-instanton core and make the action finite. However, in our case, the gravitational field a(r) is replaced by a non-gravitational scalar. Using duality and field redefinitions, these theories are shown to be equivalent to an infinite class of non-linear sigma models involving a single complex scalar. Finally, these finite action instantons are shown to induce a cosine potential to the effective potential of the associated scalar field theory.

The existence of a topologically induced cosine potential for a heretofore free scalar field may have important phenomenological and cosmological implications. For example, it has recently become clear that scalar fields with cosine potentials can play an important role in soft domain wall theories of large scale structure [7]and in inflationary cosmology [8]. The former is particularly interesting since the "signature" that soft domain walls leave on the microwave background may soon be detectable [9]. Early results show [10]that the topologically induced cosine potential discussed in this paper may give a simple and natural solution to many cosmological problems, including large scale structure and inflation. These results are discussed in the last of these lectures.

A Little Cosmology:

Consider a real manifold M of dimension n. An n-form can be written as

$$H = \frac{1}{m!} H_{\mu_1 \ldots \mu_m} dx^{\mu_1} \Lambda dx^{\mu_2} \Lambda \ldots \Lambda dx^{\mu_m} \tag{1}$$

where

$$dx^{\mu_i} \Lambda dx^{\mu_j} = -dx^{\mu_j} \Lambda dx^{\mu_i} \tag{2}$$

Note that H vanishes for $m > n$. Therefore, non-vanishing m-forms run from $0 \leq m \leq n$. The exterior derivative, d, is defined by

$$dH = \frac{1}{m!} \partial_\nu H_{\mu_1 \ldots \mu_m} dx^\nu \Lambda dx^{\mu_1} \Lambda \ldots \Lambda dx^{\mu_m} \tag{3}$$

Clearly d maps m-forms into m + 1 - forms. Note that

$$\begin{aligned} d^2 H &= \frac{1}{m!} \partial_\lambda \partial_\nu H_{\mu_1 \ldots \mu_m} dx^\lambda \Lambda dx^\nu \Lambda \ldots \\ &= -d^2 H \end{aligned} \tag{4}$$

Hence, $d^2H = 0$ for any form H. That is

$$d^2 = 0 \tag{5}$$

A form is called *closed* if

$$dH = 0 \tag{6}$$

Locally any closed m-form can be written as

$$H = d\omega \tag{7}$$

where ω is an $m - 1$ - form. If *everywhere on M*

$$H = d\omega \tag{8}$$

then H is called an *exact* form. Clearly every exact form is closed but not every closed form is exact. Denote by $Z^m(M;R)$ the set of all closed m-forms on M and by B^m(M;R) the set of all exact m-forms on M. Exact forms are trivially closed. It is the closed forms that are not exact that are topologically interesting. Therefore, consider the coset

$$H^m(M;R) = Z^m(M;R)/B^m(M;R) \tag{9}$$

This is the set of closed, non-exact, n-forms. It can be given an Abelian group structure and is called the m-th cohomology group of M. To repeat, $0 \leq m \leq n$. If $H^m(M;R) = 0$ then all closed m-form are exact and, hence, topologically trivial. If, however, $H^m(M;R) \neq 0$ then the non-zero m-forms of the group are topologically non-trivial. Finally, the elements of $H^m(M;R)$ can be obtained by integrating m-forms over topologically differing m-submanifolds called m-cycles.

D = 2 Vortices:

Let $M = R^2$ where

$$R^2 = \{(x^1, x^2)|x^1, x^2 \epsilon R\} \tag{10}$$

and consider a Euclidian metric

$$g_{\mu\nu} = \begin{pmatrix} 1 & 0 \\ 0 & 1 \end{pmatrix} \tag{11}$$

Let H = $H_\mu dx^\mu$ be any closed one-form. That is

$$\begin{aligned} dH &= \partial_\nu H_\mu dx^\nu \Lambda dx^\mu \\ &= (\epsilon^{\nu\mu}\partial_\nu H_\mu) dx^1 \Lambda dx^2 = 0 \\ &\Rightarrow \epsilon^{\nu\mu}\partial_\nu H_\mu = 0 \end{aligned} \tag{12}$$

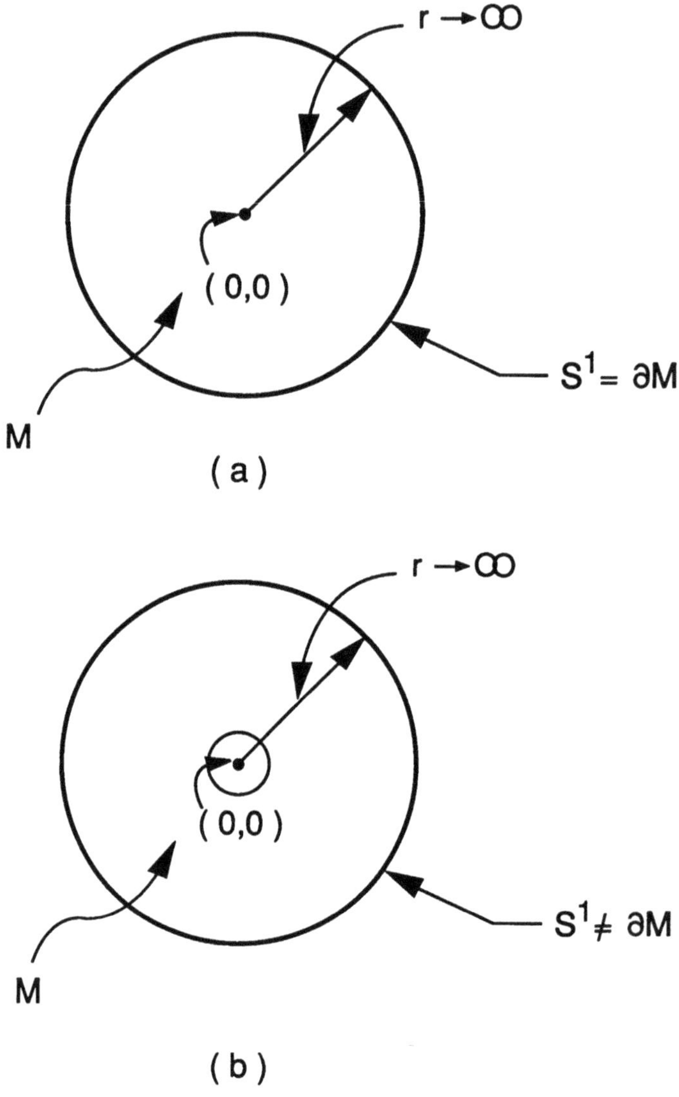

Fig. 1

The topologically non-trivial one-forms are those that are closed but not exact. These are determined from $H^1(R^2;R)$. But

$$H^1(R^2;R) = 0 \tag{13}$$

Therefore, all closed one-forms are topologically trivial. Another way to see these is to integrate any closed one-form H over the one-sphere at radial infinity, S^1, as shown in Figure 1a. Note that $S^1 = \partial M$. Hence, by Stokes theorem

$$\int_{S^1} H = \int_M dH = 0 \tag{14}$$

and $H^1(R^2;R) = 0$. We have been assuming that H is well defined everywhere on M. Now consider H well defined everywhere except at the origin. This is mathematically expressed by considering

$$M = R^2 - \{(0,0)\} \tag{15}$$

Now

$$H^1(R^2 - \{(0,0)\};R) \simeq H^1(S^1;R) = R \tag{16}$$

Therefore, there exist an infinite number of topologically non-trivial closed one-forms, indexed by $q \epsilon R$, that diverge at the origin. Another way to see this is to integrate any closed one-form H over the one-sphere at radial infinity, S^1 as shown in Figure 1b. Note that $S^1 \neq \partial M$ and therefore

$$\int_{S^1} H \neq \int_M dH = 0 \tag{17}$$

In fact

$$\int_{S^1} H = 2\pi q \tag{18}$$

where $q \epsilon R$. It is easy to give an explicit form for closed one-forms satisfying this equation. Consider

$$\hat{H}_\mu = -q\epsilon_{\mu\nu}\frac{x^\nu}{|x|^2} \tag{19}$$

Note that

$$\begin{aligned} \epsilon^{\nu\mu}\partial_\nu \hat{H}_\mu &= \epsilon^{\nu\mu}\partial_\nu(-q\epsilon_{\mu\lambda}\frac{x^\lambda}{|x|^2}) \\ &= -q\partial_\nu(\frac{x^\nu}{|x|^2}) \\ &= -\frac{2q}{|x|^2} + \frac{2q}{|x|^2} = 0 \end{aligned} \tag{20}$$

Hence, $\hat{H} = \hat{H}_\mu dx^\mu$ is a *closed* one-form. Now

$$\int_{S^1} \hat{H} = \int_{S^1} \hat{H}_\mu dx^\mu = \int_{S^1} (\hat{H}_1 x' + \hat{H}_2 dx^2) \tag{21}$$

Using polar coordinates implies that

$$\int_{S^1} \hat{H} = \int_0^{2\pi} d\theta (\hat{H}_1 \frac{dx^1}{d\theta} + \hat{H}_2 \frac{dx^2}{d\theta}) \tag{22}$$

Now

$$\begin{aligned} \hat{H}_1 &= -q\frac{x^2}{|x|^2} \\ &= -\frac{q\sin\theta}{r} \end{aligned} \tag{23}$$

and

$$\hat{H}_2 = +\frac{q\cos\theta}{r} \tag{24}$$

Also

$$\begin{aligned} \frac{dx^1}{d\theta} &= -r\sin\theta \\ \frac{dx^2}{d\theta} &= r\cos\theta \end{aligned} \tag{25}$$

Therefore

$$\begin{aligned} &\int_{S^1} \hat{H} = \int_0^{2\pi} d\theta (q\sin^2\theta + q\cos^2\theta) \\ &= q\int_0^{2\pi} d\theta \\ &= 2\pi q \end{aligned} \tag{26}$$

Hence, $\hat{H}$ is a closed one-form that is not exact. There are an infinite number of such $\hat{H}$'s labelled by $q \epsilon R$. Now

$$\frac{\partial\theta}{\partial x^1} = \frac{1}{1+u^2}\frac{du}{dx^1} = -\frac{\sin\theta}{r} \tag{27}$$

Similarly

$$\frac{\partial\theta}{\partial x^2} = \frac{\cos\theta}{r} \tag{28}$$

Therefore

$$\begin{aligned} \hat{H}_1 &= \frac{\partial q\theta}{\partial x^1} \\ \hat{H}_2 &= \frac{\partial q\theta}{\partial x^2} \end{aligned} \tag{29}$$

and

$$\hat{H} = d\hat{\phi} \tag{30}$$

where $\hat{\phi}$ is a zero-form defined by

$$\hat{\phi} = q\theta + c \tag{31}$$

and c is a constant. It looks as though $\hat{H}$ is exact, but this is not possible. Note that $\hat{\phi} = q\theta + c$ is not a continuous function on M. It is discontinuous at the positive x^1 axis. Hence, $\hat{\phi}$ is only well defined on the open submanifold $\tilde{M} = M - \{(0, x^1)/x^1 > 0\}$. Therefore, $\hat{H} = d\hat{\phi}$ on $\tilde{M}$ only and $\hat{H}$ is not exact. $\hat{H}$ (or $\hat{\phi}$) is called a vortex of charge q located at the origin.

<u>**Integer Charged Vortices:**</u>

1) $\hat{H}$ picture

$$\begin{aligned} q = +1 \Rightarrow \hat{H}_1 &= -\frac{\sin\theta}{r} \\ \hat{H}_2 &= \frac{\cos\theta}{r} \end{aligned} \tag{32}$$

Pictorially see Figure 2a.

q = - 1 (q = +1 "antivortex") $\Rightarrow$

$$\begin{aligned} \hat{H}_1 &= \frac{\sin\theta}{r} \\ \hat{H}_2 &= -\frac{\cos\theta}{r} \end{aligned} \tag{33}$$

Pictorially see Figure 2b. For higher q its the same with the vector multiplied by q.

2) $\hat{\phi}$ picture

$$\hat{\phi} = q\theta + c \tag{34}$$

It is conventional to take $c = \pi/2$. Then

$$\hat{\phi} = q\theta + \frac{\pi}{2} \tag{35}$$

$$q = +1 \Rightarrow \hat{\phi} = \theta + \frac{\pi}{2} \tag{36}$$

Pictorially see Figure 3a.

$$q = -1 \Rightarrow \hat{\phi} = -\theta + \frac{\pi}{2} \tag{37}$$

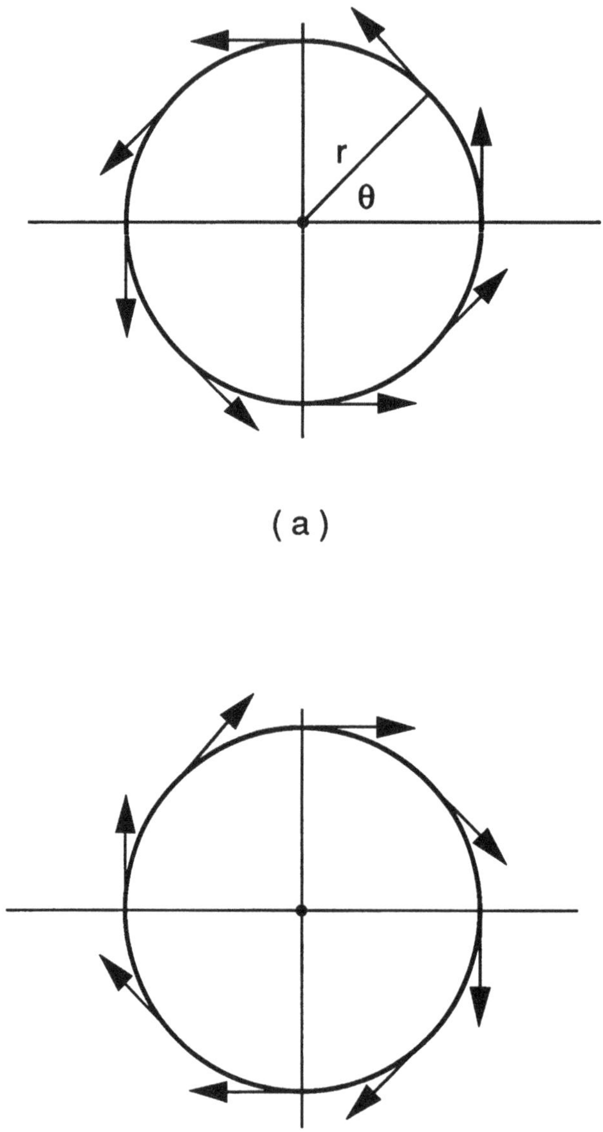

Fig. 2

Pictorially see Figure 3b.

equation of motion:

Consider

$$\hat{\phi} = q\theta + \frac{\pi}{2} \tag{38}$$

In two dimensions

$$\partial^{\mu}\partial_{\mu} = \frac{1}{r}\frac{\partial}{\partial r}(r\frac{\partial}{\partial r}) + \frac{1}{r^2}\frac{\partial^2}{\partial\theta^2} \tag{39}$$

Clearly

$$\partial^{\mu}\partial_{\mu}\hat{\phi} = 0 \tag{40}$$

Hence, the appropriate action for scalar field ϕ is

$$\begin{aligned} S &= \frac{1}{e^2}\int d^2x\sqrt{g}g^{\mu\nu}\partial_{\mu}\phi\partial_{\nu}\phi \\ &= \frac{1}{e^2}\int d^2x\sqrt{g}g^{\mu\nu}H_{\mu}H_{\nu} \\ &= \frac{1}{e^2}\int H\Lambda{}^*H \end{aligned} \tag{41}$$

where ${}^*H = (\epsilon_{\mu\nu}H^{\nu})dx^{\mu}$ is the one-form dual to H. For a single vortex

$$\begin{aligned} \hat{S}_{(1)} &= \frac{1}{e^2}\int dx^1dx^2((\hat{H}_1)^2 + (\hat{H}_2)^2) \\ &= \frac{2\pi q^2}{e^2}\int_{r_c}^{R} dr\frac{1}{r} \\ &= \frac{2\pi q^2}{e^2}ln\frac{R}{r_c} \end{aligned} \tag{42}$$

where r_c and R are small and large radius cutoffs respectively. The core action is

$$\hat{S}_c = \frac{2\pi q^2}{e^2}ln\frac{r_c}{0} \longrightarrow \infty \tag{43}$$

Is it possible to "regulate" the core action and make it finite? Introduce, in addition to closed one-form H, a real scalar field u. Let the Lagrangian be

$$\mathcal{L} = \frac{1}{2}(\partial_{\mu}u)^2 + u^2H_{\mu}H^{\mu} + v(u) \tag{44}$$

where

$$v(u) = -\frac{m^2}{2}u^2 + \frac{\lambda}{4!}u^4 + \frac{3m^4}{2\lambda} \tag{45}$$

The u equation of motion is

$$\partial^2u - 2uH_uH^u + m^2u - \frac{\lambda}{6}u^3 = 0 \tag{46}$$

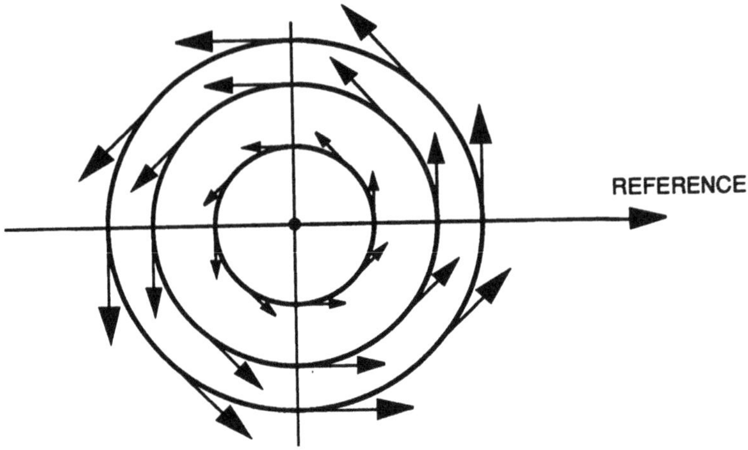

(a)

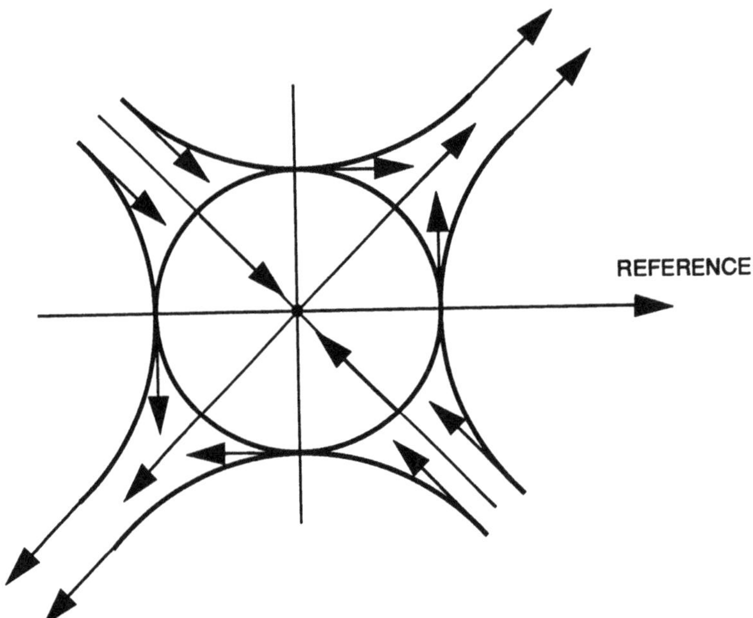

(b)

Fig. 3

Plug in $\hat{H}_u$. Then

$$\hat{H}_u \hat{H}^u = \frac{q^2}{r^2} \tag{47}$$

Assume $u = u(r)$ and use the formula

$$\partial^2 = \frac{1}{r}\frac{\partial}{\partial r}(r\frac{\partial}{\partial r}) + \frac{1}{r^2}\frac{\partial^2}{\partial \theta^2} \tag{48}$$

The u equation of motion becomes

$$u'' + \frac{1}{r}u' - (\frac{2q^2}{r^2} - m^2)u - \frac{\lambda}{6}u^3 = 0 \tag{49}$$

In the region where

$$r \ll \frac{\sqrt{2}|q|}{m} \qquad \text{region I} \tag{50}$$

one has

$$\frac{2q^2}{r^2} \gg m^2 \tag{51}$$

and the equation becomes

$$u'' + \frac{1}{r}u' - \frac{2q^2}{r^2}u = 0 \tag{52}$$

As an ansatz take

$$u' = cr^p \tag{53}$$

Then

$$\begin{aligned} &p(p-1)r^{p-2} + pr^{p-2} - 2q^2 r^{p-2} = 0 \\ &p^2 - 2q^2 = 0 \\ &p = \pm\sqrt{2}q \end{aligned} \tag{54}$$

For finite action take

$$p = +\sqrt{2}|q| \tag{55}$$

c is undetermined. Hence, in region I

$$\hat{u}_I = cr^{\sqrt{2}|q|} \tag{56}$$

Note that $u^3 \sim r^{3p} \ll r^{p-2}$ for small r and, hence, the above is a good approximation. In the region where

$$r \gg \frac{\sqrt{2}|q|}{m} \qquad \text{region III} \tag{57}$$

one has

$$\frac{2q^2}{r^2} \ll m^2 \tag{58}$$

and the equation becomes

$$m^2u - \frac{\lambda}{6}u^3 = 0 \tag{59}$$

The minimum energy solution is

$$\hat{u}_{III} =< u >= \sqrt{\frac{6}{\lambda}}m \tag{60}$$

Define

$$r_c = \frac{\sqrt{2}|q|}{m} \tag{61}$$

The complete solution is shown graphically in Figure 4a. What is c? A good approximation is defined by the graph in Figure 4b. Hence

$$\frac{c}{m}r_c^{\sqrt{2}|q|} = \sqrt{\frac{6}{\lambda}}$$

$$c = (\frac{1}{\sqrt{2}|q|})^{\sqrt{2}|q|}\sqrt{\frac{6}{\lambda}}(m)^{1+\sqrt{2}|q|} \tag{62}$$

Therefore, to a good approximation, from $0 \le r \le r_c$

$$\hat{u} = (\frac{1}{\sqrt{2}|q|})^{\sqrt{2}|q|}\sqrt{\frac{6}{\lambda}}(m)^{1+\sqrt{2}|q|}r^{\sqrt{2}|q|} \tag{63}$$

We can now compute the core action. It is given by

$$\begin{aligned} \hat{S}_c &= \int d^2x[\frac{1}{2}(\partial_\mu \hat{u})^2 + \hat{u}^2\hat{H}_\mu \hat{H}^u - \frac{m^2\hat{u}^2}{2} + \frac{\lambda\hat{u}^4}{4} + \frac{3m^4}{2\lambda}] \\ &= 2\pi\int_0^{r_c} rdr[\frac{1}{2}((\frac{1}{\sqrt{2}|q|})^{\sqrt{2}|q|-1}\sqrt{\frac{6}{\lambda}}(m)^{1+\sqrt{2}|q|}r^{\sqrt{2}|q|-1})^2 \\ &\quad +(\frac{1}{\sqrt{2}|q|})^{2\sqrt{2}|q|}\frac{6}{\lambda}(m)^{2+2\sqrt{2}|q|}r^{2\sqrt{2}|q|}\frac{q^2}{r^2} + \frac{3m^4}{2\lambda}] \\ &= \frac{3\pi q^2}{\lambda}m^2(\frac{2\sqrt{2}}{|q|} + 1) \end{aligned} \tag{64}$$

Therefore

$$\hat{S}_c = \frac{3\pi q^2}{\lambda}m^2(1 + \frac{2\sqrt{2}}{|q|})finite$$

and is naturally finite. The reason for this can be seen pictorially in Figure 5. Note that for $r \gg r_c$

$$\hat{u} \to < u >= \sqrt{\frac{6}{\lambda}}m \tag{65}$$

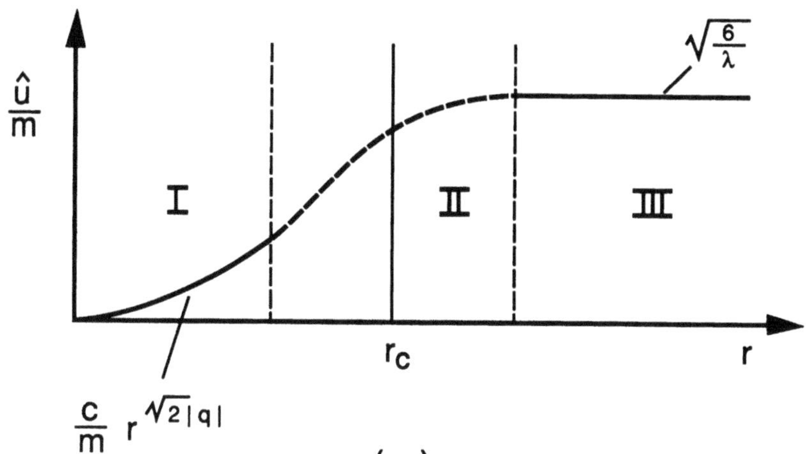

(a)

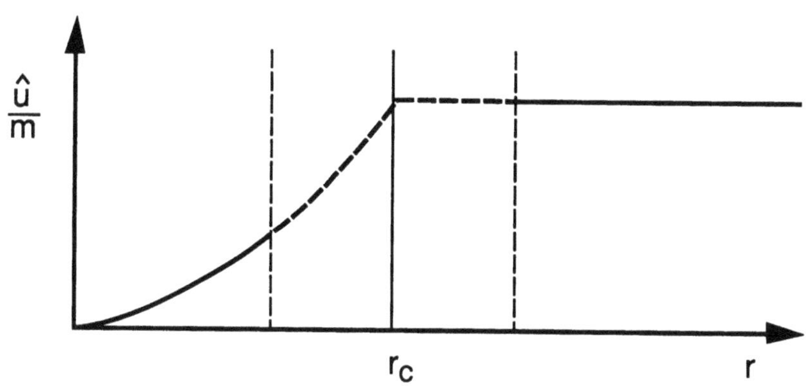

(b)

Fig. 4

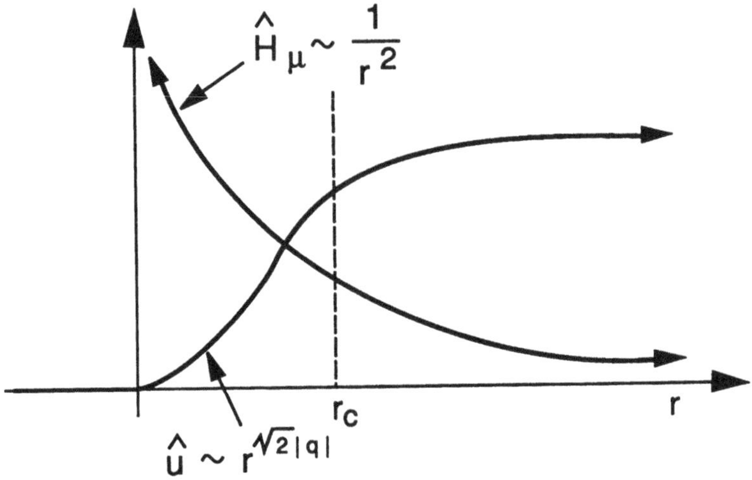

Fig. 5

Let

$$u = < u > + u' \tag{66}$$

Then

$$\mathcal{L} = \frac{1}{2}(\partial_\mu u')^2 + \frac{1}{2}(2m^2)u'^2 + ... + < u >^2 H_\mu H^\mu + O(u')H_\mu H^\mu \tag{67}$$

The mass of u' is

$$m'_u = \sqrt{2}m \tag{68}$$

Hence, for momenta and field strengths $\ll m'_u, u'$ decouples and $\mathcal{L}$ becomes

$$\mathcal{L} = \frac{6}{\lambda}m^2 H_\mu H^u \tag{69}$$

Since $\frac{6}{\lambda}m^2$ is dimensionless, define

$$\frac{1}{e^2} = \frac{6}{\lambda}m^2 \tag{70}$$

Then

$$\mathcal{L} = \frac{1}{e^2}H_\mu H^u \tag{71}$$

as in pure H_μ theory above. Henceforth, consider pure $\hat{H}$ instantons but with finite core action.

$$\hat{S}_c = \frac{\pi}{2}(\frac{q}{e})^2(1 + \frac{2\sqrt{2}}{|q|}) \tag{72}$$

Up to now we have considered $\hat{H}$ instantons centered at the origin. An $\hat{H}$ instanton of charge q centered at x_o is given by

$$\hat{H}_\mu = -q\epsilon_{\mu\nu}\frac{(x - x_0)^\nu}{|x - x_0|^2} \tag{73}$$

Note that

$$\begin{aligned} \partial_\nu ln\frac{|x - x_0|^2}{R^2} &= \frac{1}{|x - x_0|^2}\partial_\nu q_{\alpha\beta}(x - x_0)^\alpha(x - x_0)^\beta \\ &= \frac{2(x - x_0)_\nu}{|x - x_0|^2} \end{aligned} \tag{74}$$

Therefore,

$$\hat{H}_\mu = -q\epsilon_{\mu\nu}\partial^\nu ln\frac{|x - x_0|}{R} \tag{75}$$

Consider two $\hat{H}$ instantons q_1, x_1 and q_2, x_2. Then

$$\hat{H}_{(2)\mu} = -q_1\epsilon_{\mu\nu}\partial^\nu ln\frac{|x - x_1|}{R} - q_2\epsilon_{\mu\nu}\partial^\nu ln\frac{|x - x_2|}{R} \tag{76}$$

What is their action?

$$\begin{aligned}\hat{S}_{(2)} &= \frac{1}{e^2}\int d^2x \hat{H}_{(2)\mu}\hat{H}^{\mu}_{(2)} \\ &= \hat{S}_{(1)}(q_1) + \hat{S}_{(1)}(g_2) \\ &+ \frac{2q_1q_2}{e^2}\int d^2x \epsilon_{\mu\nu}\partial^{\nu} ln\frac{|x-x_1|}{R}\epsilon^{\mu\lambda}\partial_{\lambda} ln\frac{|x-x_2|}{R}\end{aligned} \tag{77}$$

The last term is

$$\begin{aligned}\frac{2q_1q_2}{e^2}\int d^2x\delta^{\lambda}_{\nu}\partial^{\nu} ln\frac{|x-x_1|}{R}\partial_{\lambda}\frac{|x-x_2|}{R} &= \frac{2q_1q_2}{e^2}\int d^2x\partial^{\mu} ln\frac{|x-x_1|}{r}\partial_{\mu} ln\frac{|x-x_2|}{R} \\ &= -\frac{2q_1q_2}{e^2}\int d^2x ln\frac{|x-x_2|}{R}\partial^2 ln\frac{|x-x_1|}{R}\end{aligned} \tag{78}$$

Now

$$\partial^2 ln\frac{|x-x_1|}{R} = 2\pi\delta(x-x_1) \tag{79}$$

Check:

$$\begin{aligned}\int d^2x\partial^2 ln\frac{|x-x_1|}{R} &= 2\pi \\ \int dS^{\mu}\partial_{\mu} ln\frac{|x-x_1|}{R} &= 2\pi \\ 2\pi r\frac{R}{r}\frac{1}{R} &= 2\pi\end{aligned} \tag{80}$$

Therefore, the last term becomes

$$\begin{aligned}-4\pi\frac{q_1q_2}{e^2}\int d^2x ln\frac{|x-x_2|}{R}\delta(x-x_1) \\ = -4\pi\frac{q_1q_2}{e^2} ln\frac{|x_1-x_2|}{R}\end{aligned} \tag{81}$$

Hence

$$\begin{aligned}\hat{S}_{(2)} = -\frac{2\pi q_1^2}{e^2} ln\frac{r_c}{R} - \frac{2\pi q_2^2}{e^2} ln\frac{r_c}{R} \\ -\frac{4\pi q_1 q_2}{e^2} ln\frac{|x_1-x_2|}{R}\end{aligned} \tag{82}$$

which we can write as

$$\hat{S}_{(2)} = -\frac{2\pi}{e^2}[\sum_{i=1}^{2} q_i^2 ln\frac{r_c}{R} + \sum_{i\neq j=1}^{2} q_iq_j ln\frac{|x_i-x_j|}{R}] \tag{83}$$

The extension to N vortices is obvious. The N-vortex is

$$\hat{H}_{(N)\mu} = -\sum_{i=1}^{N} q_i \epsilon_{\mu\nu} \partial^{\nu} ln \frac{|x - x_i|}{R} \tag{84}$$

and has action

$$\hat{S}_{(N)} = -\frac{2\pi}{e^2}[\sum_{i=1}^{N} q_i^2 ln \frac{r_c}{R} + \sum_{i \neq j = 1}^{N} q_i q_j ln \frac{|x_i - x_j|}{R}] \tag{85}$$

Note that

$$\hat{S}_{(N)} \simeq -\frac{2\pi}{e^2} \sum_{i,j=1}^{N} q_i q_j ln \frac{|x_i - x_j|}{R} \tag{86}$$

Define

$$J^N(x) = \sum_{i=1}^{N} q_i \delta(x - x_i)$$

$$G(x, y) = ln \frac{|x - y|}{R} \tag{87}$$

Then

$$\hat{S}_{(N)} = \frac{-2\pi}{e^2} \int d^2x d^2y J^N(x) G(x, y) J^N(y) \tag{88}$$

Grand Canonical Partition Function:

(1) Instanton Contribution

$$Z_{inst} = \sum_{N=0}^{\infty} \frac{z^N}{N!} \sum_{\substack{\{q_i\} \\ |q_i| = q}} (\prod_{i=1}^{N} \int \frac{d^2x_i}{\pi r_c^2}) e^{-\hat{S}_{(N)}} \tag{89}$$

where

$$z = e^{-\hat{S}_c} \tag{90}$$

Therefore

$$Z_{inst} = \sum_{i=1}^{\infty} \frac{z^N}{N!} \sum_{\substack{\{q_i\} \\ |q_i| = q}} (\prod_{i=1}^{N} \int \frac{d^2x_i}{\pi r_c^2}) e^{+\frac{2\pi}{e^2} \int d^2x d^2y J^N(x) G(x,y) J^N(y)} \tag{91}$$

Mean field theory:
Let χ be a scalar mean field. Then Z_{inst} can be written as

$$Z_{inst} = \frac{\sum_{N=0}^{\infty} \frac{z^N}{N!} \sum_{\substack{\{q_i\} \\ |q_i| = q}} (\prod_{i=1}^{N} \int \frac{d^2x_c}{\pi r_c^2}) \int [d\chi] e^{-\frac{1}{e^2} \int d^2x (\partial_\mu \chi)^2 + \frac{2\pi}{e^2} \int d^2x d^2y J^N G J^N}}{\int [d\chi] e^{-\frac{1}{e^2} \int d^2x (\partial_\mu \chi)^2}} \tag{92}$$

Let

$$\chi' = \chi + c\int d^2yG(x,y)J^N(y) \tag{93}$$

Then

$$\begin{aligned}\partial^2\chi' &= \partial^2\chi + c\int d^2y\underbrace{\partial^2 G(x,y)}_{2\pi\delta(x-y)}J^n(y)\\ &= \partial^2\chi + 2\pi cJ^N(x)\end{aligned}$$

It follows from this that

$$-\frac{1}{e^2}\int d^2x(\partial_\mu\chi)^2 - \frac{2\pi}{e^2}c^2\int d^2xd^2yJ^NGJ^N = -\int d^2x(\frac{1}{e^2}(\partial_\mu\chi')^2 + \frac{4\pi}{e^2}c\chi J^N) \tag{94}$$

Take c so that

$$-\frac{2\pi}{e^2}c^2 = \frac{2\pi}{e^2} \tag{95}$$

Therefore

$$c = i \tag{96}$$

It follows that

$$-\frac{1}{e^2}\int d^2x(\partial_\mu\chi)^2 + \frac{2\pi}{e^2}\int d^2xd^2yJGJ = -\int d^2x(\frac{1}{e^2}(\partial_\mu\chi')^2 + \frac{4\pi}{e^2}i\chi J^N) \tag{97}$$

Noting that

$$[d\chi] = [d\chi'] \tag{98}$$

implies

$$Z_{inst} = \frac{\sum_{N=0}^{\infty}\frac{z^N}{N!}\sum_{\substack{\{q_i\}\\|q_i|=q}}(\prod_{i=1}^{N}\int\frac{d^2x_i}{\pi r_c^2})\int[d\chi]e^{-\int d^2x(\frac{1}{e^2}(\partial_\mu\chi)^2+\frac{i4\pi}{e^2}\chi J^N)}}{\int[d\chi]e^{-\int d^2x\frac{1}{e^2}(\partial_\mu\chi)^2}} \tag{99}$$

Consider the numerator

$$\begin{aligned}Z^n_{inst} &= \sum_{N=0}^{\infty}\frac{z^n}{N!}\sum_{\substack{\{q_i\}\\|q_i|=q}}(\prod_{i=1}^{N}\int\frac{d^2x_i}{\pi r_c^2})\int[d\chi]e^{-\int d^2x(\frac{1}{e^2}(\partial_\mu\chi)^2+\frac{i4\pi}{e^2}\chi J^N)}\\ &= \int[d\chi]e^{-\int d^2x\frac{1}{e^2}(\partial_\mu\chi)^2}\sum_{N=0}^{\infty}\frac{z^N}{N!}\sum_{\substack{\{q_i\}\\|q_i|=q}}(\prod_{i=1}^{N}\int\frac{d^2x_i}{\pi r_c^2})\underbrace{e^{-\int d^2x\frac{i4\pi}{e^2}\chi J^N}}_{e^{-\frac{4\pi i}{e^2}\sum_{i=1}^{N}q_i\chi(x_i)}}\end{aligned}$$

$$
\begin{aligned}
&= \int [d\chi] e^{-\int d^2x \frac{1}{e^2}(\partial_\mu \chi)^2} \Big(1 + z\int \frac{d^2x_1}{\pi r_c^2} e^{-\frac{4i\pi}{e^2} q\chi(x_1)} \\
&\quad \underbrace{+\frac{z^2}{2}\int \frac{d^2x_1 d^2x_2}{\pi r_c^2} e^{-\frac{4i\pi}{e^2} q(\chi(x_1)+\chi(x_2))} + \ldots\Big)}_{(\int \frac{d^2x_1}{\pi r_c^2} e^{-\frac{4i\pi}{e^2} q\chi(x_1)})^2} \qquad (100) \\
&\quad \times\Big(1 + z\int \frac{d^2x_1}{\pi r_c^2} e^{+\frac{4i\pi}{e^2} q\chi(x_1)} + \frac{z^2}{2}\Big(\int \frac{d^2x_1}{\pi r_c^2} e^{\frac{4i\pi}{e^2} q\chi(x_1)}\Big)^2 + \ldots\Big) \\
&= \int [d\chi] e^{-\int d^2x \frac{1}{e^2}(\partial_\mu \chi)^2} e^{z\int \frac{d^2x}{\pi r_c^2} \underbrace{\left(e^{-\frac{4i\pi}{e^2} q\chi(x)} + e^{\frac{4i\pi}{e^2}\chi(x)}\right)}_{2\cos(\frac{4\pi q}{e^2}\chi)}} \\
&= \int [d\chi] e^{-\int d^2x [\frac{1}{e^2}(\partial_\mu \chi)^2 - \frac{2z}{\pi r_c^2}\cos(\frac{4\pi q}{e^2}\chi)]}
\end{aligned}
$$

Therefore

$$
Z_{inst} = \frac{\int [d\chi] e^{-\int d^2x [\frac{1}{e^2}(\partial_\mu \chi)^2 - \frac{2z}{\pi r_c^2}\cos(\frac{4\pi \chi}{e^2} x)]}}{\int [d\chi] e^{-\frac{1}{e^2}\int d^2x (\partial_\mu \chi)^2)}} \qquad (101)
$$

What about the denominator? We must include the "spin-wave" part of the partition function. For any N, expand H_μ around the N-instanton.

$$
H_\mu = \hat{H}_{(N)\mu} + H'_\mu \qquad (102)
$$

Then action

$$
S = \frac{1}{e^2}\int d^2x H_\mu H^\mu \qquad (103)
$$

becomes

$$
\begin{aligned}
S_N &= \frac{1}{e^2}\int d^2x (\hat{H}_{(N)\mu} + H'_\mu)(\hat{H}_{(N)} + H'^\mu) \\
&= \frac{1}{e^2}\int d^2x \hat{H}_{(N)\mu}\hat{H}^\mu_{(N)} + \frac{1}{e^2}\int d^2x H'_\mu H^{\mu\prime} \qquad (104) \\
&+ \frac{2}{e^2}\int d^2x \hat{H}_{(N)\mu} H'^\mu
\end{aligned}
$$

Now H_μ is closed and, hence, $H'^\mu = \partial^\mu \phi'$. Therefore

$$
\frac{2}{e^2}\int d^2x \hat{H}_{(N)\mu}\partial^\mu \phi' = -\frac{2}{e^2}\int d^2x (\partial^\mu \hat{H}_{(N)\mu})\phi' \qquad (105)
$$

Now

$$
\begin{aligned}
&\hat{H}_{(1)\mu} = -q_i \epsilon_{\mu\nu}\partial^\nu ln\frac{|x - x_i|}{R} \\
&\partial^\mu \hat{H}_{(1)\mu} = 0 \qquad (106) \\
&\Rightarrow \partial^\mu \hat{H}_{(N)\mu} = 0
\end{aligned}
$$

Therefore, expression (105) vanishes. Hence

$$S_{(N)} = \hat{S}_{(N)} + \frac{1}{e^2}\int d^2x H'_\mu H^{\mu'} \tag{107}$$

The *full* partition function has $e^{-S_{(N)}}$ and not $e^{-\hat{S}_{(N)}}$.

$$\int [d\phi']e^{-S_{(N)}} = e^{-\hat{S}_{(N)}} \underbrace{\int [d\phi']e^{-\frac{1}{e^2}\int d^2x(\partial_\mu\phi')^2}}_{\text{same factor for every N}} . \tag{108}$$

Therefore

$$\begin{aligned} Z_T &= Z_{inst}\int [d\phi']e^{-\frac{1}{e^2}\int d^2x(\partial_\mu\phi')^2} \\ &= \frac{\int [d\chi]e^{-\int d^2\chi[\frac{1}{e^2}(\partial_\mu\chi)^2 - \frac{2z}{\pi r_c^2}\cos(\frac{4\pi q\chi}{e^2})]}\int [d\phi']e^{-\frac{1}{e^2}\int d^2x(\partial_\mu\phi')^2}}{\int [d\chi]e^{-\frac{1}{e^2}\int d^2x(\partial_\mu\chi)^2}} \end{aligned} \tag{109}$$

It follows that the total partition function is

$$Z_T = \int [d\chi]e^{-S_{eff}} \tag{110}$$

where

$$S_{eff} = \int d^2x[\frac{1}{e^2}(\partial_\mu\chi) - \frac{2z}{\pi r_c^2}\cos(\frac{4\pi q}{e^2}\chi)] \tag{111}$$

and

$$\begin{aligned} r_c &= \frac{\sqrt{2}|q|}{m} \\ z &= e^{-\frac{3\pi q^2}{\lambda}m^2(1+\frac{2\sqrt{2}}{|q|})} \end{aligned} \tag{112}$$

Put the kinetic energy in standard form by defining

$$\chi' = \frac{\sqrt{2}}{e}\chi \tag{113}$$

Therefore, dropping the prime,

$$S_{eff} = \int d^2x[\frac{1}{2}(\partial_\mu\chi)^2 - \frac{2z}{\pi r_c^2}\cos(\frac{2\sqrt{2}\pi}{e}q\chi)] \tag{114}$$

Application to String Theory

The bosonic string worldsheet action is, in general,

$$S = \frac{1}{4\pi}\int d^2x(G_{\alpha\beta}(X)\partial^\mu X^\alpha\partial_\mu X^\beta + B_{\alpha\beta}(X)\epsilon_{\mu\nu}\partial^\mu X^\alpha\partial^\nu X^\beta \tag{115}$$

Consider a one-dimensional string compactified on a circle of radius R. Then

1.

$$\begin{aligned} &\alpha = \beta = 1 \\ &\Rightarrow B_{\alpha\beta} = 0 \end{aligned} \tag{116}$$

2.

$$\begin{aligned} &ds^2 = R^2 d\theta^2 \\ &\Rightarrow G(x) = R^2 \end{aligned} \tag{117}$$

Therefore

$$S = \frac{R^2}{4\pi} \int d^2x (\partial_\mu X)^2 \tag{118}$$

S is in exactly the same form as the previous ϕ action where

$$\frac{1}{e^2} = \frac{R^2}{4\pi} \tag{119}$$

Since X is an "angular" variable, it admits vortices. Note that any vortices with integer q are allowed. These vortices induce an effective action given by

$$S_{eff} = \int d^2x [\frac{R^2}{4\pi}(\partial_\mu X)^2 - \frac{2z}{\pi\epsilon^2}\cos(qR^2X)] \tag{120}$$

Note that r_c has been replaced by infinitesimal cutoff ϵ. This is because string theory does not contain the regulating u scalar field and one must simply cutoff the vortex core, as discussed previously. Also z is now given by

$$z = e^{-\frac{3\pi m^4}{2\lambda}\epsilon^2} \tag{121}$$

where one uses of the constant part of the core action cutoff at radius ϵ. Defining

$$X' = RX \tag{122}$$

and then dropping the prime gives

$$S_{eff} = \int d^2x [\frac{1}{4\pi}(\partial_\mu X)^2 - \frac{2z}{\pi\epsilon^2}\cos(qRX)] \tag{123}$$

Now go to complex coordinates

$$\begin{aligned} z &= \frac{1}{\sqrt{2}}(x^1 + ix^2) \\ \bar{z} &= \frac{1}{\sqrt{2}}(x^1 - ix^2) \end{aligned} \tag{124}$$

Note that

$$\begin{aligned} ds^2 &= dx^{12} + dx^{22} \\ &\rightarrow 2dzd\bar{z} \end{aligned} \tag{125}$$

and, hence,

$$g_{\mu\nu} \begin{pmatrix} 0 & 1 \\ 1 & 0 \end{pmatrix} \tag{126}$$

Also

$$\int d^2x \rightarrow \int d^2z \tag{127}$$

and

$$\begin{aligned} &(\partial_\mu X)^2 = g^{\mu\nu}\partial_\mu X \partial_\nu X \\ &\rightarrow 2\partial_{\bar{z}} X \partial_z X \end{aligned} \tag{128}$$

Therefore

$$S_{eff} = \int d^2z [\frac{1}{2\pi}\partial_{\bar{z}} X \partial_z X - \frac{2z}{\pi\epsilon^2}\cos(qRX)] \tag{129}$$

Treating the cosine term as a perturbation, the free equation of motion is

$$\partial_{\bar{z}}\partial_z X = 0 \tag{130}$$

which has the solution

$$X(z,\bar{z}) = X(z) + \bar{X}(\bar{z}) \tag{131}$$

The free stress energy tensor is given by

$$\begin{aligned} T(z) &= -\frac{1}{2} : (\partial_z X)^2 : \\ \bar{T}(z) &= -\frac{1}{2} : (\partial_{\bar{z}} X)^2 : \end{aligned} \tag{132}$$

The cosine interaction given above is a "classical" result. To do conformal field theory we must turn it into a normal ordered operator. This is done using the formula

$$\cos(\alpha X(z,\bar{z})) =: \cos(\alpha X(z,\bar{z})) : e^{-\lim_{z'\to z}\frac{\alpha}{2}<X(z'\bar{z}'),X(z,\bar{z})>} \tag{133}$$

With the above normalization

$$\begin{aligned} < X(z')X(z) > &= -ln(z'-z) \\ < \bar{X}(\bar{z}')\bar{X}(\bar{z}) > &= -ln(\bar{z}'-\bar{z}) \end{aligned} \tag{134}$$

therefore

$$< X(z',\bar{z}')X(z,\bar{z}) >= -2ln|z-z'| \tag{135}$$

and

$$\lim_{z' \to z} < X(z', \bar{z}')X(z, \bar{z}) >= -2ln\epsilon \tag{136}$$

It follows that

$$\begin{aligned} \cos(\alpha X(z,\bar{z})) &=: \cos(\alpha X(z,\bar{z})) : e^{+\alpha^2 \ln \epsilon} \\ &=: \cos(\alpha X(z,\bar{z})) : \epsilon^{\alpha^2} \end{aligned} \tag{137}$$

In our case

$$\alpha = qR \tag{138}$$

and,hence,

$$S_{eff} = \int d^2 z [\frac{1}{2\pi} \partial_{\bar{z}} X \partial_z X - \frac{2z}{\pi} \epsilon^{\alpha^2 - 2} : \cos(\alpha X) :] \tag{139}$$

What is the conformal dimension of the cosine operator?

$$\begin{aligned} : \cos(\alpha X) : &= \frac{1}{2}(: e^{i\alpha X} : + : e^{-i\alpha X} :) \\ &= \frac{1}{2}(: e^{i\alpha(X(z)+\bar{X}(\bar{z}))} : + e^{-i\alpha(X(z)+\bar{X}(\bar{z}))}) :) \\ &= \frac{1}{2}(: e^{i\alpha X(z)} :: e^{i\alpha \bar{X}(\bar{z})} : + : e^{-i\alpha X(z)} :: e^{-i\alpha \bar{X}(\bar{z})} :) \end{aligned} \tag{140}$$

Now, in general,

$$T(z) : e^{i\alpha X(z')} := \frac{\alpha^2/2}{z - z'} e^{i\alpha X(z')} + ... \tag{141}$$

where $\alpha^2/2$ is the z conformal dimension. Similarly for $\bar{T}(\bar{z})$. Noting that

$$< X(z)\bar{X}(\bar{z}') >= 0 \tag{142}$$

implies that $: \cos(\alpha X) :$ has conformal dimension

$$(\frac{\alpha^2}{2}, \frac{\alpha^2}{2}) \tag{143}$$

It follows that the cosine term in the action is conformally invariant if and only if

$$\begin{aligned} \frac{\alpha^2}{2} &= 1 \\ \Rightarrow \alpha &= \sqrt{2} \\ \Rightarrow R_c &= \frac{\sqrt{2}}{q} \end{aligned} \tag{144}$$

What happens for other values of R? Note that the cosine coupling is

$$C = \frac{2z}{\pi} e^{\alpha^2 - 2} \tag{145}$$

As $\epsilon \to 0$ then $z \to 1$. Therefore, for $\alpha^2 > 2$

$$C \stackrel{\epsilon \to 0}{\longrightarrow} 0 \tag{146}$$

However, for $\alpha^2 < 2$

$$C \stackrel{\epsilon \to 0}{\longrightarrow} \infty \tag{147}$$

We conclude that for $R > \frac{\sqrt{2}}{q}(\alpha^2/2 > 1)$ the cosine coupling is driven to zero and the cosine is an "irrelevant" operator. Such string theories are conformally invariant and stable. However, for $R < \frac{\sqrt{2}}{q}(\alpha^2/2 < 1)$ the cosine coupling is large and the cosine is a "relevant" operator. Such string theories are unstable and must be driven to some other conformal field theory, not that of a free string on a circle.

D = 4 H - Instantons:

Consider

$$M = R^4 - \{(0,0,0,0)\} \tag{148}$$

Note that

$$H^4(R^4 - \{(0,0,0,0)\}) \simeq H^3(S^3; R) = R \tag{149}$$

Hence, there exists an infinite number of topologically non-trivial closed three-forms, indexed by $q \epsilon R$, that diverge at the origin. These three forms satisfy

$$\int_{S^3} H = 2\pi q \tag{150}$$

where $q \epsilon R$. It is easy to give an explicit form for closed three-forms satisfying this equation. It is

$$\hat{H}_{\mu\nu\lambda} = \frac{q}{\pi f} \epsilon_{\mu\nu\lambda\gamma} \frac{x^\gamma}{|x|^4} \tag{151}$$

where f is an arbitrary mass parameter. It is straightforward to show that such $\hat{H}'s$ are closed and satisfy the above equation. Since they are closed, these $\hat{H}'s$ can be written locally as

$$\hat{H} = d\hat{B} \tag{152}$$

where $\hat{B}$ is a two-form, but we will not compute $\hat{B}$ explicitly in these lectures. Once again, the action is given by

$$\begin{aligned} S &= \frac{1}{e^2} \int H \Lambda^* H \\ &= \frac{1}{3! e^2} \int d^4x \sqrt{g} H_{\mu\nu\lambda} H^{\mu\nu\lambda} \end{aligned} \tag{153}$$

What is the action for an H-instanton of charge q?

$$\begin{aligned}
\hat{S}_{(1)} &= \frac{1}{3!e^2}\int d^4x\sqrt{g}\frac{q^2}{\pi^2 f^2}\epsilon_{\mu\nu\lambda\gamma}\frac{x^\gamma}{|x|^4}\epsilon^{\mu\nu\lambda\delta}\frac{x_\delta}{|x|^4} \\
&= \frac{q^2}{e^2\pi^2 f^2}\int d^4x\sqrt{g}\frac{1}{|x|^6} \\
&= \frac{2q^2}{e^2 f^2}(\frac{1}{r_c^2}-\frac{1}{R^2}) \\
&\xrightarrow{R\to\infty} \frac{2q^2}{e^2 f^2}\frac{1}{r_c^2}
\end{aligned} \tag{154}$$

Pictorially see Figure 6. The core action is

$$\hat{S}_c = \frac{2q^2}{e^2 f^2}(\frac{1}{0}-\frac{1}{r_c^2}) \to \infty \tag{155}$$

Is it possible to regulate the core action and make it finite? Introduce in addition to the closed form H a real scalar field u. Let the Lagrangian be

$$\begin{aligned}
\mathcal{L} &= \frac{1}{2}(\partial_\mu u)^2 + Bu^2 H_{\mu\nu\lambda}H^{\mu\nu\lambda} + V(u) \\
V(u) &= -\frac{m^2}{2}u^2 + \frac{\lambda}{4!}u^4 + \frac{3m^4}{2\lambda}
\end{aligned} \tag{156}$$

where $B > 0$. The u equation of motion is

$$\partial^2 u - 2BuH_{\mu\nu\lambda}H^{\mu\nu\lambda} + m^2 u - \frac{\lambda}{6}u^3 = 0 \tag{157}$$

Plug in $\hat{H}_{\mu\nu\lambda}$

$$\hat{H}_{\mu\nu\lambda}\hat{H}^{\mu\nu\lambda} = \frac{6q^2}{\pi^2 f^2}\frac{1}{r^6} \tag{158}$$

Assuming $u = u(r)$ implies

$$u'' + \frac{3u'}{r} - (\frac{c}{r^6} - m^2)u - \frac{\lambda u^3}{6} = 0 \tag{159}$$

where

$$c = 12B\frac{q^2}{\pi^2 f^2} \tag{160}$$

The core radius is defined by

$$\begin{aligned}
\frac{c}{r_c^6} &= m^2 \\
\Rightarrow r_c &= \frac{c^{1/6}}{m^{1/3}}
\end{aligned} \tag{161}$$

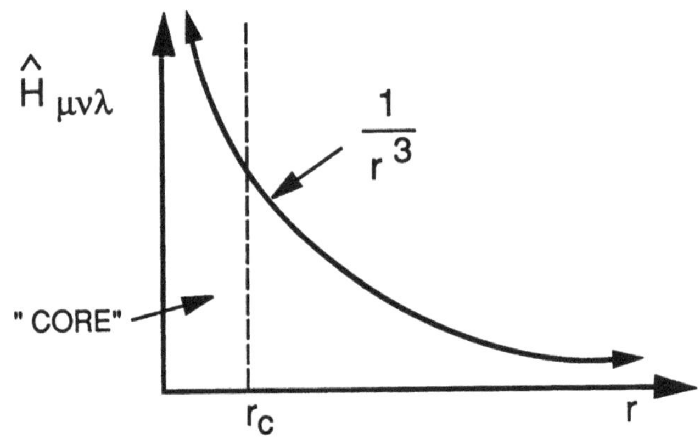

Fig. 6

In the region where

$$r << r_c \qquad \text{region I} \tag{162}$$

the equation becomes

$$u'' + \frac{3u'}{r} - \frac{cu}{r^6} = 0 \tag{163}$$

which has the solution

$$\hat{u}_I = ae^{-\frac{\sqrt{c}}{2r^2}} \tag{164}$$

for arbitrary a. In the region where

$$r >> r_c \qquad \text{region III} \tag{165}$$

The equation of motion becomes

$$m^2u - \frac{\lambda}{6}u^3 = 0 \tag{166}$$

which has as its solution

$$\hat{u}_{III} = < u > = \sqrt{\frac{6}{\lambda}}m \tag{167}$$

We have been unable to find an analytic solution of the equation of motion for the region $r \simeq r_c$. Graphically see Figure 7a. What is a? A good approximation is that a is such that $\hat{u}_I$ matches onto $\hat{u}_{III}$. Then

$$a = \sqrt{\frac{6}{\lambda}}m \tag{168}$$

Therefore, to a good approximation, for $0 \leq r \leq r_c$

$$\hat{u} = \sqrt{\frac{6}{\lambda}}me^{-\frac{\sqrt{c}}{2r^2}} \tag{169}$$

We can now compute the core action. It is

$$\hat{S}_c = 4\pi^2 \int_0^{r_c} r^3 dr (\frac{1}{2}(\hat{u}')^2 + \frac{c\hat{u}^2}{r^6} - \frac{m^2}{2}\hat{u}^2 + \frac{\lambda\hat{u}^4}{4!} + \frac{3m^4}{2\lambda}) \tag{170}$$

After some calculation the result is

$$\hat{S}_c = \frac{3\pi^2 m^2}{2\lambda}(8\sqrt{c}e^{-(m^2\sqrt{c})^{1/3}} + (mc)^{2/3}) \tag{171}$$

which is finite. Graphically, the reason is shown in Figure 7b. Note that for $r \gg r_c$

$$\hat{u} \to < u > = \sqrt{\frac{6}{\lambda}}m \tag{172}$$

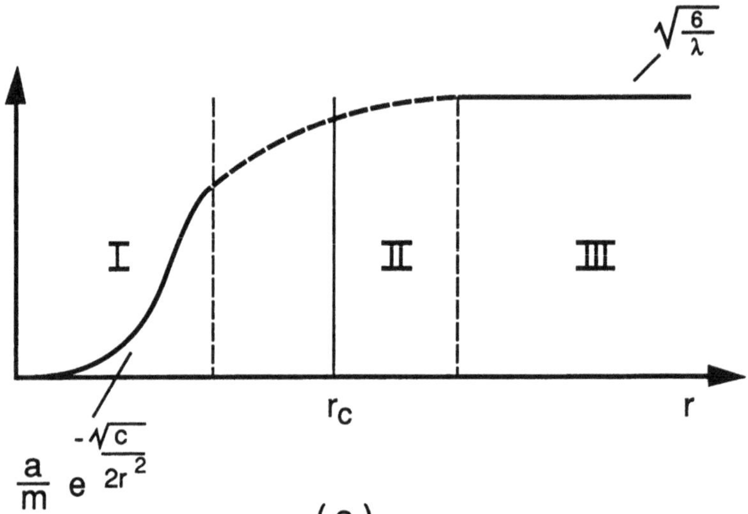
$\sqrt{\frac{6}{\lambda}}$
I
II
III
r_c
r
$\frac{a}{m} e^{-\frac{\sqrt{c}}{2r^2}}$

(a)

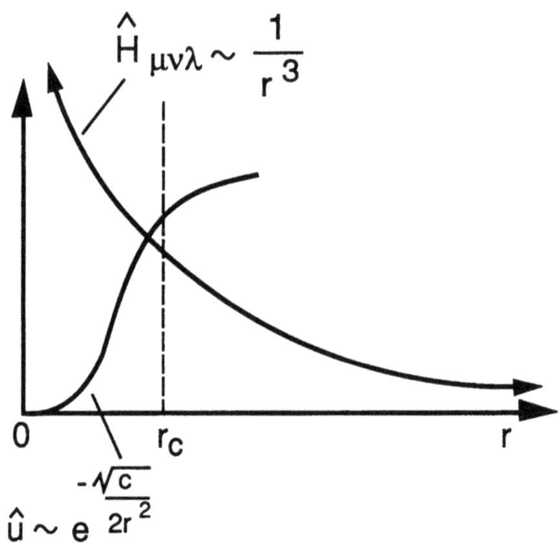
$\hat{H}_{\mu\nu\lambda} \sim \frac{1}{r^3}$
0
r_c
r
$\hat{u} \sim e^{-\frac{\sqrt{c}}{2r^2}}$

(b)

Fig. 7

Let

$$u = < u > + u' \tag{173}$$

Then

$$\begin{aligned}\mathcal{L} &= \frac{1}{2}(\partial_\mu u')^2 + \frac{1}{2}(2m^2)u'^2 + \ldots \\ &+ B < u >^2 H_{\mu\nu\lambda}H^{\mu\nu\lambda} + O(u')H_{\mu\nu\lambda}H^{\mu\nu\lambda}\end{aligned} \tag{174}$$

The mass of u' is

$$m_{\mu'} = \sqrt{2}m \tag{175}$$

For momenta and field strengths $<< m_{\mu'}, u'$ decouples and $\mathcal{L}$ becomes

$$\mathcal{L} = \frac{B6}{\lambda}m^2 H_{\mu\nu\lambda}H^{\mu\nu\lambda} \tag{176}$$

Since $\frac{B6}{\lambda}m^2$ is dimensionless define

$$\frac{1}{3!e^2} = \frac{B6m^2}{\lambda} \tag{177}$$

Then

$$\mathcal{L} = \frac{1}{3!e^2}H_{\mu\nu\lambda}H^{\mu\nu\lambda} \tag{178}$$

as in the pure $H_{\mu\nu\lambda}$ theory above. Therefore, we, henceforth, consider pure $\hat{H}$ instantons but with the finite core action given above. Before computing the effective action due to the H-instantons, let us briefly discuss the dual description to the theory. Recall that

$$\mathcal{L} = \frac{1}{2}(\partial_\mu u)^2 + Bu^2 H_{\mu\nu\lambda}H^{\mu\nu\lambda} + V(u) \tag{179}$$

We now perform a duality transformation on $H_{\mu\nu\lambda}$. The relevant result is

$$\int [dB_{\mu\nu}]e^{-\int d^4x Bu^2 H_{\mu\nu\lambda}H^{\mu\nu\lambda}} = \int [d\phi]e^{-\int d^4x \frac{1}{24Bu^2}(\partial_\mu\phi)^2} \tag{180}$$

where ϕ is a real scalar field. It follows that the $u, H_{\mu\nu\lambda}$ theory with the above Lagrangian is equivalent, both classically and quantum mechanically, to the u, ϕ theory with the Lagrangian

$$\mathcal{L} = \frac{1}{2}(\partial_\mu u)^2 + \frac{1}{24Bu^2}(\partial_\mu\phi)^2 + V(u) \tag{181}$$

We can write this in a simple form by introducing the complex scalar field Φ defined by

$$\Phi = \frac{1}{2}(\frac{1}{4\sqrt{3B}}u^2 + i\phi) \tag{182}$$

In terms of Φ the Lagrangian is given by

$$\begin{aligned} \mathcal{L} &= \frac{1/2\sqrt{3b}}{(\Phi+\Phi^\dagger)}\partial_\mu\Phi\partial^\mu\Phi^\dagger + V(u) \\ u^2 &= \frac{1}{\sqrt{3B}}(\Phi+\Phi^\dagger) \end{aligned} \tag{183}$$

Hence, this theory is a non-linear sigma model of a simple form. The expression for a single H-instanton of charge $\pm q$ located at x_o is

$$\hat{H}_{\mu\nu\lambda} = \frac{\pm q}{\pi f}\epsilon_{\mu\nu\lambda\gamma}\frac{(x-x_0)^\gamma}{|x-x_0|^4} \tag{184}$$

Its self action is

$$\hat{S}_{(1)} = \frac{2q^2}{e^2 f^2}\frac{1}{r_c^2} \tag{185}$$

The N-instanton is given by

$$\hat{H}_{\mu\nu\lambda(N)} = \sum_{i=1}^{N}\frac{q_i}{\pi f}\epsilon_{\mu\nu\lambda\gamma}\frac{(x-x_0)^\gamma}{|x-x_0|^4} \tag{186}$$

where $q_i = \pm q$ for all i. The action for the N-instanton is found to be

$$\hat{S}_{(N)} = \frac{2}{e^2 f^2}[\sum_{i=1}^{N} q_i^2\frac{1}{r_c^2} + \sum_{i\neq j=1}^{N} q_i q_j\frac{1}{|x_i-x_j|^2}] \tag{187}$$

This can be written as

$$\hat{S}_{(N)} \simeq \frac{2}{e^2 f^2}\sum_{i=1}^{N} q_i q_j\frac{1}{|x_i-x_j|^2} \tag{188}$$

Define

$$\begin{aligned} J^N(x) &= \sum_{i=1}^{N} q_i\delta(x-x_i) \\ G(x,y) &= \frac{1}{|x-y|^2} \end{aligned} \tag{189}$$

Therefore

$$\hat{S}_{(N)} = \frac{2}{e^2 f^2}\int d^4x d^4y J^N(x)G(x,y)J^N(y) \tag{190}$$

The grand canonical partition function for the instantons is

$$\begin{aligned} Z_{inst} &= \sum_{N=0}^{\infty}\frac{1}{N!}z^N \sum_{\substack{\{q_i\} \\ |q_i|=q}} (\prod_{i=1}^{N}\int\frac{d^4x}{\pi^2 r_c^4})e^{-\hat{S}_{(N)}} \\ &= \sum_{N=0}^{\infty}\frac{1}{N!}z^N \sum_{\substack{\{q_i\} \\ |q_i|=q}} (\prod_{i=1}^{N}\int\frac{d^4x}{\pi^2 r_c^4})e^{-\frac{2}{e^2f^2}\int d^4x d^4y J^N(x)G(x,y)J^N(y)} \end{aligned} \tag{191}$$

Using the method of mean fields, it follows that

$$Z_{inst} = \frac{\sum_{N=0}^{\infty} \frac{1}{N!} z^N \sum_{\substack{\{q_i\} \\ |q_i| = q}} (\prod_{i=1}^{N} \int \frac{d^4x}{\pi^2 r_c^4}) \int [d\chi] e^{-S_M^{(N)}}}{\int [d\chi] e^{-\frac{e^2}{4} \int d^4x (\partial_\mu \chi)^2}} \tag{192}$$

where

$$S_M^{(N)} = \int d^4x (\frac{e^2}{4} (\partial_\mu \chi)^2 - \frac{16\pi}{e^2} \frac{i\chi}{f} J^{(N)}) \tag{193}$$

and χ is the scalar mean field. This series can be resummed to give

$$Z_{inst} = \frac{\int [d\chi] e^{-S_{eff}}}{\int [d\chi] e^{-\frac{e^2}{4} \int d^4x (\partial_\mu \chi)^2}} \tag{194}$$

where

$$S_{eff} = \int d^4x (\frac{e^2}{4} (\partial_\mu \chi)^2 - \frac{2z}{\pi^2 r_c^4} \cos[\frac{4\pi q}{f} \chi]) \tag{195}$$

The complete partition function must include the contribution of the $H_{\mu\nu\lambda}$ spin-wave defined by

$$H'_{\mu\nu\lambda} = H_{\mu\nu\lambda} - \hat{H}_{\mu\nu\lambda(N)} \tag{196}$$

for any N. Noting that

$$\partial^\mu \hat{H}_{\mu\nu\lambda} = 0 \tag{197}$$

and that $\hat{H}_{\mu\nu\lambda}$ falls off as r^{-3} at infinity, it follows that

$$S_{(N)} = \hat{S}_{(N)} + \frac{1}{3! e^2} \int d^4x H'_{\mu\nu\lambda} H'^{\mu\nu\lambda} \tag{198}$$

for every N. Therefore, the complete partition function is given by

$$Z_T = Z_{inst} Z_o \tag{199}$$

where

$$Z_o = \int [dB'_{\mu\nu}] e^{-\frac{1}{3! e^2} \int d^4x H'_{\mu\nu\lambda} H'^{\mu\nu\lambda}} \tag{200}$$

Using the duality transformation, Z_o can be written as

$$Z_o = \int [d\phi'] e^{-\frac{e^2}{4} \int d^4x (\partial_\mu \phi')^2} \tag{201}$$

which cancels the denominator in Z_{inst}. Therefore

$$\begin{aligned} Z_T &= \int [d\chi] e^{-S_{eff}} \\ S_{eff} &= \int d^4x (\frac{e^2}{4} (\partial_\mu \chi)^2 - \frac{2z}{\pi^2 r_c^4} \cos[\frac{4\pi q}{f} \chi]) \end{aligned} \tag{202}$$

where fugacity $z = e^{-\hat{S}_c}$ and

$$\begin{aligned} r_c &= \frac{c^{1/6}}{m^{1/3}} \\ \hat{S}_c &= \frac{3\pi^2 m^2}{2\lambda}(8\sqrt{c}e^{-(m^2\sqrt{c})^{1/3}} + (mc)^{2/3}) \end{aligned} \tag{203}$$

Finally, it is easy to show that, by adjusting the constant term in the u field potential energy, S_{eff} can be put in the form

$$S_{eff} = \int d^4x(\frac{e^2}{4}(\partial_\mu\chi)^2 + \frac{2z}{\pi^2 r_c^4}[1 \pm \cos(\frac{4\pi q}{f}\chi)])$$

or, defining $\chi' = \frac{e}{\sqrt{2}}\chi$ and dropping the prime

$$S_{eff} = \int d^4x(\frac{1}{2}(\partial_\mu\chi)^2 + \frac{2z}{\pi^2 r_c^4}[1 \pm \cos(\frac{4\sqrt{2}q}{ef}\chi)]) \tag{204}$$

For the q = 1 case, we find that the completed partition function for the scalar mean field, χ, is given by

$$Z = \int [d\chi]e^{-S_{eff}} \tag{205}$$

where

$$S_{eff} = \int d^4x(\frac{1}{2}(\partial_\mu\chi)^2 + \frac{2z}{\pi^2 r_c^4}(1 \pm \cos[4\sqrt{2}\pi\frac{\chi}{ef}])) \tag{206}$$

If, for simplicity, we assume

$$\frac{\sqrt{\lambda}m}{ef} \ll 22 \tag{207}$$

then $\hat{S}_c$ reduces to

$$S_c = 4\sqrt{\frac{3}{\lambda}\frac{m}{ef}}e^{-(\frac{1}{\pi}\sqrt{\frac{\lambda}{3}}\frac{m}{ef})^{1/3}} \tag{208}$$

where we have used (203) and the fact that $c = \lambda(\frac{q}{\pi emf})^2$. The instanton core radius, r_c, is given by

$$r_c = (\frac{1}{\pi}\sqrt{\frac{\lambda}{3}}\frac{1}{em^2 f})^{1/3} \tag{209}$$

even if constraint (207) is not imposed. The effective scalar theory defined by action (206) can potentially play an important role in the understanding of various aspects of inflationary cosmology, domain walls and large scale structure, as we now show.

Inflation:

It has recently been shown [8], that an acceptable inflationary universe scenario can be achieved using a single scalar field χ with potential energy

$$V(\chi) = \Lambda^4(1 \pm \cos[\frac{\chi}{M}]) \tag{210}$$

as an inflaton, as long as $M \sim O(M_p)$ and $\Lambda \sim O(M_{GUT})$ where M_p is the Planck mass and $M_{GUT} \sim 10^{15} GeV$. In [8], the χ field was taken to be a pseudo-Nambu-Goldstone boson and the cosine potential was assumed to arise in the usual way from instantons in the non-Abelian gauge fields which couple to χ. Parameter Λ is the scale at which the gauge coupling becomes strong. Since $\Lambda \sim 10^{15} GeV$, one must introduce a new non-Abelian gauge group beyond the $SU(3)_c \times SU(2)_L \times U(1)$ gauge group of the standard model. Although this is possible, it represents a major departure from conventional models. In this talk, we suggest that an equally satisfactory inflationary scenario can arise using the scalar mean field χ with action (206) introduced above. One need only demand that

$$\frac{ef}{4\sqrt{2}\pi} = M_P \tag{211}$$

$$(\frac{2z}{\pi^2 r_c^4})^{1/4} = M_{GUT} \tag{212}$$

and all the conclusions of [8], except those pertaining to reheating, remain valid. It is not hard to show that if we restrict e and f to satisfy (211) then (212) will be valid if we choose, for example

$$\begin{aligned} m &\sim M_p \\ \lambda &\sim 10^{-4} - 10^{-5} \end{aligned} \tag{213}$$

Note that these parameters satisfy constraint (207), as they must. Also, expanding the cosine potential as a power series around $\chi = 0$, we find that the mass of the χ field is $m_\chi \sim 10^{-8} M_p$. The mass ofthe u' field is $m_{u'} \sim \sqrt{2} M_p$. Therefore, $m_{u'} \gg m_\chi$and the assumption that the u' field decouples is valid. To complete thescenario, it is necessary to give a mechanism for reheating. A detaileddiscussion of the coupling of the scalar mean field χ to quarks and leptons is beyond the scope of these lectures. Suffice it here to say that these couplings contain a term of the form

$$\mathcal{L}_{eff} = \frac{1}{f'} \partial_\mu \chi (\bar{t}_L \gamma^5 \gamma^u t_L + \bar{t}_R \gamma^5 \gamma^\mu t_R) + \ldots \tag{214}$$

where t_L and t_R are the left and right chiral top quark spinor fields and f' is a new mass parameter unrelated to m or f. The reheating temperature is given by

$$T_{RH} = (\frac{45}{4\pi^3 g^*})^{1/4} min[\sqrt{H(\phi_2) M_p}, \sqrt{\Gamma M_p}] \tag{215}$$

where g^* is the number of light relativistic degrees of freedom, $H(\phi_2)$ is the Hubble parameter evaluated at the point in field space where the slow roll part of the potential ends, and Γ is the decay rate of χ into two top quarks. Γ is easily computed from (11) and is found to be

$$\Gamma \sim (\frac{m_t}{f'})^2 10^{-8} M_p \tag{216}$$

where m_t is the top quark mass and we have used the fact that $m_\chi \sim 10^{-8} M_p$. For parameters satisfying (212) and (213) we find that $H(\phi_2) \sim 10^{-8} M_p$. Assuming, for example, that $f' \geq m_t$ then $\sqrt{\Gamma M_p} < \sqrt{H(\phi_2) M_p}$ and T_{RH} is given by

$$T_{RH} \sim (\frac{m_t}{f'}) 10^{-3} M_p \tag{217}$$

where we have used the fact that in the standard model $g^* = 106.75$. It follows that one can choose f' so that $0 < T_{RH} \leq 10^{16} GeV$, which is more than adequate. We emphasize that this scenario arises naturally from the instantons in the $u, B_{\mu\nu}$ fields themselves and does not require new gauge groups or other modifications of the standard model.

Domain Walls and Large Scale Structure:

For several years there has been interest in theories in which large scale cosmological structure is induced by domain walls. Recently, to avoid predicting unobserved large fluctuations in the photon microwave background, it has been suggested [7] that such theories should have a phase transition in the scalar potential energy that occurs after photon decoupling. Domain walls form only after this late time phase transition and, hence, only minimally perturb the homogeneity of the microwave background. General principles for such theories were given, and a tentative particle physics model based on pseudo-Goldstone bosons was proposed. In these lectures, we present a new class of theories in which domain walls naturally form after photon decoupling and many of the generic results of [7] are maintained, but in which no late time phase transition of the potential takes place. Instead, we assume that a temperature independent potential for a weakly self-interacting real scalar field, χ, with at least two degenerate minima, is present in the theory, and that the early universe was dominated by an epoch of inflation unrelated to this scalar field. As discussed in [11], the Hawking temperature [12] present during inflation does not produce fluctuations of the microwave background in excess of the exponential limit as long as the Hubble parameter during inflation, H_i satisfies

$$H_i \leq 10^{14} GeV \tag{218}$$

However, as shown in [13, 14] the Hawking temperature does substantially effect the amplitude probability distribution, P, of the quasiclassical vacuum χ field. The probability distribution is defined by

$$< (\chi - \bar{\chi})^2 >= \int_{-\infty}^{\infty} d\chi P(\chi, l/l_c)\chi^2 \tag{219}$$

where, at a specific time to be discussed below, $l_c = \frac{1}{3}\pi r_H, r_H$ being the horizon size, l is the length scale over which one is observing, $l \geq l_c$ and $\bar{\chi}$ is the vacuum expectation value of χ. P is easily computed for a non-interacting scalar field χ, and is found to be

$$P(\chi, l/l_c) = \sqrt{\frac{2\pi}{\sigma^2}} e^{-\frac{(\chi-\bar{\chi})^2}{2\sigma^2}} \tag{220}$$

where

$$\sigma = \frac{H_i}{2\pi}\sqrt{ln(l/l_c)} \tag{221}$$

and $\bar{\chi}$ is arbitrary. Using (220), it is straightforward to compute the vacuum expectation value of the free Hamiltonian density of $\chi, \mathcal{H}$. We find that

$$< \mathcal{H} > \simeq \frac{H_i^2}{4l_c^2(z)} \tag{222}$$

Note that l_c is a function of redshift variable z. We now allow χ to acquire a potential energy of the form

$$V(\chi) = cM^4(1 + \cos\frac{\chi}{M}) \tag{223}$$

where c and M have mass dimensions zero and one respectively. We could have chosen any potential with at least two degenerate absolute minima for our scenario to work. We discuss (223) because it is exactly of the same form as the instanton generated potential in (206). Define $\hat{z}$ to be the redshift at which

$$\frac{H_i^2}{4l_c^2(\hat{z})} = 2m_\chi^2 M^2 \tag{224}$$

where $m_\chi = \sqrt{c}M$ is the χ mass and $2m_\chi^2 M^2$ is the maximum height of the potential. For $z > \hat{z}$ the energy of the system is dominated by the Hawking fluctuations, the potential is insignificant and domain walls do not form. For $z < \hat{z}$ the potential energy dominates and domain wall formation becomes possible. The equation of motion for rolling down the potential is

$$\ddot{\chi} + 3H\dot{\chi} + m_x^2 M \sin\frac{\chi}{M} = 0 \tag{225}$$

where $3H\dot{\chi}$ acts as a damping term. Undamped rolling can occur for $z < z^*$, where redshift z^* is defined by

$$3H(z^*) = m_\chi \tag{226}$$

It is at z^*, and not at $\hat{z}$, when the system can freely roll down the potential and form domain walls. To keep the domain walls from generating large inhomogeneities in the microwave background, z^* must be chosen to satisfy $z^* < z^\gamma \simeq 1100$ where z^γ is the redshift at photon decoupling. It is the physics at and around z^* that, in our scenario, replace the late time phase transition of [7]. Writing m_χ as $m_\chi = \frac{2}{\gamma}(Mpc)^{-1}$, and using the fact that the present inverse Hubble parameter $H_o^{-1} \simeq 3000 Mpc$, it follows from (226) that

$$z^* = (\frac{2000}{\gamma})^{2/3} - 1 \qquad (227)$$

Note that it is the size of the horizon at $z = z^*$ that is relevant to the calculation of l_c. The exact relation is $l_c(z^*) = \frac{1}{3}\pi r_H(z^*) = \pi\gamma Mpc$. Using (222), it is not hard to show that residual Hawking fluctuations will not significantly effect rolling down the potential as long as

$$\frac{M}{H_i} \geq .16 \qquad (228)$$

This bound also insures that $z^* < \hat{z}$, as it must be. We will see below that parameters c and M have values far below those required in [8] for slow roll inflation. Hence, domain wall formation proceeds rapidly for $z \leq z^*$ and there is no new inflation. Furthermore, the natural width σ of distribution P is such that only domain walls consisting of a single kink between nearest neighbor minima of the cosine potential are statistically likely to form. The width of such a domain wall is given by

$$r_\omega = \gamma Mpc \qquad (229)$$

whereas the energy density stored in the domain wall is

$$\varphi_\omega^c = 6.65 \times 10^{-58}(\frac{M}{\gamma})^2(eV)^4 \qquad (230)$$

It is essential to note that the wall thickness and energy density do not redshift. The density perturbation amplitude at redshift $z < z^*$ is defined to be $(\frac{\delta\varphi}{\varphi})(z) = \frac{\varphi_\omega^c}{\varphi_m(z)}$, where φ_m is the energy density of matter. It can also be shown that

$$(\frac{\delta\varphi}{\varphi})(z) = \frac{8.04 \times 10^{-30}(1 + z^* + (1+z)^3)}{(1+z)^3(1+z^*)}(\frac{M}{\gamma})^2 \qquad (231)$$

where M is given in Gev. Structure forms rapidly for redshifts at or below z_1 defined by

$$(\frac{\delta\varphi}{\varphi})(z_1) \simeq 1 \qquad (232)$$

It follows from (231) and (232) that

$$\frac{M}{\gamma} = (\frac{1.24 \times 10^{29}(1+z_1)^3(1+z^*)}{1 + z^* + (1+z)^3})^{1/2} \qquad (233)$$

A good measure of the relative width of the amplitude probability distribution to the size of vacuum expectation value in the cosine potential is $\alpha = \frac{\sigma}{\pi M}$. Using (221), it follows that

$$\alpha = \frac{1}{2\pi^2}(\frac{H_i}{M})\sqrt{ln(l/l_c)} \tag{234}$$

Note that α is a function of l/l_c. It is simplest to evaluate (234) at $z = z^*$ since, at that redshift, $l_c(z^*) = \pi\gamma Mpc$. The smallest value of $l(z^*)$ for which domain wall formation should be reasonably probable is that value, denoted by $\hat{l}(z^*)$, which scales up to the present horizon size at $z = 0$. That is

$$\hat{l}(z^*) = \frac{6000}{1 + z^*}Mpc \tag{235}$$

It follows that the value of α at $\hat{z}$, denoted $\hat{\alpha}$ is

$$\hat{\alpha} = \frac{1}{2\pi^2}(\frac{H_i}{M})\sqrt{\ln(\frac{6000}{(1+z^*)\pi\gamma})} \tag{236}$$

Comparing $P(\chi, \hat{l}/l_c)$ with the cosine potential, it is clear that one gets significant domain wall formation when the field at which the distribution is centered, $\bar{\chi}$, satisfies $-\sigma \le \bar{\chi} \le \sigma$. The probability that $\bar{\chi}$ is so located is clearly given $\mathcal{P} = 100\hat{\alpha}(\%)$. The distribution of the domain walls is a complicated problem in percolation theory [15] that is beyond the scope of these lectures. In putting z^* one can, using (227), and (229), predict the domain wall thickness r_ω. Recall that $z^* < z^\gamma$. Similarly, in putting $z_1 < z^*$ and using (233) and (236), one can predict the probability $\mathcal{P}$. We list here some examples. Take $H_i = 10^{14}GeV$, thus saturating the upper bound of (218).
a) For $z^* = 1000, z_1 = .3$ we find $r_\omega = 60Kpc$ and $\mathcal{P} = 30\%$. b) for $z^* = 353, z_1 = .3$ we find $r_\omega = 300Kpc$ and $\mathcal{P} = 6\%$. c) for $z^* = 200, z_1 = .3$ we find $r_\omega = 700Kpc$ and $\mathcal{P} = 2.3\%$. Finally, d) for $z^* = 158, z_1 = .3$ we find $r_\omega = 1Mpc$ and $\mathcal{P} = 1.5\%$. We conclude that, in our scenario, it is more likely to form domain walls of thickness $r_\omega \simeq O(100Kpc)$, though megaparsec thick walls are quite possible. Furthermore,we find that for $z_1 = 3$ all of the above cases have, in general, extremely tiny probability $\mathcal{P}$. Therefore, in our scenario, it is very unlikely to see structure at redshifts $z \ge 1$. We would like to point out that the idea that quantum fluctuations produced in a light scalar field during inflation could lead to domain walls at a later epoch, was explored recently for the case of the Pecci-Quinn axion [4].

Finally, the above scenario stands on its own and does not require motivation from particle physics. However, it is interesting to note that this scenario arises naturally in the particle physics theories described by effective Lagrangian (206). For example, it follows from (207) and (208) that case a) above will occur if we choose $\lambda = .33, m = 10^{16}GeV$ and ef$= 5.5 \times 10^{14}GeV$. Similarly case d) will occur with the choice $\lambda = .14, m = 10^{17}GeV$ and ef $= 9.2 \times 10^{15}GeV$. Thus instantons in antisymmetric tensor fields may lead to a simple and natural explanation of large scale structure.

References

1. J. Frolich in "Renormalization Theory: Proceedings of the NATO Advanced Study Institute", Erice 1975, ed. G. Velo and A. S. Wightman (Reidel, Dordrecht/Boston); S. Samuel, Phys. Reb. **D18**, 1916 (1978).

2. A. M. Polyakov, Nucl. Phys. **B120**, 429 (1977); see also "Gauge Fields and Strings" by A. M. Polyakov, Harwood Academic Publishers, 1987.

3. R. Rohm and E. Witten, Annals of Physics **170**, 454 (1986).

4. S. W. Hawking, Phys. Lett. **B195**, 337 (1987); Phys. Rev. **D37**, 904 (1988); G. V. Lavrelashvili, V. A. Rubakov, and P.-G. Tinyakov, JETP Lett. **46**, 167 (1987); S. B. Giddings and A. Strominger, Nucl. Phys., **B306**, 890 (1988); **B307**, 854 (1988); S. Coleman, Nucl. Phys. **B307**, 867 (1988); K. Lee, Phys. Rev. Lett. **61**, 263 (1988); R. Myers, Phys. Rev. **D38**, 1327 (1988); L. F. Abbott and M. B. Wise, Nucl. Phys. **B325**, 687 (1989).

5. S.-J. Rey, "The Confinine Phase of Superstrings and Axiomatic Strings", UCSB TH-89/23,

6. B. Ovrut and S. Thomas, University of Pennsylvania Preprint preprint UPR-0465T, 1991.

7. C. T. Hill, D. N. Schramm, and J. N. Fry, Comments Nucl. Part. Phys. **19**, 25 (1989).

8. K. Freese, J. A. Frieman and A. V. Olinto, Fermilab-Pub-90/177-A (1990).

9. G. Goetz and D. Notzold Fermilab-Pub 149-A (1990); M. S. Turner, R. Watkins and L. Widrow, Fermilab-Pub-90/185-A (1990).

10. B. Ovrut and S. Thomas, University of Pennsylvania Preprint UPR-0450T, 1990; UPR-0455T, 1991.

11. D. H. Lyth, Phys. Lett. **147B**, 403 (1984); **150B**, 465 (1985); **246B**, 359 (1990); S. W. Hawking, Phys. Lett. **150B**, 339 (1985); see also reference [3]for a discussion of this point.

12. G. W. Gibbons and S. W. Hawking, Phys. Rev. **D15**, 2752 (1977).

13. A. A. Starobinsky, Phys. Lett. **117B**, 175 (1982) also in "Current topics in field theory, quantum gravity and strings", Lectures notes in Physics ed. H. J. de Vega and N. Sanchez, vol. 246 (Springer-Heidelberg, 1986) p. 107; A. Goncharov and A. Linde, Sov Phys. JETP **65**, 635 (1987); S.-J. Rey, Nucl. Phys. **B284**, 706 (1987).

14. A. Linde and D. H. Lyth, Phys. Lett. **246B**, 353 (1990); D. H. Lyth 'Axions and Inflation' University of Lancaster preprint (1991).

15. See, for example, the review by D. Stauffer, Physics Reports **54**, 1 (1979).

ERNEST MA

UNIFIED DESCRIPTION OF QUARK AND LEPTON MASS MATRICES

Ernest Ma
University of California, Riverside, CA 92521

ABSTRACT

Prompted by the present empirical information regarding quark masses and mixing angles, I propose an S_3XZ_3 discrete symmetry from which two successful relationships are derived. This model is then extended to include leptons, resulting in two qualitatively different solutions for a possible 17 keV neutrino.

1. Demise of the Fritzsch Ansatz

Assuming that there are 6 quarks (d, s, b of charge $-1/3$ and u, c, t of charge 2/3), the two $3X3$ quark mass matrices are in general given by

$$M_d = U_1 V_{KM} \begin{bmatrix} m_d & 0 & 0 \\ 0 & m_s & 0 \\ 0 & 0 & m_b \end{bmatrix} U_2^\dagger , \tag{1}$$

and

$$M_u = U_1 \begin{bmatrix} m_u & 0 & 0 \\ 0 & m_c & 0 \\ 0 & 0 & m_t \end{bmatrix} U_3^\dagger . \tag{2}$$

In the standard $SU(2)$ X $U(1)$ electroweak gauge model, only the quark masses and the charged-current mixing matrix V_{KM} are measurable (the Kobayashi-Maskawa matrix). The unitary matrices $U_{1,2,3}$ are completely arbitrary, and yet it is believed that M_d and M_u may have simple forms if only clever choices for $U_{1,2,3}$ are made. A very attractive scheme was proposed by Fritzsch[1], who assumed that M_u and M_d are Hermitian, i.e. $U_1 = U_3$ and $U_2 = U_1 V_{KM}$, and are of the form

$$M_d = \begin{bmatrix} 0 & A & 0 \\ A & 0 & B \\ 0 & B & C \end{bmatrix} , \tag{3}$$

and

$$M_u = \begin{bmatrix} 0 & D & 0 \\ D^* & 0 & E \\ 0 & E^* & F \end{bmatrix} . \tag{4}$$

The parameters A and B have been rendered real by redefining the two relative phases between the 3 left-handed quark doublets $(u,\ d)_L$, $(c,\ s)_L$, and $(t,\ b)_L$. We then have 6 real parameters and 2 phases in Eqs. (3) and (4) to determine the 6 quark masses and the 3 angles and 1 phase of V_{KM}. This means that we can express the latter 4 quantities in terms of the 6 masses and the 2 original phases. In particular,[2]

$$V_{us} \simeq (m_d/m_s)^{\frac{1}{2}} - (m_u/m_c)^{\frac{1}{2}} e^{i\sigma} , \tag{5}$$

$$V_{cb} \simeq -(m_s/m_b)^{\frac{1}{2}} + (m_c/m_t)^{\frac{1}{2}} e^{i(\tau-\sigma)} , \tag{6}$$

$$V_{ub} \simeq (m_d/m_b)^{\frac{1}{2}} - (m_u/m_t)^{\frac{1}{2}} e^{i\tau} + (m_s m_u/m_b m_c)^{\frac{1}{2}} e^{i\sigma} . \tag{7}$$

The constraint on V_{us} is easily satisfied and it was in fact the motivation for Eqs. (3) and (4) in the first place. The constraint on V_{cb} is only marginally satisfied because $|V_{cb}| = 0.044 \pm 0.009$ whereas $(m_s/m_b)^{\frac{1}{2}} \simeq 0.182$ and $(m_c/m_t)^{\frac{1}{2}} < 0.123$ from the present experimental limit of $m_t > 89\ GeV$. The clincher is Eq. (7) because $|V_{ub}|$ is now measured to be 0.004 ± 0.002 whereas $(m_d/m_b)^{\frac{1}{2}} \simeq 0.041$ cannot be significantly reduced by $(m_u/m_t)^{\frac{1}{2}} < 0.008$ and $(m_s m_u/m_b m_c)^{\frac{1}{2}} \simeq 0.011$. Hence the Fritzsch ansatz is far from agreement with present data.

2. Two Empirical Observations

Let M_u be diagonal in Eq. (2), i.e. $U_1 = U_3 = 1$, then

$$M_d = V_{KM} \begin{bmatrix} m_d & 0 & 0 \\ 0 & m_s & 0 \\ 0 & 0 & m_b \end{bmatrix} U_2^\dagger . \tag{8}$$

If we now choose $U_2 = V_{KM}$, then

$$(M_d)_{ij} = \sum_k m_k V_{ik} V_{jk}^* \tag{9}$$

is essentially entirely known. In particular,

$$(M_d)_{11} = m_d|V_{ud}|^2 + m_s|V_{us}|^2 + m_b|V_{ub}|^2. \tag{10}$$

Since the last term is numerically negligible, we obtain the desirable relationship $|V_{us}| \simeq (m_d/m_s)^{\frac{1}{2}}$ for $|V_{ud}| \simeq 1$ if $(M_d)_{11} = 0$. Note that a change in sign for the ratio m_d/m_s is required, but that can be done trivially by a γ_5 phase transformation of the quark fields.

Consider now

$$M_d M_d^\dagger = V_{KM} \begin{bmatrix} m_d^2 & 0 & 0 \\ 0 & m_s^2 & 0 \\ 0 & 0 & m_b^2 \end{bmatrix} V_{KM}^\dagger \, . \tag{11}$$

It is indepedent of U_2 and again all its entries are more or less known. In particular, the only possibly zero entry is

$$\begin{aligned} (M_d M_d^\dagger)_{12} &= m_d^2 V_{ud} V_{cd}^* + m_s^2 V_{us} V_{cs}^* + m_b^2 V_{ub} V_{cb}^* \\ &= (m_s^2 - m_d^2) V_{us} V_{cs}^* + (m_b^2 - m_d^2) V_{ub} V_{cb}^* . \end{aligned} \tag{12}$$

Numerically,

$$\left| \frac{V_{ub} V_{cb}}{V_{us} V_{cs}} \right|^{\frac{1}{2}} \simeq 0.03 \pm 0.01 \, , \tag{13}$$

whereas

$$\frac{m_s}{m_b} \simeq \frac{150 \; MeV}{5 \; GeV} = 0.03. \tag{14}$$

Hence $(M_d M_d^\dagger)_{12} = 0$ is possibility, from which the relationship

$$\frac{m_s^2}{m_b^2} \simeq - \frac{V_{ub} V_{cb}}{V_{us}} \tag{15}$$

can be derived, assuming $V_{cs} \simeq 1$.

Going back to M_d, we can use the arbitrary U_2 to get it in the form

$$M_d = \begin{bmatrix} 0 & X & 0 \\ X & X & X \\ 0 & X & X \end{bmatrix} , \tag{16}$$

where X denotes a nonzero entry and 0 a zero one. To obtain $(M_d M_d^\dagger)_{12} = 0$, we need $(M_d)_{22} = 0$. Also, $(M_d)_{12} = (M_d)_{21}$ is desired for $|V_{us}| \simeq (m_d/m_s)^{\frac{1}{2}}$. Hence the form

$$M_d = \begin{bmatrix} 0 & a & 0 \\ a & 0 & b \\ 0 & c & d \end{bmatrix} \tag{17}$$

is what we are aiming for.

3. S_3XZ_3 Model of Quarks

To maintain the form of M_d, a symmetry is required, otherwise the zero entries of Eq. (17) are not protected from large quantum corrections. I propose therefore an S_3XZ_3 discrete symmetry, according to which the quarks and leptons transform as follows.[3,4,5] Let 1, ω, and ω^2 denote the elements of Z_3 with $\omega^3 = 1$, then $[(u,d)_L,\ (c,s)_L]$ and $[(\nu_1,l_1)_L,\ (\nu_2,l_2)_L]$ transform as $(2,\omega)$ under S_3XZ_3; $(c,u)_R$, $(s,d)_R$, $(\nu_2,\nu_1)_R$, $(l_2,l_1)_R$ as $(2,\omega^2)$; and $(t,b)_L$, t_R, b_R, $(\nu_3,l_3)_L$, ν_{3R}, l_{3R} as $(1,1)$. There are 4 scalar doublets: (ϕ_1^+,ϕ_1^0) transforms as $(1,1)$; (ϕ_2^+,ϕ_2^0) as $(1,\omega^2)$; and $[(\eta_1^+,\eta_1^0),\ (\eta_2^+,\eta_2^0)]$ as $(2,\omega)$. There are also 2 scalar singlets in 2 possible arrangements: (I) (χ_2^0,χ_1^0) transforms as $(2,\omega^2)$ and χ_3^0 as $(1,\omega^2)$; $(II)(\chi_1^0,\chi_2^0)$ as $(2,\omega)$ and χ_3^0 as $(1,1)$. Furthermore, ν_i and l_i have lepton number assignment $L = 1$, χ_i have $L = -2$, and all others have $L = 0$.

As a result, the quark mass matrices are given by

$$M_u = \begin{bmatrix} f_1 < \bar{\eta}_1^0 > & 0 & 0 \\ 0 & f_1 < \bar{\eta}_2^0 > & 0 \\ 0 & 0 & f_2 < \bar{\phi}_1^0 > \end{bmatrix}, \tag{18}$$

and

$$M_d = \begin{bmatrix} 0 & f_3 < \phi_2^0 > & f_4 < \eta_1^0 > \\ f_3 < \phi_2^0 > & 0 & f_4 < \eta_2^0 > \\ f_5 < \eta_1^0 > & f_5 < \eta_2^0 > & f_6 < \phi_1^0 > \end{bmatrix}. \tag{19}$$

Note that $(M_d)_{12} = (M_d)_{21}$ is possible without any left-right interchange symmetry hrere because of the non-Abelian nature of S_3. Also, note that m_u and m_c come from the decomposition $2X2X2 \rightarrow 1$ in S_3. Now since $< \eta_1^0 > / < \eta_2^0 > = m_u/m_c$ from Eq. (18), I can rewrite Eq. (19) as

$$M_d = \begin{bmatrix} 0 & a & \xi b \\ a & 0 & b \\ \xi c & c & d \end{bmatrix}, \tag{20}$$

where a, b, c, d, are chosen real and ξ is complex with $|\xi| = m_u/m_c$. This is not quite Eq. (17), but since $|\xi| \simeq 4X10^{-3}$ is very small, it is a very good approximation. I find $m_b \simeq (c^2+d^2)^{\frac{1}{2}}$, $m_s \simeq bc/(c^2+d^2)^{\frac{1}{2}}$, $m_d \simeq a^2d/bc$, $V_{cb} \simeq bd/(c^2+d^2)$, $V_{ub} \simeq ac/(c^2+d^2)$, and $V_{us} \simeq -ad/bc$. Hence Eq. (15) is obtained as well as

$$|V_{us}| \simeq \left(\frac{m_d}{m_s}\right)^{\frac{1}{2}} \left[1 + \frac{|V_{ub}|}{|V_{us}V_{cb}|}\right]^{-\frac{1}{4}}, \tag{21}$$

which favors smaller values of $|V_{ub}/V_{cb}|$ and m_s/m_d within their allowed ranges.

The dependence on ξ shows up to first approximation only in the parameter J which is a measure of CP nonconservation[6],

$$J \simeq -V_{us}V_{cb}^2\left[1 + \frac{|V_{ub}|}{|V_{us}V_{cb}|}\right] Im\xi$$
$$< 5X10^{-6}\ , \tag{22}$$

but it is too small by an order of magnitude to account for the parameter ϵ in the $K-\bar{K}$ system. On the other hand, ϵ has contributions from the extended scalar sector of this model, and a correct value for it can be obtained. In fact, there are terms here such as $\phi_1^\dagger\phi_2$ and $\phi_i^\dagger\eta_j$ which break S_3XZ_3 softly and may also be useful for understanding the baryon asymmetry of the Universe.[7]

4. The Lepton Sector

The charged leptons have a mass matrix just like Eq. (20) but with different values of a, b, c, d. The neutrinos have a diagonal Dirac mass matrix just like M_u of Eq. (18):

$$M_\nu = \begin{bmatrix} \xi A & 0 & 0 \\ 0 & A & 0 \\ 0 & 0 & B \end{bmatrix}. \tag{23}$$

The singlet neutrinos ν_{iR} also have a Majorana mass matrix from their couplings to the $\chi_i's$. As it turns out, given the constraint of S_3XZ_3, there are only 2 forms for this mass matrix to be of rank two, which is a desirable way[8] to accommodate a possible 17 keV neutrino[9,10] as a pseudo-Dirac neutrino to avoid conflict with the experimental limit on neutrinoless double beta decay.[11] With the assignment of Model I, I get[4]

$$M'_\nu = \begin{bmatrix} C & E & 0 \\ E & D & 0 \\ 0 & 0 & 0 \end{bmatrix}, \tag{24}$$

whereas in Model II, I get[5]

$$M'_\nu = \begin{bmatrix} 0 & 0 & C \\ 0 & 0 & D \\ C & D & E \end{bmatrix}. \tag{25}$$

In Model I, ν_3 is a Dirac neutrino of mass B ($\simeq$ 17 keV) and the others form a $4X4$ mass matrix with eigenvalues $\pm\ E\ +\ (C+D)/2$, A^2C/E^2, and

$- \xi^2 A^2/C$, in the approximation $A \ll C, D < E$. The last two mass eigenstates are mostly mixtures of ν_{1L} and ν_{2L} with mixing angle $\simeq \xi E/C$. If the $\nu_e - \nu_3$ overlap is 0.092 as indicated by the experiment of Hime and jelley,[9] then ν_3 is mostly ν_τ, so l_3 is mostly τ. I now use the freedom of redefining $(e, \mu, \tau)_R$ to rewrite the charged-lepton mass matrix as

$$M_l \simeq \begin{bmatrix} \xi b & -\xi a & a \\ b & a & \xi^* a/c \\ d & 0 & c \end{bmatrix}, \tag{26}$$

so that for $b^2, d^2 \ll a^2 \ll c^2$, it is easily seen that $m_\tau \simeq c$, $m_\mu \simeq a$, $m_e \simeq |ad/c - 2\xi b|$, $V_{e\tau} \simeq a/c = 0.06$, $V_{\mu\tau} \simeq \xi^* a/c$, and $V_{e\mu} \simeq -\xi + \xi E/C$. Note that the $\xi E/C$ component of $V_{e\mu}$ is from the neutrino mass matrix and may be as large as 0.1. In that case, if $A \simeq 50$ keV, then $m(\nu_\mu) \simeq 10^{-3}$ eV, $m(\nu_e) \simeq 10^{-5}$ eV, and $sin^2(2\theta)\,\Delta\, m^2 \simeq 4X10^{-8}\ eV^2$, which is good[12] for the MSW explanation[13] of the solar neutrino problem. On the other hand, the value of 0.06 for $V_{e\tau}$ is actually its upper limit in this model, corresponding to the limit $b^2 \ll a^2$. Hence this model can be ruled out if the present experimental indication of $V_{e\tau} \simeq 0.09$ is confirmed.

The spontaneous breaking of lepton number in this model implies the existence of a massless Goldstone boson called the Majoron,[14] but it has no tree-level coupling to ν_3. The decay of ν_3 to $\nu_{1,2}$ and Majoron occurs only as a loop-induced process and has a lifetime of about 10^4 years. Note also that the Majoron here is an $SU(2)$ singlet, so it does not couple to the Z boson and does not contribute to its invisible width.

In Model II, the eigenvalues of the $6X6$ neutrino mass matrix are $\pm\,(C^2 + D^2 + E^2/4)^{\frac{1}{2}} + \mathrm{E}/2$, $\pm\, AC(C^2 + D^2)^{-\frac{1}{2}} + A^2D^2E(C^2+D^2)^{-2}/2$, and $\pm\,\xi AB/C + \xi^2 A^2 E C^{-2}/2$, in the approximation $A^2, B^2 \ll C^2, D^2, E^2$. As it turns out, an MSW solution in this model requires E also to be much less than $(C^2 + D^2)^{\frac{1}{2}}$, hence there are 3 pseudo-Dirac neutrinos: a very massive one perhaps of order 10^2 GeV formed by $(C\nu_{1R} + D\nu_{2R})\,(C^2+D^2)^{-\frac{1}{2}}$ and $\bar{\nu}_{3R}$, a not-so-massive one of 17 keV formed by $\nu_{2L} - (\xi D/C)\nu_{1L}$ and $(D\nu_{1R} - C\nu_{2R})\,(C^2+D^2)^{-\frac{1}{2}}$, and a light one perhaps of order 10^{-3} eV formed by $\nu_{1L} + (\xi D/C)\nu_{2L}$ and $\bar{\nu}_{3L}$. This means that ν_{2L} is now mostly ν_τ and l_{2L} has to be mostly τ_L.

Consider

$$M_l M_l^\dagger = V \begin{bmatrix} m_e^2 & 0 & 0 \\ 0 & m_\mu^2 & 0 \\ 0 & 0 & m_\tau^2 \end{bmatrix} V^\dagger . \tag{27}$$

Let $b^2 \gg a^2, c^2, d^2$, then $m_\tau \simeq b$, $m_\mu \simeq (c^2 + a^2)^{\frac{1}{2}}$, $m_e \simeq |a^2 d -$

$2\xi abc|/b(c^2+a^2)^{\frac{1}{2}}$, and

$$V = \begin{bmatrix} cos\theta & sin\theta & \xi \\ (d/b)sin\theta & -(d/b)cos\theta & 1 \\ -sin\theta & cos\theta & d/b \end{bmatrix}, \quad (28)$$

where $tan\theta \equiv a/c$. The ν_e and ν_μ overlaps with the 17 keV neutrino are respectively $-(\xi D/C)cos\theta + (d/b)sin\theta$ and $(\xi D/C)sin\theta + (d/b)cos\theta$. If the latter is set equal to zero to avoid $\nu_\mu - \nu_\tau$ oscillations, then ξ D/C $cos\theta \simeq$ 0.092 is needed to account for the present 17 keV data. The formula for m_e implies that $sin\theta \simeq 0.3$, hence C/D $\simeq$ 0.043 and $A \simeq$ 400 keV. If $B \simeq 10\,keV$, $C \simeq 10\,GeV$, and $E \simeq 0.85\,GeV$, then $m(\nu_{e,\mu}) \simeq 1.5X10^{-3}$ eV, and $sin^2(2\theta)\Delta m^2$ is again about $4X10^{-8}\ eV^2$. The 17 keV neutrino now decays into $\nu_{e,\mu}$ and Majoron at tree level with a lifetime of about $3X10^{-3}s$.

The mass difference squared between the 2 mass eigenstates which make up the 17 keV pseudo-Dirac neutrino of this model is of order 1 to 10^2 eV^2. This means that at the age of about 1s in the early universe, they oscillate very rapidly into each other and would count as two relativistic degrees of freedom in its expansion rate during this epoch of nucleosynthesis. However, their effective decay lifetime is about 0.2s, which may be short enough to avoid their being in full thermal equilibrium with the other particles. Hence the effective number of neutrinos should be somewhere between 2 and 4, in possible agreement with the present astrophysical upper bound[15] of 3.3. On the other hand, there is also an upper bound of the order 10 keV on the mass of a Dirac or pseudo-Dirac neutrino with an inert partner from supernova dynamics.[16] This may be alleviated here because the population of ν_τ in the core of the supernova is likely to be suppressed.

5. Conclusion

The S_3XZ_3 discrete symmetry allows 2 derivable and successful relationships in the quark sector, i.e. Eqs. (15) and (21). It also allows 2 solutions with a 17 keV pseudo-Dirac neutrino with a significant overlap with ν_e.

6. Acknowledgements

I thank the organizers, especially Jose′ Nieves, for their hospitality in Puerto Rico. This work was supported in part by the U.S. Department of Energy under Contract No. DE-AT03-87ER40327.

7. References

1. H. Fritzsch, Phys. Lett. 73B, 317 (1978).
2. X.-G. He and W.-S. Hou, Phys. Rev. D41, 1517 (1990).

3. E. Ma, Phys. Rev. D43, R2761 (1991).
4. E. Ma, Univ. of Calif., Riverside Report No. UCRHEP-T68 (Feb 1991).
5. T.V. Duong and E. Ma, Univ. of Calif., Riverside Report No. UCRHEP-T74 (Apr 1991).
6. C. Jarlskog, Phys. Rev. Lett. 55, 1039 (1985).
7. N. Turok, These Proceedings.
8. S.L. Glashow, Phys. Lett. B256, 218 (1991).
9. J.J. Simpson and A. Hime, Phys. Rev. D39, 1825 (1989); A. Hime and J.J. Simpson, ibid. 39, 1837 (1989); A. Hime and N.A. Jelley, Phys. Lett. B257, 441 (1991).
10. B. Sur et al., Phys. Rev. Lett 66, 2444 (1991).
11. Particle Data Group, J.J. Hernandez et al., Phys. Lett. B239, 1 (1990).
12. J.N. Bahcall and H.A. Bethe, Phys. Rev. Lett. 65, 2233 (1990).
13. S.P. Mikheyev and A. Yu. Smirnov, Yad. Fiz. 42, 1441 (1985) [Sov. J. Nucl. Phys. 42, 913 (1985)], and Nuovo Cim. 9C, 17 (1986); L. Wolfenstein, Phys. Rev. D17, 2369 (1978).
14. Y. Chikashige, R.N. Mohapatra, and R.D. Peccei, Phys. Lett. 98B, 265 (1981).
15. K.A. Olive et al., Phys. Lett. B236, 454 (1990).
16. R. Gandhi and A. Burrows, Phys. Lett. B246, 149 (1990).

ANTHONY ZEE

From Semionics to Topological Fluids*

A. ZEE

Institute for Theoretical Physics
& Department of Physics
University of California
Santa Barbara, California 93106-4030

TABLE OF CONTENTS

I. A PARABLE

I would like to begin with a parable. Once upon a time, a particle physicist decided that he would like to dabble in condensed matter physics. He thought that as a first try he should try to understand spontaneous symmetry breaking and Goldstone's theorem, particularly since that was one of the first things Professor Swieca had taught him many, many years ago. It's all simple stuff, child's play, he thought. First, write down

$$\mathcal{L} = \partial_\mu \Phi^\dagger \partial^\mu \Phi - V(\Phi^\dagger \Phi) \tag{1.1}$$

with V the usual double well potential. Set

$$\Phi = v e^{i\alpha}. \tag{1.2}$$

The Lagrangian reduces to

$$\mathcal{L} = v^2 (\partial_\mu \alpha)^2 \tag{1.3}$$

and there it is, the Goldstone boson α.

Well, our particle physicist is smart enough to know that condensed matter physics is non-relativistic, and so he writes down a non-relativistic Lagrangian

$$\mathcal{L} = i\phi^\dagger \partial_0 \phi - \frac{1}{2M} \partial_i \phi^\dagger \partial_i \phi - V(\phi^\dagger \phi) \tag{1.4}$$

instead of (1.1). There is manifestly a $U(1)$ symmetry. He breaks the symmetry by setting

$$\phi = v e^{i\alpha} \tag{1.5}$$

just as in (1.2). Now the Lagrangian becomes

$$\mathcal{L} = -v^2 (\partial_i \alpha)^2 \tag{1.6}$$

with the equation of motion

$$\partial_i^2 \alpha = 0. \tag{1.7}$$

Oops, what is this? It's not even a propagating degree of freedom. Where is the Goldstone boson?

Our physicist shows this puzzling result to some condensed matter postdocs. After throwing in some red herrings and mumbling something about charges and non-relativistic reduction and superfluids, they managed to confuse our physicist even further.

The resolution of this apparent puzzle is simple. In the non-relativistic case, we must keep the radial excitation or the "Higgs field" h. In other words, instead of (1.5) we must set

$$\phi = (v + h)e^{i\alpha} \tag{1.8}$$

so that the Lagrangian becomes

$$\mathcal{L} = -vh\partial_0\alpha - \frac{v^2}{2M}(\partial_i\alpha)^2 - \frac{1}{2}m_h^2h^2 + \ldots \tag{1.9}$$

where we keep terms up to quadratic order in the fields and dropped total derivatives. (Here m_h^2 is just a symbol used to denote the second derivative of V at its minimum.) If we now integrate out the heavy field h, we obtain

$$\mathcal{L} = \frac{v^2}{2m_h^2}(\partial_0\alpha)^2 - \frac{v^2}{2M}(\partial_i\alpha)^2. \tag{1.10}$$

(Note that this result holds in the limit of low energy and momentum when the Higgs propagator becomes $-i/m_h^2 + \ldots$.) We have found the Goldstone boson! Furthermore, we have determined its speed of propagation to be $m_h/M^{1/2}$.

When we substituted in (1.5) we had effectively assumed that the double well potential is infinitely steep, that is $m_h \to \infty$. The Goldstone boson was still here except that its velocity was infinite and had thus disappeared from the spectrum for $k \neq 0$.

Why is it that we are allowed to go to this limit in the relativistic case? Because we have Lorentz invariance! The velocity of a linearly dispersing mode, if such a mode exists, is guaranteed by symmetry reasons to be equal to 1 in suitable units. Those condensed guys don't have Lorentz invariance to help them! The absence of Lorentz invariance is the root cause of many headaches, messy formulas, yukky calculations, and other unpleasant things, but on the other hand, it also allows for interesting physics otherwise not allowed by Lorentz invariance.

The moral is that the relativistic mind working on non-relativistic physics must be ever vigilant in order to avoid the many pitfalls that lie in wait.

Note incidentally that we even get, with considerable less labor, the classic Bogoliubov result[1] for the velocity of sound in the superfluid.

To see this, we note that the number density n is related to the "Higgs mass" m_h by the standard formula

$$n \equiv v^2 = m_h^2/\lambda \tag{1.11}$$

where λ is the quartic coupling in the potential V and measures the interaction between the bosons in Bogoliubov's formulation.

Thus, eliminating m_h we have the speed of propagation

$$c = (\lambda n/M)^{1/2} \tag{1.12}$$

The learned reader would notice that we have reproduced Bogoliubov's result without ever doing a Bogoliubov rotation!

(Let us recall schematically how Bogoliubov did it. In the boson Hamiltonian $H = \sum(k^2/2M)a^+a + \sum ga^+a^+aa$, he replaced the operator a_o by the c-number $\sqrt{n}$, thus obtaining something like

$$H = \sum(k^2/2M)a^+a + gn\sum\left(a_k^+a_k + a_ka_{-k} + a_k^+a_{-k}^+\right) \tag{1.13}$$

which he diagonalized by doing the rotation named after him. His coupling g is related to our λ by some kinematic factor.)

As another example, consider the same theory gauged. so that (1.1) becomes

$$\mathcal{L} = (\partial_\mu + iA_\mu)\Phi^\dagger(\partial^\mu - iA^\mu)\Phi - V(\Phi^\dagger\Phi) - \frac{1}{4}F_{\mu\nu}^2. \qquad (1.14)$$

Our relativistic physicist recalls instantly the result of Coleman and Weinberg[2] that to one loop order the effect of the gauge field is to simply modify V to $V + \Delta V$, but the qualitative double well shape of V is unchanged. (In a suitable gauge, the Feynman diagram calculation involves only the gauge field propagating in the loop.) In the non-relativistic case, in contrast, the Lagrangian contains a term linear in A_0

$$\mathcal{L} = i\phi^\dagger(\partial_0 - iA_0)\phi + \ldots \qquad (1.15)$$

Upon integrating out A_0 we obtain a non-local interaction instantaneous in time between $\phi^\dagger\phi(\vec{x},t)$ and $\phi^\dagger\phi(\vec{y},t)$. This is of course just the Coulomb interaction but our particle physicist in his infinite sophistication has long forgotten such elementary triviality as the Coulomb interaction!

II. THE HUBBARD MODEL

In these lectures I am going to talk about a possible theory of high temperature superconductivity based on the possibility that the excitations in high temperature superconductors obey fractional statistics, a theory I like to call semionics. In the last year or so, this highly speculative subject has suddenly become fashionable and sometimes it seems to me that every particle theorist with time on his or her hands is rushing into this field. They would write down a relativistic field theory containing the Chern-Simons term and declare that they have solved high temperature superconductivity or at least they end their papers with some pieties like "We hope that our considerations here will prove to be relevant to high temperature superconductivity."

I would like to emphasize here that the subject consists of two parts. In the first part, you have to get from the microscopic solid state physics to some kind of long distance field theory in which the excitations may turn out to be semions. In the second part, you have to show that a liquid of semions is indeed a superfluid, and a superconductor when the semions are treated as charged. In many respects, the first part is the hard part, and at the moment much less well established than the second part.

Quite aside from any possible relevance to high temperature superconductivity, the subject I am going to talk about here is a part of a fascinating area of physics dealing with strongly correlated quantum electron systems in two dimensions. Examples of such systems include the Hall fluid and the highly frustrated antiferromagnet. It appears that these systems may exhibit striking universality properties. The long distance physics of these systems may be described by the so-called topological field theories.

I find it convenient to organize this subject by asking four "big questions". First, under what circumstances does the ground state of a commonly accepted model Hamiltonian spontaneously violate time reversal and parity? The theory to be described here will be largely based on assuming that T and P are broken in high temperature superconductors. We know, of course, that if I give you a piece of superconducting material, the underlying Coulomb interaction of the electrons certainly does not violate time reversal and parity. These invariances have to be dynamically and spontaneously violated. So the first question is whether this is possible? Second, how do we describe the dynamics of the excitations about the ground state? What are their quantum numbers and in particular what are their statistics? What is the long distance effective theory? The third question is how does a gas or liquid of fractional statistics particles behave? Is it a superfluid and does it superconduct if these particles are charged? And of course, most important is the question whether there are experimental tests that we can invent to test our scenario.

Before I go on I have to give you a lightning review of high temperature superconducting materials. All you have to know is that these materials have a layered structure so that the electrons are largely confined to moving in planes. Most of the theoretical work on the subject assumes as an approximation that the electrons live in a 2 dimensional world. The natural question to ask is whether there is some physics that is generically different in 2+1 dimensional spacetime. Maybe this new phenomenon is associated with the fact that the electrons are forced to move in a 2 dimensional world.

Very crudely speaking, theoretical work on high temperature superconductivity follows two main currents. The vast majority of researchers in this field assume that there is no need for a drastically new physical concept and that more or less the same constructs that have been so successful in condensed matter physics can be applied. This majority view could well be right. But it's sort of a somewhat boring point of view, at least in my opinion and in the opinion of a small minority of theorists who have gone down another path by suggesting that perhaps new theoretical constructs peculiar to 2 dimensional systems are necessary.[3]

Let us begin with some of the model Hamiltonians typically considered by condensed matter physicists, such as the Hubbard model and the t-J model. These models are defined on a lattice and hence are not even rotation invariant, let alone Lorentz invariant. Where is the Lorentz invariant field theory going to come from? We will see, in due time, but first let us define the model.

The Hamiltonian for the Hubbard model consists of two terms

$$H = -t \sum_{\langle ij \rangle} c^{\dagger}_{i\alpha} c_{j\alpha} + h.c. + U \sum_{i} c^{\dagger}_{i\uparrow} c_{i\uparrow} c^{\dagger}_{i\downarrow} c_{i\downarrow} \tag{2.1}$$

The operator $c_{i\alpha}$ annihilates an electron of spin α ($\alpha = \uparrow$ or $\downarrow$) at site i on a lattice (which we will take for the purposes of these lectures to be square and two-dimensional). The symbol $\langle ij \rangle$ indicates that i and j are nearest neighbors. Thus the first term describes the hopping of electrons from site i to site j. The second term tells us that if two electrons

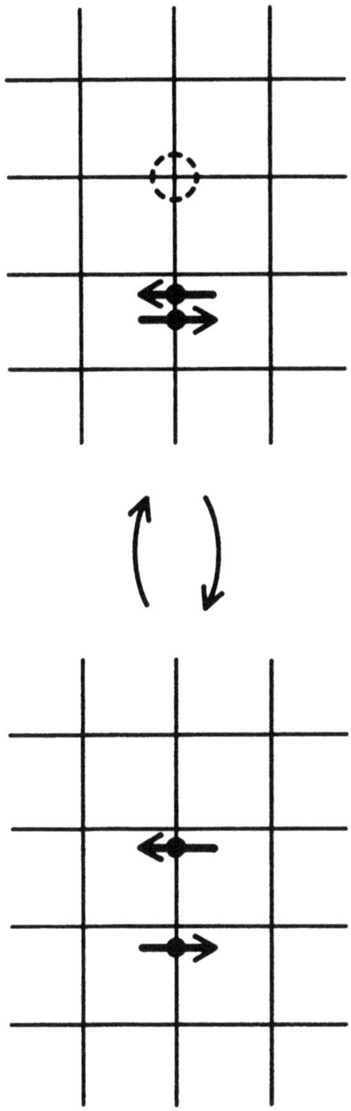

FIGURE 1

are on the same site, the energy goes up by U. (Note that for two electrons to occupy the same site they must have opposite spins, by the Pauli exclusion principle.)

We also have to specify how many electrons are there in the system. It is natural to start with the half-filled case, in which the number of electrons N_e (half of them with spin up and half of them with spin down) is equal to the number of sites on the lattice N. We can then dope the system by removing electrons (keeping the number of spin up electrons equal to the number of spin down electrons). The density of holes is defined as $\delta = (N - N_e)/N$.

While the Hubbard model is easy enough to write down, it has challenged the ingenuity of condensed matter physicists for decades. Two limits suggest themselves: $U/4t$ small (weak coupling) and $U/4t$ large (strong coupling). The literature bifurcates according to whether you believe U is large or small compared to $4t$. Schrieffer and his collaborators[4] have suggested that the physics for small $U/4t$ and for large $U/4t$ may interpolate smoothly into each other. Here we will concentrate on the large $U/4t$ limit.

Let us start with $\delta = 0$ so that the number of electrons is equal to the number of sites. As $U/4t$ tends to infinity, obviously the energy of the system would be the lowest possible if there is only one electron on each site. Let us now ask perturbatively what the effect of a small t might be. The electrons are now allowed to hop around. Focus on two neighboring electrons. If their spins are opposite, then one electron can hop onto the site occupied by the other (so that on one site we have two electrons with opposite spins). See Fig. (1). Through this virtual process of the electron hopping back and forth, the energy is lowered by the $\sim t^2/U$ "delocalization energy" as is easily estimated by second order perturbation theory. In contrast, if the two neighboring electrons have the same spin, this back-and-forth hopping is forbidden by the Pauli exclusion principle. Thus, an antiferromagnetic arrangement of spins is favored energetically. More formally, the Hubbard Hamiltonian at half-filling reduces in the large U limit to the Heisenberg antiferromagnet

$$H = J \sum_{\langle ij \rangle} \vec{S}_i \vec{S}_j \tag{2.2}$$

where $J = 4t^2/U$ is positive.

The Heisenberg antiferromagnet is manifestly an insulator with only spin degrees of freedom. In real life, high T_c superconducting materials are also insulators before they are doped with holes. As is well-known by now, they become superconducting at a finite concentration of holes (Fig. 2). The critical concentration is typically about $x_c \sim 0.05$ where x denotes the number of holes per site. The Néel order disappears for $x_c \sim 0.03$. Thus, we have to ask what happens when the model described here is doped with holes, *i.e.*, when electrons are removed.

The assignment for theorists is thus easily stated: Show that when the Hubbard model is doped with holes it superconducts.

Clearly, holes tend to destroy the antiferromagnetic order. At the most naive level, the presence of holes allow the electrons to move about. The spin order in the electrons is destroyed by "churning". A more precise formulation is that holes generate non-nearest neighbor antiferromagnetic coupling. Focus on two electrons neighboring on a hole, as shown in Fig. (3). One electron can now hop onto the site occupied by the other electron "directly". Repeating the same argument as before, we now conclude that electrons that are next nearest neighbors to each other would prefer to have their spins aligned antiferromagnetically.

Consider the Hamiltonian

$$H = J_1 \sum_{nn} \vec{S}_i \vec{S}_j + J_2 \sum_{nnn} \vec{S}_i \vec{S}_j + J_3 \sum_{nnnn} \vec{S}_i \vec{S}_j + \ldots \tag{2.3}$$

The symbols $nn, nnn, nnnn$, and so on under the various summation symbols indicate that the sites i and j are to be nearest neighbors, next nearest neighbors, next next nearest

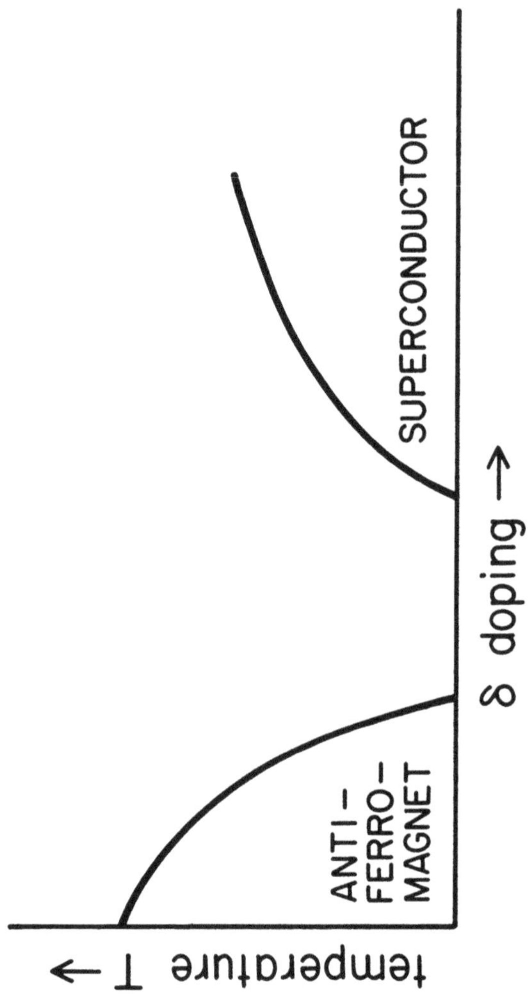

FIGURE 2

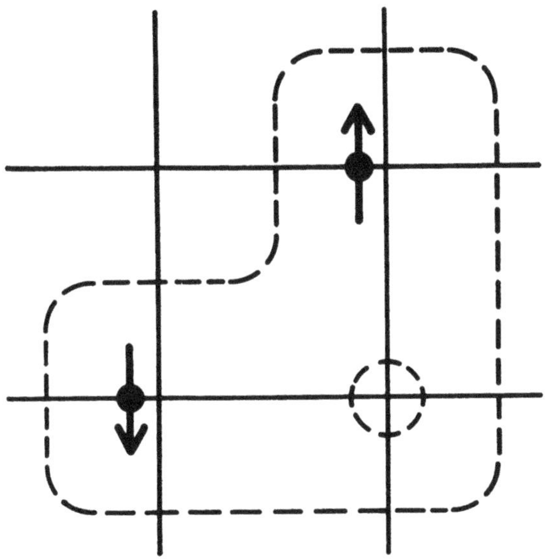

FIGURE 3

neighbors, and so on. Just as in real life, frustration ensues when conflicting demands are replaced on a system. For instance, two next neighbor spins are told by the J_1 coupling to point in the same direction but by the J_2 coupling to point in the opposite direction. Thus, this Hamiltonian is known as the frustrated Heisenberg antiferromagnet.

People have argued that a Hubbard model with a small but finite density of holes may be "equivalent" to a frustrated Heisenberg antiferromagnet, with one spin per site. The word "equivalent" of course requires a great deal of explanation. To begin with, the two models do not even have the same number of electrons. What is meant, or rather, hoped for, is that the two models belong to the same "universality class" in the sense that various relevant spin-spin correlation function may be (at least qualitatively) the same. ("Universality class" is of course one of the most abused terms in physics behind which people hide a great deal of ignorance). What we are talking about will become clearer later.

The crucial issue may then be stated as follows: what does the ground state of the frustrated Heisenberg Hamiltonian look like? Is it still antiferromagnetic, or is it, at least for some range of parameters, a "quantum disordered spin liquid"?

The exact solution of the frustrated Heisenberg Hamiltonian is of course a formidable problem. Instead, people have been trying to guess what the ground state may look like. In particular, the resonating valence bond (RVB) approach[5,6] of Anderson and others represents an attempt to write down the ground state by an inspired guess. In general, it has proved rather difficult to describe this putative quantum disordered spin liquid state.

Fortunately, instead of tackling this difficult task, we can ask a clear cut and binary question about the ground state: Does it violate T and P invariances? The answer to this question provides a binary classification of effective Hamiltonians.

III. A STATE IN SEARCH OF A HAMILTONIAN

In school courses on quantum mechanics, we learned to solve for the ground state, *etc.*, once we are given various specific Hamiltonians. The strongly correlated Hamiltonians we have encountered here are much too difficult to solve exactly. If we do not want to do numerical work, all we can hope for is some sort of variational or mean field approach whereby we hope to capture some of the essential physics implied by the Hamiltonian.

In modern condensed matter physics, people often work in a direction opposite to what we learned in school: people describe or characterize the quantum state of a system without specifying the Hamiltonian. A classic example of this is Laughlin's theory of the quantum Hall fluid,[7] a two dimensional system of spinless electrons moving in an external magnetic field. The important physics is that the electrons move in Larmor-Landau orbits and obey fermi statistics. Thus, each electron takes up a finite amount of room and because of fermi statistics refuses to occupy the same room with another electron. This is in essence the reason why the Hall fluid is incompressible. To attack this problem Laughlin does not solve some model Hamiltonian. Rather, he writes down a many-body wave function for the electrons that incorporates fermi statistics and at the same time can be decomposed into a linear combination of single-body Landau wave functions. This wave function, precisely because it incorporates the essential physics, turns out to have very low energy for a whole class of Hamiltonians. In Laughlin's theory, there is no need to specify the detailed form of the interaction between the electrons, as long as it has some suitable general properties such as being short ranged.

For the spin Hamiltonians we have encountered here, it is not even clear *a priori* how the ground state can be written down in a neat and closed form by the stroke of a pen. Rather, to describe the ground state we should begin by characterizing its order. In effect, we already do this in elementary discussions. For instance, when we propose that a state is anti-ferromagnetic, we are saying that the staggered spin-spin correlation function

$$(-)^{i-j}\left\langle \vec{S}_i \cdot \vec{S}_j \right\rangle \tag{3.1}$$

does not vanish as the distance between the two sites $i-j$ tends to infinity. We can describe other kinds of order by specifying how fast this correlation vanishes at infinity, according to a power law or an exponential decay law, for instance.

Anderson and his collaborators[5,6] first proposed that the ground state of highly frustrated Heisenberg Hamiltonians is a spin liquid, in which the spins are not antiferromagnetically ordered and in which translation invariance is maintained. Later, the so-called chiral spin state, in which T and P are spontaneously violated, was proposed.[8]

We will follow here the development leading to the construction of the chiral spin state. As we will see, the story as it is now told in the literature is far from completely satisfactory and may even be pure fiction in some places.

Let us proceed to construct the chiral spin state. Although the whole point is that we do not have to specify the Hamiltonian, it is pedagogically helpful to keep some Hamiltonians in mind lest the discussion becomes too vague and unanchored. Take for example the generic frustrated Heisenberg spin Hamiltonian

$$H = \sum_{\langle ij \rangle} J_{ij} \vec{S}_i \cdot \vec{S}_j \tag{3.2}$$

The notation J_{ij} is used to incorporate compactly the couplings J_1, J_2, *etc.* and does not imply that translation invariance is explicitly broken. The spin-spin interaction may be rearranged by the Fierz identity:

$$\vec{S}_i \cdot \vec{S}_j = (c_i^\dagger \vec{\sigma} c_i)(c_j^\dagger \vec{\sigma} c_j) = -2(c_i^\dagger c_j)(c_j^\dagger c_i) - (c_i^\dagger c_i)(c_j^\dagger c_j) + c_i^\dagger c_i \tag{3.3}$$

For a spin system (so that the number operator on each site $c_i^\dagger c_i$ is equal to 1), the Hamiltonian in (3.2) is thus equivalent to

$$H = -2 \sum_{\langle ij \rangle} J_{ij} (c_i^\dagger c_j)(c_j^\dagger c_i) \tag{3.4}$$

Traditionally, order in spin systems is described in terms of the spin operators $\vec{S}_i$. One motivation for rearranging is that some type of order may be more naturally described in terms of the creation $c_i^\dagger$ and annihilation c_i operators. In particular, define the hopping operator $\chi_{ij} \equiv c_i^\dagger c_j$ which causes an electron to hop from site j to site i. Then we can consider a "flux order" whereby we specify the behavior of the expectation of a product of hopping operators in the ground state $|0\rangle$:

$$\langle 0|\chi_{jk} \cdots \chi_{li}\chi_{ij}|0\rangle_\Gamma \sim e^{ibA_\Gamma} \tag{3.5}$$

The product of the hopping operators is taken around a loop Γ and A_Γ denotes the area enclosed by the loop. In (3.5) we omit writing an overall real factor, if there is any. By conjugating (3.5), we see clearly that this order violates parity P and time reversal T, if $b \neq 0$.

The flux order is realized if

$$\left\langle 0|c_i^\dagger c_j|0\right\rangle \equiv (\text{real overall factor})U_{ij} \tag{3.6}$$

where U_{ij} is a complex number of magnitude 1 and such that the product of the U_{ij}'s around a square plaquette

$$\prod_{\square} U_{ij} = e^{i\Phi} \tag{3.7}$$

is not equal to 1. The flux Φ through a plaquette characterizes the order.

In the mean field approximation, we would then study the single-body Hamiltonian

$$H = -\sum_{\langle ij\rangle} U_{ij} c_j^\dagger c_i$$

Notice that at this level the Hamiltonian is really the sum of two uncoupled pieces, describing the hopping of spin up electron and of spin down electrons separately. Thus, we have effectively a problem of a single spinless electron hopping in a prescribed background U_{ij}.

The Hamiltonian in (3.4) has a $U(1)$ gauge invariance under which $c_i \to e^{i\theta_i}c_i$. Indeed, the spin Hamiltonian in (3.2) is obviously invariant since $\vec{S}_i$ is unchanged under this transformation. (Note that the Hubbard model breaks this invariance with the t term; this is an approximate symmetry that emerges at half-filling for $U \gg t$.)

Affleck and Marston,[9] Kotliar,[10] and Ioffe and Larkin[11] initiated the study of flux order by taking $\Phi = \pi$. We can always choose a gauge in which $U_{iy} = 1$ and $U_{ix} = (-)^{i_y}$. Henceforth we ignore irrelevant overall factors. Here we write U_{ix} for $U_{i+x,i}$, U_{iy} for $U_{i+y,i}$, and denote the coordinates on the lattice by $i = (i_x, i_y)$. Our lattice spacing is fixed to be 1 and i_x, i_y are integers. The energy spectrum of the single-body hopping problem is then determined by

$$H = -\sum \left\{ c^\dagger_{i+x} c_i (-)^{i_y} + c^\dagger_{i+y} c_i + h.c. \right\} \tag{3.8}$$

(We write $i+x$ for (i_x+1, i_y) *etc.*) The operators c_i and $c^\dagger_i$ no longer carry any spin label. We Fourier-transform to momentum space $c_i = \frac{1}{\sqrt{N}} \sum_k c_k e^{-iki}$ where as usual $k = \frac{2\pi}{N} l$ with $l = (l_x, l_y)$ with l_x, l_y integers and k ranges over the Brillouin zone $-\pi \le k_x \le \pi$, $-\pi \le k_y \le \pi$. We suppress the two dimensional character of k, l, i, *etc.* We also trust the reader to distinguish between the site location i and the imaginary unit i. We obtain

$$H = -\left\{ \sum c^\dagger_{k+w} c_k e^{ik_x} + c^\dagger_k c_k e^{ik_y} + h.c. \right\}. \tag{3.9}$$

Note that the hopping from i to $i+\mu$ ($\mu = x$ or y) in real space is associated with e^{ik_μ} in momentum space because of the phase mismatch between c_i and $c^\dagger_{i+\mu}$. A position dependence in U_{ij} transfers momentum. Thus, here $U_{ix} = (-)^{i_y} = e^{i\pi i_y} = e^{iwi}$ with $w = (0, \pi)$ and the "gauge background" kicks in a momentum w. This is crucial. In contrast, consider what would happen if there is no flux order and $U_{ix} = U_{iy} = 1$. Instead of (3.9) we would obtain

$$H = -\sum \left\{ c^\dagger_k c_k e^{ik_x} + c^\dagger_k c_k e^{ik_y} + h.c. \right\}. \tag{3.10}$$

The energy eigenvalues can be read off instantaneously to be

$$E(k) = -(\cos k_x + \cos k_y) \tag{3.11}$$

a result derived in Feynman's "freshman lectures" or in Chapter 1 of solid state physics texts.

The energy spectrum (3.11) has lines of zeroes given by $\pm k_x \pm k_y = \pm\pi$ (uncorrelated sign) as indicated in Fig. (4). At half filling, by symmetry between $E \leftrightarrow -E$ it is obvious that electrons are filled up to these lines in the Brillouin zone. Thus, the Fermi surface is a square.

Let us return to the case of π flux. To diagonalize this Hamiltonian in (3.9), we shift the dummy summation variable by $k \to k + w$, thus obtaining

$$H = -\left\{\sum c_k^\dagger c_{k+w} e^{ik_x} - c_{k+w}^\dagger c_{k+w} e^{ik_y} + h.c.\right\} \tag{3.12}$$

Note the minus sign. The Hamiltonians in (3.9) and (3.12) are identical. Let us add them and divide by 2 to obtain

$$H = -\sum{}'(c_k^\dagger c_{k+w}^\dagger) \mathcal{H} \begin{pmatrix} c_k \\ c_{k+w} \end{pmatrix} \tag{3.13}$$

where

$$\mathcal{H} = \begin{pmatrix} \cos k_y & \cos k_x \\ \cos k_x & -\cos k_y \end{pmatrix}. \tag{3.14}$$

In (3.13), the summation over k should be restricted to half of the Brillouin zone if one wants to be pedantic. The prime on the summation symbol serves as a reminder. Equivalently, we can recognize (3.13) as defining for a given k a two-site hopping problem in momentum space: a particle hope from k to $k + w$ and with another hop gets back to $k + 2w = k$.

The eigenvalues are

$$E = \pm\sqrt{\cos^2 k_x + \cos^2 k_y}. \tag{3.15}$$

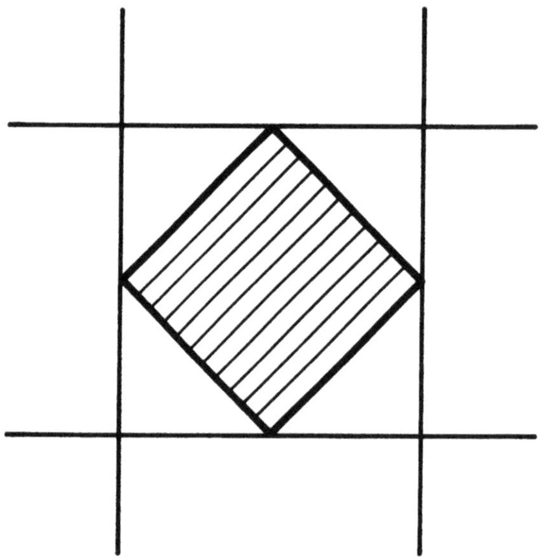

FIGURE 4

The spectrum has been drastically altered from (3.11). Here $E = 0$ only at isolated points, namely $k = k_* = \left(\frac{\pi}{2}, \frac{\pi}{2}\right)$ and $k = k'_* = \left(-\frac{\pi}{2}, \frac{\pi}{2}\right)$ and equivalent points. The lines of zero in Fig.(4) has been squeezed into isolated zeroes, which I will call Dirac zeroes for reasons that will become clear presently.

Let us expand $\mathcal{H}$ around one of the Dirac zeroes, say $(\pi/2, \pi/2)$. For $k = k_* + q$, q small, we have

$$\mathcal{H} = -q_y\tau_3 - q_x\tau_1 \tag{3.16}$$

Multiplying by τ_2, we see that the energy and momentum are related by

$$\gamma_0\mathcal{H} - q_x\gamma_x - q_y\gamma_y = 0 \tag{3.17}$$

where we have defined

$$\gamma_0 = \tau_2, \quad \gamma_x = i\tau_3, \quad \gamma_y = -i\tau_1. \tag{3.18}$$

Remarkably, in a system that is not even rotationally invariant, we find that the $E \sim 0$ excitations satisfy the massless Dirac equation!

The next crucial step was taken by Wen, Wilczek, and Zee[8] who proposed that diagonal correlations such as $\left\langle c^\dagger_{i+x+y}c_i\right\rangle$ may also be present. The imagined that such next nearest neighbor correlations should be generated by the frustrating next nearest neighbor spin-spin coupling J_2 in (3.2). In general, we can always simply invite ourselves to consider such correlations. Whether or not we will produce an energetically favored state is of course the bottom line issue to be addressed later.

Suppose then that we add to the Hamiltonian (3.8) a term

$$\Delta H = \beta\sum_i c^\dagger_{i+x+y}c_iU_{i_{x+y}} + h.c. \tag{3.19}$$

(The overall real constant β reflects the relative magnitude of the couplings J_2 to J_1.) What should we take for U_{ix+y}?

Since the flux through a square plaquette is π, we may want the flux through half of a plaquette, namely a triangle (travelling clockwise $i \to i+x+y \to i+x$) to be $\pi/2$, that is $U_{ix+y}(-)^{i_y} = e^{i\pi/2}$ (see Fig.(5)) and thus

$$U_{ix+y} = i(-)^{i_y} \tag{3.20}$$

Having gone through the discussion given earlier we can write down ΔH in momentum space immediately:

$$\Delta H = \beta \sum c^{\dagger}_{k+w} c_k i e^{i(k_x+k_y)} + h.c. \tag{3.21}$$

After shifting, we see that we obtain a term

$$\beta c^{\dagger}_{k+w} c_k (+i) e^{-i(k_x+k_y)} \tag{3.21$'$}$$

to go with the term in (3.21). Under hermitean conjugation the i flips sign while the shift of k by w produces another sign, so that we again obtain a cosine upon adding the two terms. Thus,

$$\Delta \mathcal{H} = \begin{pmatrix} 0 & -i\beta\cos(k_x+k_y) \\ i\beta\cos(k_x+k_y) & 0 \end{pmatrix} \tag{3.22}$$

(Again, we ignore irrelevant overall real factors such as factors of 2.) The fact that we hop to our next nearest neighbor rather than the nearest neighbor produces $\cos(k_x+k_y)$ rather than $\cos k_x$ or $\cos k_y$. This is crucial, because then $\Delta\mathcal{H}$ is just a constant near a Dirac zero, say $(\frac{\pi}{2}, \frac{\pi}{2})$, so that we now have, instead of (3.16)

$$\mathcal{H} = -q_y\tau_3 - q_x\tau_1 - m\tau_2 \tag{3.23}$$

where m is 2β or something like that.

We obtain a massive Dirac fermion! In condensed matter language, a gap opens in the energy spectrum.

Why is this important? Let us recall that we want to construct the ground state of some spin Hamiltonian. We have now determined the energy spectrum of a single-body

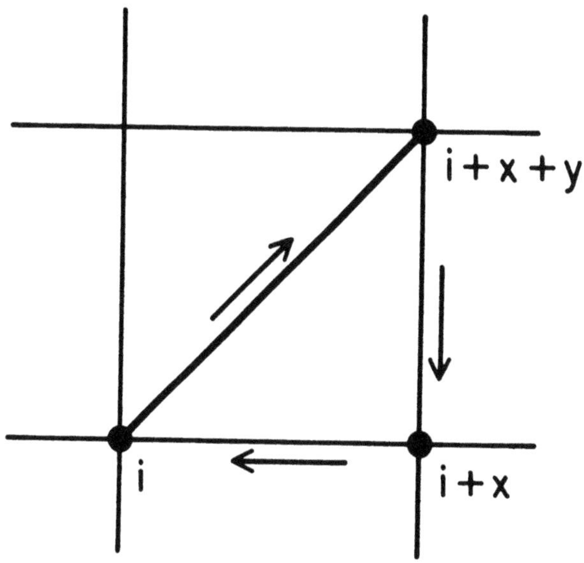

FIGURE 5

hopping problem. For non-interacting electrons we can then construct a many-body state by filling the lower band (or Dirac sea in field theory language) in accordance with Pauli exclusion. We have to do this separately for the spin up electrons and the spin down electrons, thus obtaining something like

$$|\Psi\rangle = \mathcal{P}|\Psi \uparrow\rangle \otimes |\Psi \downarrow\rangle \tag{3.24}$$

as a candidate ground state. This state is known as the chiral spin state.

Since $|\Psi \uparrow\rangle$ and $|\Psi \downarrow\rangle$ are constructed completely independently, in the product state $|\Psi \uparrow\rangle \otimes |\Psi \downarrow\rangle$ there would be non-zero amplitudes for a spin up electron and a spin down electron at the same site. (Note that we have already guaranteed that no more than one spin up electron occupies a site.) If we want to have a variational ground state for a spin Hamiltonian (which by definition has only one electron per site), we have to project with the Gutzwiller projector

$$\mathcal{P} = \prod_i (1 - n_{i\uparrow} n_{i\downarrow}) \tag{3.25}$$

It is easy enough to write down curly $\mathcal{P}$ in (3.24) but the actual projection is of course a mess. In this case, Zou and Laughlin[12] have shown that by a minor miracle the projected state $|\Psi\rangle$ is actually relatively simple. (The "miracle" is summarized by saying that the Gutzwiller projected Wen-Wilczek-Zee state is the same as the Kalmeyer-Laughlin state.[13]) We will come back to this projection.

Generically, the energy of the Dirac sea is lowered when a gap opens. Thus, we might expect the chiral spin state to be an attractive candidate for the ground state of some spin Hamiltonians. The appearance of an energy gap is crucial.

The low energy dynamics of the excitations around the ground state is thus governed by the Dirac Lagrangian

$$\mathcal{L} = \sum \bar{\psi}(i\partial\!\!\!/ - m)\psi \tag{3.26}$$

For simplicity of presentation, I have only spoken of the hopping from i to $i+x+y$ (and back of course). In general, we would expect also hopping from i to $i+x-y$:

$$\Delta H = \beta' \sum c^{\dagger}_{i+x-y} c_i U_{i,x-y} \tag{3.27}$$

I will leave it to the reader as an exercise to work this out. There is only one slight subtlety: we want the flux through the triangles ($i \to i+y \to i+x$) to be the same as the flux through the triangles ($i \to i+x+y \to i+x$) and thus we require

$$U_{i,x-y} = -ie^{i\pi/2} \tag{3.28}$$

This extra (-1) relative to (3.20) is crucial. The reader can work out that effectively the $\beta \cos(k_x + k_y)$ term in $\Delta \mathcal{H}$ in (3.22) is replaced by $\beta \cos(k_x + k_y) + \beta' \cos(k_x - k_y)$. We expect β and β' to be comparable and perhaps even exactly equal. Were the $(-)$ sign absent in (3.29), the two cosines would subtract rather than add, cancelling out the energy gap.

With these diagonal correlations it is clear that T and P are broken. (To see whether or not a symmetry is broken in a gauge theory, it is of course important to look at transformation of gauge invariant quantities. We have to verify that U_{ij} and the transformed U_{ij} are not related by a gauge transformation $U'_{ij} = W_i^{\dagger} U_{ij} W_j$.) Field theorists have long known that in (2+1)-dimensions the Dirac mass violates T and P. Thus, we could not possibly obtain a massive Dirac equation in the flux phase with π flux through each plaquette since the microscopic lattice physics does not violate T and P ($e^{i\pi} = e^{-i\pi}$). Here under time reversal $e^{i\pi/2} \to e^{-i\pi/2}$, which is certainly not equal to $e^{+i\pi/2}$. Similarly, under parity, a triangle with flux $e^{i\pi/2}$ is reflected into a triangle with flux $e^{-i\pi/2}$.

To see explicitly the violation of T and P, let us form the gauge invariant quantity

$$Pl_{ijk} = \langle 0 | \chi_{ij} \chi_{jk} \chi_{ki} | 0 \rangle \tag{3.28}$$

where i, j, k are the vertices of an elementary triangle on the lattice. By explicit computation, we can verify that

$$\mathcal{P}l_{ijk} - \mathcal{P}l_{ikj} = -\frac{i}{2}\left\langle 0|\vec{S}_i \cdot (\vec{S}_j \times \vec{S}_k)|0\right\rangle \qquad (3.30)$$

A non-zero expectation value of the triple spin product $\vec{S}_i(\vec{S}_j \times \vec{S}_k)$ manifestly violates T and P.

The chiral spin state is inspired by a mean field approximation. According to more-or-less standard arguments, the mean field approximation is expected to be exact in some sort of large N limit, where N is the number of spin components. Since N is actually equal to 2, we have no *a priori* reason to expect the mean field approximation to be reliable. The fluctuation of $\left\langle 0|c_i^\dagger c_j|0\right\rangle$ around its mean value may well be enormous. (In quantum chromodynamics, however, the large N approximation seems to be qualitatively correct, but there N is as large as 3! Is 2 a lot less than 3?)

Of course, given a spin Hamiltonian we can simply present the chiral spin state as the candidate ground state. To determine whether or not the chiral spin state is indeed the ground state numerical work is indispensible. We argued earlier that the chiral spin state may well be the ground state because of the energy gap in the excitation spectrum. The trouble with this argument is that the Gutzwiller projection $\mathcal{P}$ in (3.24) may well wash out this hoped for energetic advantage. The unprojected state $|\Psi \uparrow\rangle \otimes |\Psi \downarrow\rangle$ in (3.24) is constructed by putting down electrons with up and down spins in a totally uncorrelated fashion. Thus, the amplitude of having one electron per site in $|\Psi \uparrow\rangle \otimes |\Psi \downarrow\rangle$ may be exponentially small in the number of sites. The Gutzwiller projection keeps only a tiny component of $|\Psi \uparrow \otimes |\Psi \downarrow\rangle$. In this case, Gutzwiller may well have thrown the baby out with the bathwater, as they say.

Devotees of the chiral spin state could still argue that the solid state physics of high temperature superconductors is so complicated that we really do not know what a reasonable Hamiltonian for the system may be. In the space of all reasonable Hamiltonians, there may well be, hopefully, some Hamiltonians that would favor the chiral spin state. As I said earlier, the present discussion may be best regarded as motivating a binary classification of the literature, according to whether T and P are violated or not.

IV. GAUGE INVARIANCE AND THE APPEARANCE OF SEMIONS

Having now studied the ground state, let us consider deviations from the ground state. Denote the U_{ij}'s we have used thus far by U^o_{ij}. Consider phase fluctuations $U_{ij} = U^o_{ij} e^{ia_{ij}}$. (We can also consider fluctuations in the magnitude of U_{ij} but these clearly correspond to high energy fluctuations because they change the overall energy scale of the Hamiltonian.) Under a gauge transformation $c_i \to e^{i\theta_i} c_i$, the Hamiltonian is now unchanged if we take $a_{ij} \to a_{ij} + \theta_i - \theta_j$. By gauge invariance, we can argue immediately that in the long distance limit the effective Lagrangian (3.26) is now replaced by

$$\mathcal{L} = \sum \bar{\psi}(i\partial\!\!\!/ - m - a\!\!\!/)\psi \tag{4.1}$$

where $a_\mu(x)$ is a gauge potential. (The phase fluctuations a_{ij} go into the spatial components of $a_\mu(x)$ as usual. The time component $a_0(x)$ emerges in a treatment more careful than the one given here. It originates in a Lagrange multiplier term $\sum_i a_{0i}(c_i^\dagger c_i - 1)$ we have to add in order to enforce the constraint of one particle per site.)

Here we see again that the mass gap is crucial. In the π flux phase, the long distance theory

$$\mathcal{L} = \sum \bar{\psi}(i\partial\!\!\!/ - a\!\!\!/)\psi \tag{4.2}$$

is infrared singular. In other words, the long distance fluctuations are large and uncontrollable and so the mean field approach is *a fortiori* not trustworthy.

With a mass gap, the theory is not infrared singular. While the mean field approach is still questionable, it is at least not clearly wrong. Before the Gutzwiller projection, the spin up electrons form a uniform fluid, and similarly the spin down electrons. The gap discourages excitation from the lower band to the higher band, thus causing the density to deviate from uniformity. Thus, the gap suppresses both charge and spin degree of freedom.

Since the fermions are massive, at energies low compared to m, we can integrate out the fermions to obtain an effective theory expressed solely in terms of the gauge potential a_μ. We have to calculate a simple one loop Feynman diagram (see Fig. (6)). The relevant integral is proportional to

$$\int d^3p \ \mathrm{tr} \left[\gamma_\mu \frac{1}{\not p + \not q - m} \gamma_\nu \frac{1}{\not p - m}\right] \tag{4.3}$$

We are only interested in the low energy long distance physics and thus we may expand in powers of momentum. Set q_λ to zero after differentiating with respect to q_λ to obtain

$$\int \frac{d^3p}{(p^2-m^2)^3} \ \mathrm{tr} \ \left[\gamma_\mu(\not p + m)\gamma_\lambda(\not p + m)\gamma_\nu(\not p + m)\right] \tag{4.3}$$

I will leave it to the reader to evaluate this integral in full. Let us simply look at one piece of the integral, say the piece coming from the term in the trace proportional to m^3, thus

$$m^3 \int \frac{d^3p}{(p^2-m^2)^3} \epsilon_{\mu\nu\lambda} \tag{4.4}$$

First, we see that the antisymmetric symbol appears: in (2+1)-dimensional spacetime $\mathrm{tr}\gamma_\mu\gamma_\nu\gamma_\lambda$ is proportional to $\epsilon_{\mu\nu\lambda}$. Thus, in the effective Lagrangian we obtain the Chern-Simons term $\epsilon^{\mu\nu\lambda} a_\mu \partial_\nu a_\lambda$. Incidentally, the appearance of the ϵ symbol indicates that T and P are manifestly violated. Second, by dimensional analysis, we see that the integral in (4.4) is up to a numerical constant equal to $1/m^3$. But be careful, the integral depends only on m^2 and doesn't know about the sign of m. The correct answer involves $1/|m|^3$, not $1/m^3$. Thus, the coefficient of the Chern-Simons term is equal to $m^3/|m|^3 = m/|m| =$ sign of m, up to a numerical constant.

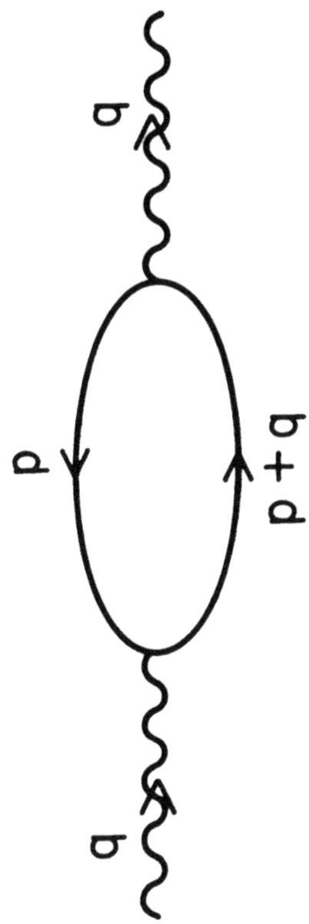

FIGURE 6

Now I have to tell you what the summation symbol in (3.26) and (4.1) means, if you don't already know. We actually have to sum over four species of Dirac fermions: there are two Dirac zeroes in the Brillouin zone and the up and down spin electrons separately leads to the Hamiltonian in (3.8). Note that particle physicists would call what the condensed types call spin isospin. Indeed, the Lagrangian in (4.1) exhibits a global $SU(4)$ symmetry, of which an $SU(2)$ subgroup describes spin (isospin).

At this point we should be worried. The coefficient of the Chern-Simons term is proportional to the sign of m and thus the contributions of the four different fermion fields to the Chern-Simons term may cancel. Indeed, recall that the mass term comes from a term like $\cos(k_x+k_y)$, which takes on opposite signs at the two Dirac zeroes $k_* = (\pi/2, \pi/2)$ and $(-\pi/2, \pi/2)$. It looks like we are not going to get a Chern-Simons term. But wait, we may still be okay. Recall that $\mathcal{H}$ contains a term like $(\cos k_x)\tau_1$. Thus, at one Dirac zero, τ_1 is used to construct the gamma matrices, while at the other, $(-\tau_1)$ is used. In other words, the representations of the gamma matrices at the two Dirac zeroes are conjugate of each other. The contributions of the four Dirac fields thus add and we obtain for the effective Lagrangian

$$\mathcal{L} = 4\left(\frac{1}{8\pi}\right)\varepsilon^{\mu\nu\lambda}a_\mu\partial_\nu a_\lambda \tag{4.5}$$

By now it is well known that particles coupled to a gauge potential with Chern-Simons dynamics acquire fractional statistics.[14,15] Specifically, with the Lagrangian

$$\mathcal{L} = 2\alpha\epsilon^{\mu\nu\lambda}a_\mu\partial_\nu a_\lambda + a_\mu j^\mu \tag{4.6}$$

a particle carrying unit charge $q = \int d^2x j_0 = 1$ has statistics defined by[16]

$$\theta = \frac{1}{8\alpha} \tag{4.7}$$

Thus, the induced Chern-Simons term in (4.5) will endow the excitations in the system with semion statistics ($\theta = \pi/2$). The word semion, half-Latin and half-Greek in origin, is perhaps particularly appropriate as the name for a particle whose statistics lies half way between fermi and bose statistics. Note that careful accounting practices, according to which we are required to sum over four fermion fields in (4.1), are crucial in obtaining semion statistics.

Ishikawa and his collaborators[17,18] have given an elegant representation of the Feynman diagrams contributing to the coefficient of the Chern-Simons term. (Their argument was actually given in the context of developing a rigorous theory of the integer quantum Hall effect. We will adapt and simplify their arguments for our rather limited purposes here.) The antisymmetric part of the Feynman integral in (4.3) may be written as

$$\int d^3p\ \varepsilon^{\mu\nu\lambda}\mathrm{tr}\left[\frac{\partial S^{-1}}{\partial p_\mu}S\frac{\partial S^{-1}}{\partial p_\nu}S\frac{\partial S^{-1}}{\partial p_\lambda}S\right] \tag{4.8}$$

where the inverse propagator

$$S^{-1}(p) = \not{p} + m \tag{4.9}$$

and

$$\frac{\partial S^{-1}}{\partial p_\mu} = \gamma_\mu \tag{4.10}$$

The mathematically well-schooled reader would immediately recognize this integral as some sort of homotopic invariant. Indeed, introducing the standard (Cartan-Maurer) form

$$dS^{-1}S \equiv \frac{\partial S^{-1}}{\partial p_\mu}S\,dp^\mu \tag{4.11}$$

we have the integral

$$\int \mathrm{tr}(dS^{-1}S)^3 \tag{4.12}$$

The Feynman integral in (4.3) is no garden-variety a-dime-a-dozen integral: it is a very special integral.

Before we conclude that we have a homotopic invariant, we have to deal with several mathematical details. In the context of the original problem, the integration region over

d^2p is a Brillouin zone and thus a torus in mathematics. The integration over p_0 is over the band. Thus, we are integrating over some sort of "pinched torus" rather than the three-sphere S^3. Ishikawa *et al.* had to invoke some mathematical theorem asserting that it is homotopically equivalent to integrate over S^3 (assuming we have already performed some sort of Euclidean continuation). Furthermore, in Euclidean space we recognize that

$$S^{-1}(p) = ip_a\tau_a + m \tag{4.13}$$

(where we sum over the three Pauli matrices) is up to an irrelevant overall factor an element of $SU(2)$ (since $(SS^\dagger)^{-1} = p^2 + m^2$). For further details and a more general treatment than given here, see.Ref. 18

I include this mathematical, and admittedly sketchy interlude in order to indicate how a logical gap in the original presentátion of Wen, Wilczek, and Zee may be filled. They had simply jumped from the lattice discussion to the continuum Feynman integral in (4.3) without further ado. Having identified this integral as a homotopic invariant, we can now see that the detailed shape of the band structure does not matter. (For instance, we may homotopically modify $S^{-1}(p)$ to have the form $ip_a\tau_a\alpha(p) + \beta(p)$. For reasonably smooth functions α and β the value of the integral would not change.)

The massive fermions we integrated over are known as spinons. To determine the electric charge of a spinon, we couple in the electromagnetic gauge potential by the usual substitution $c_i^\dagger c_j \to c_i^\dagger c_j e^{i\int_i^j d\vec{x}\cdot\vec{A}}$. We also have to remember to couple A_0 to the neutralizing positive background charge density ρ_B. Thus, the effective Lagrangian becomes

$$\mathcal{L} = \bar{\psi}(i\partial\!\!\!/ - m - (a\!\!\!/ - A\!\!\!/))\psi - A_0\rho_B \tag{4.14}$$

In the ground state, that is, the field theory vacuum, there is of course no net charge

$$\frac{\delta\mathcal{L}}{\delta A_0} = \rho_B - \langle 0|\bar{\psi}\gamma_0\psi|0\rangle = 0 \tag{4.15}$$

Now excite a spinon from the lower band to the upper band. Let the spinon and the anti-spinon move far apart from each other, just like a separating electron positron pair produced from the vacuum. The electromagnetic current associated with the spinon is given by

$$J_\mu = \frac{\delta S}{\delta A_\mu} = \frac{\delta S}{\delta a_\mu} = 0 \tag{4.16}$$

The first equality is just the definition of the electromagnetic current, the second equality comes from the fact that $\mathcal{L}$ depends on $(a-A)_\mu$, and the third equality represents the equation of motion for a_μ. We thus learn that the spinon does not carry any electromagnetic current. The spinon represents a spin excitation and does not carry any charge.

In this formalism, we can introduce holes as external test particles. It is easy to see that they have charge but no spin. We have thus obtained in an extraordinarily simple way the phenomenon of spin charge separatism as originally discussed by Anderson and collaborators[6] and by Kivelson, Rokshar, and Sethna.[19]

We should also check that the gauge field is massive. In our effective field theory treatment, we keep systematically only the leading long distance terms. Higher dimensional terms, less important at long distances, also appear of course. By gauge invariance, the next possible term in (4.5) is the Maxwell term $f_{\mu\nu}f^{\mu\nu}$ with some coupling g^2 determined by microscopic physics. With the Lagrangian

$$\mathcal{L} = \frac{k}{4\pi}\epsilon a\partial a - \frac{1}{2g^2}f^2 \tag{4.17}$$

we can easily compute the mass of the gauge field by counting powers of derivative or momentum. We denote the coefficient of the Chern-Simons term generically by k. The Chern-Simons term has one derivative while the Maxwell term has two. In momentum space, the inverse propagator thus has the form $k\epsilon q + \frac{1}{g^2}q^2$ where we do not bother with numerical factors and the Lorentz structure. The propagator thus has a pole at $q^2 \sim (g^2k)^2$.

Here we are faced with an apparent paradox (which turns out to have a rather instructive resolution). If the gauge field is massive and thus short ranged, how can it mediate fractional statistics, which is after all an infinitely long ranged phenomenon? The answer is that from the form of the inverse propagator given above it is clear that the propagator also has a pole at $q = 0$. To avoid clutter, let us scale out in (4.17) a factor of g^2, and define $m = g^2k/4\pi$. Thus, we will determine the propagator for the Lagrangian $\mathcal{L} = m\epsilon a\partial a - f^2/2$. We find, after some arithmetic, that the propagator is

$$\frac{1}{(q^2 - m^2)}\left[\left(-g_{\mu\nu} + \frac{q_\mu q_\nu}{q^2}\right) - im\frac{\epsilon_{\mu\nu\lambda}q_\lambda}{q^2}\right] + \gamma\frac{q_\mu q_\nu}{q^4} \tag{4.18}$$

Here γ is some gauge parameter and as usual, the terms proportional to $q_\mu q_\nu$ are irrelevant. We note that the term proportional to $g_{\mu\nu}$ has a pole only at $q^2 = m^2$. It is the pole at $q = 0$ that appears in the term proportional to $\epsilon_{\mu\nu\lambda}$ that propagates phase information (*i.e.*, fractional statistics) all the way to infinity.

This story is unsatisfactory in several respects. It is far from clear that the presence of a finite density of holes is really "equivalent" to frustration. What is hoped for is that in the phase diagram, with doping on one axis and frustration on the other, there may be a broad region belonging to the same universality class. See Fig. (7). Even if this is true, the chiral spin state may not be the true ground state of the frustrated Heisenberg antiferromagnet. From numerical work, there is some indirect evidence that the Heisenberg antiferromagnet, no matter how frustrated, may still have some kind of spin ordered ground state (although the order may be exceedingly complicated). Finally, the holes play two separate roles. They are supposed to generate the frustration, but once the frustration appears, we forget about the holes and deal with a spin model with no holes. After a gap opens in the spinon spectrum and a Chern-Simons term appears, the holes are then treated individually as externally introduced test particles. Fukuyama and his collaborators have given an improved treatment that addresses some of these points.[20]

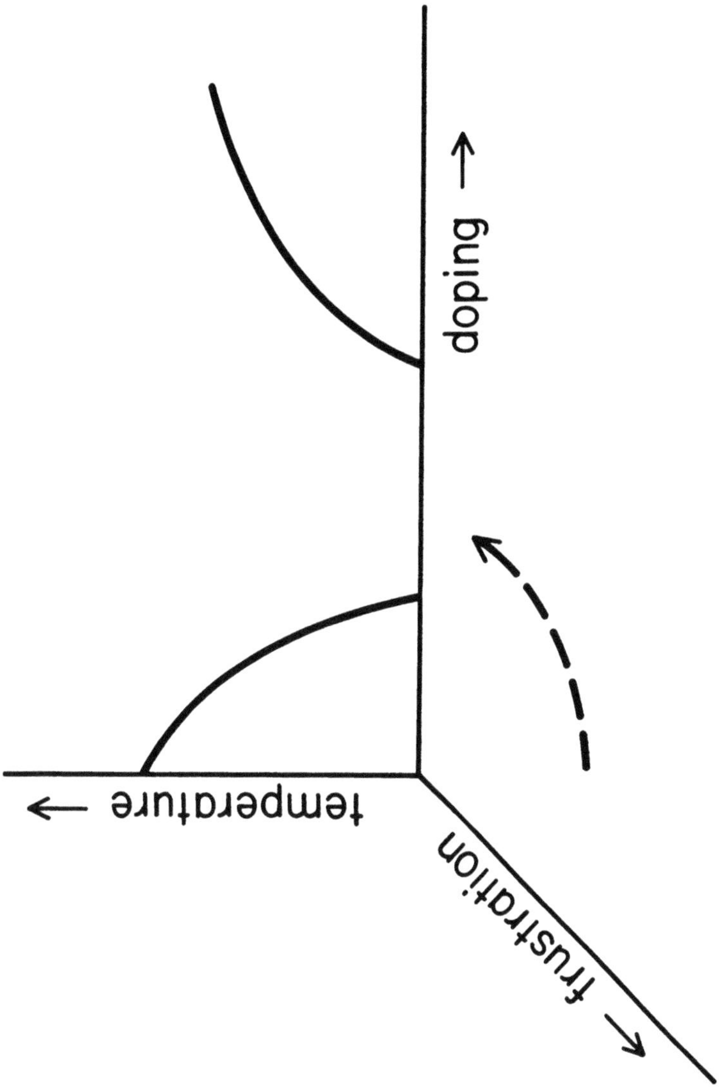

FIGURE 7

Let us mention an immediate generalization: we can consider the state with flux per plaquette equal to $\Phi = 2\pi(p/q)$. This may be achieved with $U_{iy} = 1$ and $U_{ix} = e^{i2\pi(p/q)i_y}$. (Our previous discussion in section III. corresponds to $q = 2$.) Going through the same steps as before, we now obtain a Hamiltonian of exactly the same form as the Hamiltonian in (3.9) but with the crucial exception that the vector w is now equal to $(0, 2\pi p/q)$, the momentum transferred by U_{ix}. We thus have a q-site hopping problem in momentum space. The corresponding $\mathcal{H}$ (see (3.14)) is now a q by q matrix, thus giving us q energy bands. As before, the crucial question is how many Dirac zeroes there are at $E = 0$. It turns out that there are q Dirac zeroes for q even, but none for q odd. Actually, this result is guaranteed by an index theorem, as was discussed by Kohmoto[21] and by Wen and Zee.[22]

The consequence of having q Dirac zeroes rather than 2 is easy to work out: the coefficient of the Chern-Simons term in (4.5) is scaled up by $q/2$ and thus the statistics angle of the excitations around this generalized flux state is scaled down by $q/2$ so that

$$\theta = (\pi/2)(2/q) = \pi/q \tag{4.19}$$

As we will see below, when we construct a theory of superconductivity based on the chiral spin state, the quantum of magnetic flux in the superconductor turns out to have the canonical value of $2\pi/2e$. (This is basically because when two semions pair they form a boson. See sections VII and XI below.) It takes, however, more than two of such $\theta = \pi/q$ excitations to form a boson. Consequently, the quantum of magnetic flux would be smaller than $2\pi/2e$! Experimentally, the quantum of magnetic flux in high temperature superconductors has been measured to have the canonical value of $2\pi/2e$, thus ruling out these $q > 2$ states. It would be fun if exotic superconductors with magnetic flux quantum less than $2\pi/2e$ are ever discovered.

There are a number of other generalizations that can be pursued. We merely mention them here. Non-abelian U_{ij}'s may be considered, leading to non-abelian flux states with many interesting properties.[23,24,25,26] The flux states discussed here on two dimensional lattices may be generalized to three dimension.[27,28,29,30] Also, staggered flux, in which the flux varies from plaquette to plaquette, may be considered, but the corresponding states are typically "not nice" from the field theoretic point of view. Finally, random flux states may also be considered.[31] Perhaps the most important generalization is the incommensurate flux state,[32] which we now discuss.

In the treatment of (Ref. 8), we fix the flux per plaquette to be π and fill half of the available states. In effect, we treat the density of holes as infinitesimal. Physically, it seems that the flux per plaquette would adjust itself to the presence of the holes.

Suppose we dope the system with holes until the filling factor ν decreases from $\frac{1}{2}$ to 47/100, for example. With a flux of π per plaquette, there are two bands as we have seen. The lower band is then not quite completely filled. On the other hand, if the flux per plaquette is correspondingly decreased to $(2\pi)47/100$, there are now 100 bands. With the filling factor $\nu = 47/100$ the lower 47 bands are filled. Energetically, it may thus be particularly favorable for the flux per plaquette $\phi/2\pi$ to be equal to ν. This is known as the commensurate flux phase.[32] Indeed, for a flux of $\phi = 2\pi p/q$, the gap between the pth band and the $(p+1)$th band is particularly large. (In Hofstadter's butterfly pattern,[33] this gap is part of the wing.)

Note that in this approach, there is no need to assume an additional diagonal correlation since T and P are already broken. Note also that as stated earlier the middle two bands touch at 100 places (!) in our example, but within the present context we don't care. We are only interested in the gap between 47th and the 48th bands.

A striking result is that in the commensurate flux phase the excitation is always a semion.[34]

Heuristically, this result may be understood by noting that the flux per particle in the commensurate flux phase is the same as the flux per particle in the π-flux chiral spin state. More rigorously, we can invoke a theorem from the theory of the Hall fluid. The contribution of each band to the Hall conductance is (in some suitable units) equal to some crazy integer, but when all these contributions are added up, the total turns to be equal to 2. The Hall conductance is directly related to the coefficient of the Chern-Simons term. The theorem that guarantees the Hall conductance to be equal to 2 says equivalently that the statistics of the excitation in the commensurate flux phase is always a semion. Another argument is that we can imagine continuing "quasi-adiabatically" from the chiral spin state, gradually changing the flux and the diagonal correlations to zero only at the end. Since the gap never closes, and since the coefficient of the Chern-Simons term is a topological invariant (see the discussion below in section XI.), we conclude that the excitations remain semionic. For more recent work on the incommensurate flux phase, see Zou, Lèvy, and Laughlin.[35]

In this context, let us also mention the $t-J$ model, defined by the Hamiltonian

$$H = \sum_{\langle ij \rangle} \left\{ -t(1-n_{j,-\alpha})c^{+}_{j\alpha}c_{i\alpha}\left(1-n_{i,-\alpha}\right) + J\vec{S}_i\vec{S}_j \right\} \tag{4.20}$$

This Hamiltonian describes an antiferromagnet doped with holes. The rather complicated looking first term simply tells us that for an electron of spin α (with $\alpha =$ up or down) to hop onto site j, the site j cannot be already occupied by an electron of spin $(-\alpha)$. Of course, Pauli already told us that a spin up electron cannot hop onto a site already occupied by another spin up electron. The constraint here is much more stringent and amounts to decreeing that there can never be two particles on any given site. This Hamiltonian, with $J \sim t^2/U$, may thus be derived from the Hubbard model in the limit $U/t \to \infty$. However, the $t-J$ model, with J not necessarily small compared to t, is a perfectly legitimate model in its own right. Indeed, some condensed matter physicists consider the $t-J$ model to be a more accurate representation of the real system than the Hubbard model.

Unfortunately, the $t-J$ model is no easier to solve than the Hubbard model. (Indeed, even for $J=0$, the $t-J$ model is non-trivial.[36]) The constraint

$$\sum_{\alpha} c_{i\alpha}^{+} c_{i\alpha} \leq 1, \tag{4.21}$$

that there can only be 0 or 1 electron on each site, is real nasty. In physics, inequality is much more difficult to deal with than equality. The slave boson formalism was invented precisely to replace the inequality in our constraint by an equality. We write

$$c_{i\alpha} = b_i^{\dagger} f_{i\alpha} \tag{4.22}$$

where $f_{i\alpha}$ is a fermion annihilation operator and $b_i^{\dagger}$ is a boson creation operator. The physical idea is that a vacant site (sometimes confusingly referred to in literature as a hole) may be thought of as occupied by a fictitious "slave" boson. We are told by (4.21) that when we annihilate an electron we also create a slave boson. The inequality constraint (4.20) is then traded in for an equality constraint

$$b_i^{\dagger} b_i + \sum_{\alpha} f_{i\alpha}^{\dagger} f_{i\alpha} = 1 \tag{4.23}$$

The boson is the fermion's slave.

The $t-J$ model may be treated in the mean field approximation more or less following the same line of logic that led us to (3.6) and subsequent equations. One assumes expectation value for $\left\langle b_i^{\dagger} b_j \right\rangle$ and $\left\langle \sum_{\alpha} f_{i\alpha}^{\dagger} f_{j\alpha} \right\rangle$. With flux order, we obtain a gauge theory with the gauge field now interacting with a fermion and a boson field.[20]

In the context of the $t-J$ model, it is the J term that favors flux order. The heuristic reason is just that when rearranged according to the Fierz identity (see (3.4)) the spin-spin interaction corresponds to "coherent" motion of the electrons: when an electron hops from site i to site j, another electron hops from site j to site i. One may imagine that it is this sort of coherent hopping that generates flux. On the other hand, the t-term alone (that is, without the all important constraint (4.21)) would tell the electrons to move in independent Bloch waves, certainly without any flux order. Phenomenologically, it appears that $t \sim 0.45$ ev and $J \sim 0.13$ ev and so the t-term may dominate the J-term, in which case flux order may not be favored.

V. NON-LINEAR σ-MODEL

An important first step in elucidating the nature of a condensed matter system is the determination of the quantum numbers of the elementary excitations or quasiparticles. A long distance effective field theory approach is particularly suitable since the most essential quantum numbers, such as charge and fractional statistics, can be deduced by studying the physics at long distances. In this respect, field theory has an important role to play in condensed matter physics. Also, a field theory treatment often accommodates analytic calculation more readily than a lattice model.

I believe that the field theory treatment to be given below captures the essence of the long distance physics we are interested in. By its very nature, a field theory discussion misses the details of the short distance physics which is simply absorbed into a cut-off. Of course, even in a lattice model, the interaction on the lattice scale already represents an approximation, often rather drastic, of the true underlying physics.

Using the coherent state formalism and evaluating the path integral representation of the partition function carefully,[37] one can show that Heisenberg spin Hamiltonians may be represented by the non-linear σ-model:[38,39]

$$\mathcal{L} = \frac{2}{g^2} \left(\partial_\mu \vec{n}\right) \left(\partial^\mu \vec{n}\right) \tag{5.1}$$

with the constraint

$$\vec{n}^2(x) = 1 \tag{5.2}$$

and the spacetime coordinates $x^\mu, \mu = 0,1,2$. The order parameter $\vec{n}$ represents an expectation value of the staggered spin operator $(-)^i \vec{S}_i$. In proceeding to this effective field theory description, one has implicitly assumed that the spin-spin correlation is larger than the lattice spacing. In this case, over the length scale of a few lattice spacing there is antiferromagnetic ordering and that an effective field or local spin order parameter $\vec{n}(x)$ can be defined. This implicit assumption appears as an assertion about smoothness in the path integral formalism.

Actually, the representation of the long distance physics of the Heisenberg spin Hamiltonians by the nonlinear σ-model is quite general. As soon as we recognize that the spin rotation symmetry is spontaneously broken by the ground state, it follows that there is a Nambu-Goldstone mode, namely the spin wave. On general grounds, we know that for small $\vec{k}$ the spin wave satisfies the linear dispersion relation $\omega^2 \simeq \vec{k}^2$ in an antiferromagnet and the quadratic dispersion relation $\omega \simeq \vec{k}^2$ in a ferromagnet. (This is consistent with the time reversal transformation T. Ferromagnetic ordering violates T and hence the spin wave dispersion should not be invariant under $\omega \to -\omega$. On the other hand, antiferromagnetic ordering is invariant under T followed by a translation through one lattice spacing.) Thus, with a suitable choice of units to set the spin wave velocity to unity, we may describe the long wavelength physics of an antiferromagnetic system by a "relativistic" field theory.

The non-linear σ model in (2+1)-dimensions was studied in the present context by Wilczek and Zee.[15] A crucial feature of this model follows from the existence of a topologically non-trivial map from $S^3 \to S^2$, the Hopf map. The existence of the Hopf map appears somewhat mysterious to a "practical" physicist. (Indeed, he may even conjecture that since the maps $S^1 \to S^0$ and $S^2 \to S^1$ are topologically trivial, all maps of higher dimensional spheres into lower dimensional spheres are trivial.) In fact, the fundamental Hopf map is nothing more mysterious than Pauli spin. Let

$$\vec{n} = z^\dagger \vec{\sigma} z, \tag{5.3}$$

with

$$z = \begin{pmatrix} z_1 \\ z_2 \end{pmatrix} \tag{5.4}$$

a doublet of two complex numbers. The constraint $z^\dagger z = 1$ implies $\vec{n}^2 = 1$. With this constraint the four real components contained in z define the sphere S^3, while the three components contained in $\vec{n}$ define S^2. The relation in (5.3) then defines a map from $S^3 \to S^2$.

With the boundary condition $\vec{n}(x) \to n_0 =$ some fixed unit vector as $x \to \infty$, Euclidean spacetime is compactified into S^3. The Hopf map may be defined in general as $n(z(x))$ where $z(x)$ defines a map of spacetime S^3 onto the S^3 on which z lives. The map is characterized by a topological degree, namely the number of times the S^3 of spacetime "wraps around" the target S^3. This map is then followed by a map from the S^3 on which z lives to the S^2 on which n lives, as defined in (5.3).

The existence of the Hopf map implies that the action for the system may be augmented by a topological term

$$S = \int d^3x \frac{2}{g^2}(\partial_\mu \vec{n})^2 + 2\pi n \mathcal{H} \tag{5.5}$$

where $\mathcal{H}$ is the topological invariant associated with the map. $\mathcal{H}$ is equal to the integer k characterizing the topological degree of the map measuring how many times the S^3 of Euclidean spacetime is "wrapped around" S^2. (For the lowest non-trivial map, $\mathcal{H} = 1$.) The invariant $\mathcal{H}$ may be written as a double integral over an integrand which is a non-local function of n. When written in terms of z, it is a single integral over a local function:[40]

$$\mathcal{H} = \frac{1}{4\pi^2} \int d^3x \epsilon^{\mu\nu\lambda} (z^\dagger \partial_\mu z)(\partial_\nu z^\dagger \partial_\lambda z) \tag{5.6}$$

(You may recognize the integrand in (5.6) simply as an infinitesimal "area" element on S^3 by evaluating the integrand around the "north pole" $z \simeq \begin{pmatrix} 1 + iw_1 \\ w_2 + iw_3 \end{pmatrix}$ for w_i infinitesimal. Thus the integral suitably normalized defines the degree of the map $z(x)$. By construction, the degree of the map $n(z(x))$ is clearly equal to the degree of the map $z(x)$.)

In the action (5.5), the coefficient of $\mathcal{H}$ is quantized to be $2\pi n$, with n an integer (obviously not to be confused with the field $\vec{n}(x)$). The path integral is modified by a factor $e^{i2\pi nk} = 1$, that is, not modified at all. The reason for requiring the coefficient of $\mathcal{H}$ to be quantized is that $\mathcal{H}$ involves the ϵ symbol and thus would violate T and P unless its contribution to the path integral is nil.

We are now ready to discuss the physics described by this action. Our treatment[41] will be non-rigorous and based on the mean-field approximation.

Simple physical considerations suggest that there may be a disordered or spin liquid phase. A fluctuation costs an amount in action of order $(1/g^2)$. Thus, at large coupling fluctuations should be so inexpensive that they occur freely and the system becomes disordered. At small coupling, the system is in an ordered phase with $\langle\vec{n}(x)\rangle = \vec{n}_0$. Note the standard Heisenberg antiferromagnet has only a parameter J setting the overall energy scale at zero temperature. The coupling g^2 measures the tendency to disorder as given by the frustration introduced by non-nearest neighbor couplings.

The philosophy here is to start from the Néel ordered phase and to move towards the order-disorder transition. Because of the assumption of a spin correlation length over which $\vec{n}(x)$ can be defined, the field theory description may not be valid far into the disordered phase. In other words, we start with a theory with a Nambu-Goldstone boson and we show that it disappears as a coupling increases past some critical value.

The action in (5.5) without the Hopf term was treated in the late 70's by various field theorists.[42] Here we essentially follow their treatment but focus on the effect of the Hopf term.

Using (5.3) we may rewrite the Lagrangian corresponding to (5.5) as

$$\mathcal{L} = \frac{2}{g^2}\left[\partial_\mu z^\dagger \partial^\mu z + z^\dagger \partial_\mu z z^\dagger \partial^\mu z\right] + \frac{n}{2\pi}\epsilon^{\mu\nu\lambda} z^\dagger \partial_\mu z \partial_\nu z^\dagger \partial_\lambda z \tag{5.7}$$

We have apparently traded in two degrees of freedom (three components of $\vec{n}$ minus one constraint) for three degrees of freedom (four components of z minus one constraint). In fact, the z-field represents only two degrees of freedom since an overall phase transformation

$$z(x) \rightarrow e^{i\alpha(x)} z(x) \tag{5.8}$$

does not change $\vec{n}(x)$ and hence the Lagrangian. It is easy to verify that the Lagrangian in (5.7) is invariant under this local transformation, as it must. This gauge invariance can be explicitly displayed by rewriting the Lagrangian as

$$\mathcal{L} = \frac{2}{g^2} |(\partial_\mu - ia_\mu) z|^2 + \frac{n}{4\pi} \epsilon^{\mu\nu\lambda} a_\mu \partial_\nu a_\lambda + \frac{g^2 n^2}{(4\pi)^2} f_{\mu\nu}^2 + \ldots . \tag{5.9}$$

Since the Lagrangian depends on a_μ up to quadratic order only, we can eliminate a_μ and verify that the Lagrangian in (5.7) is recovered. Under the transformation (5.8) we have

$$a_\mu \rightarrow a_\mu + \partial_\mu \alpha. \tag{5.10}$$

The transformation from $\vec{n}$ to z in (5.3) apparently makes it easier to see the physics involved. In a sense, z is like a square root of $\vec{n}$. The introduction of z allows us to probe the possibility that a fraction of the quantum numbers carried by $\vec{n}$ is exposed.

At this point, the constraint $z^\dagger z = 1$ is incorporated by adding to the Lagrangian a term $-\lambda(z^\dagger z - 1)$ where λ is a Lagrange multiplier field. It can be shown[42] that in the mean field approximation the expectation value $\langle \lambda \rangle \neq 0$ for g^2 large enough. We replace the field $\lambda(x)$ in the Lagrangian by a constant λ equal to the expectation value of λ. Thus, the effective theory of the disordered phase is given by

$$\mathcal{L}_{\text{eff}} = \frac{2}{g^2} |(\partial_\mu - ia_\mu) z|^2 - \lambda z^\dagger z - \frac{n}{2\pi} a_\mu \partial_\nu a_\lambda \epsilon^{\mu\nu\lambda} - \frac{1}{\gamma} f_{\mu\nu}^2 \tag{5.11}$$

where z is an effective field no longer subject to a constraint on its magnitude. The precise value of γ does not concern us here. The z quantum acquires a mass and thus the spin-spin correlation function becomes short ranged. We have in effect traded in the constraint for a mass.

To summarize, we find that in the disordered phase the field theory may have many possible vacua labelled by an integer n. Note that in the ordered phase the integer n is hidden. For $n = 0$ the gauge field is massless and mediates a logarithmic confining interaction between z and z^*.

For $n \neq 0$, T and P are violated due to the Chern-Simons term. The gauge field becomes massive and thus the z and $z^\dagger$ quanta are de-confined. They are the spinons and anti-spinons of our earlier discussion. Their statistics may be read off from the coefficient of the Chern-Simons term (see (4.7)): they carry fractional statistics with $\theta = \pi/2n$. For $n = 1$, the spinon is a semion.

This field theoretic discussion also gives a particularly nice and clean demonstration of spin-charge separation.[6,19] The z-quantum carries spin (isospin) but not charge.

The treatment given here for the half-filled antiferromagnet may be extended to include holes.[43] To the effective Lagrangian in (5.11) we add the terms

$$\chi^\dagger(i\partial_0 + a_0\sigma_3)\chi + \frac{1}{2M}\chi^\dagger(i\partial_i + a_i\sigma_3)^2\chi \tag{5.12}$$

where χ is a non-relativistic fermionic field describing the dressed holes. χ carries charge but not spin and may be thought of as a hole bound with a z quantum. Because of its gauge coupling to a_μ, it acquires a fractional statistics given by $\theta = \pi + \frac{\pi}{2n}$.

How trustworthy is the mean field approximation? Wen and I[44] have attempted to obtain some feel for this question by studying field theories similar to the ones discussed here in (1+1)-dimensional spacetime non-Abelian bosonization and other field theoretic techniques are then available to solve some of these theories more or less exactly, or at least without invoking the mean field approximation. Our general impression from these studies is that the results of the mean field approximation appears to be reliable as far as quantum numbers are concerned.

VI. DUALITY

As is well known, the nonlinear σ-model contains solitons or skyrmions. In particular, it was shown in Ref. 15 that these skyrmions can be quantized in such a way that they have fractional statistics.

In Ref. 45 we showed that the statistics of the skyrmion is related to the statistics of the z quantum by the reciprocal relation

$$\frac{\theta_{\text{skyrmion}}}{\pi} = \frac{\pi}{\theta_z}. \tag{6.1}$$

This result may also be obtained by direct computation.

With this duality relation, it is easy to understand the value of the fractional statistics derived in the approaches mentioned in the last two sections. The fractional statistics $\theta_z = \frac{\pi}{2n}$ carried by the z-quantum just corresponds to the soliton in the dual picture being quantized with the statistics of the boson. (According to (6.1), the value $\theta_z = \frac{\pi}{2n}$ corresponds to $\theta_{\text{skyrmion}} = 2\pi n$.)

This duality relation indicates that fractional statistics occurs generically, either in the particle or in the soliton sector. The inverse duality relation discussed here is mathematically reminiscent of the duality between the electric charge of the electron and the magnetic charge of the magnetic monopole.

I believe that the duality derivation for $\theta = \pi/2n$ given here may be more general than the derivations mentioned earlier and which all rely on the mean field approximation. The duality structure may be made rigorous by formulating it on a lattice[46,47,48,49] and appears to have a deep mathematical origin.

In the preceeding section we see that the nonlinear σ-model exhibits an ordered phase at small coupling (in which $\langle z \rangle \neq 0$) and a disordered phase at large coupling (in which $\langle z \rangle = 0$). Wen and I have suggested in Ref. 45 a duality picture in which we claimed that the disordered phase corresponds to a state in which skyrmions condense into the vacuum. This is rather plausible since around each skyrmion the z-field is twisting around. If the quantum ground state consists of a condensate of skyrmions, the z-field should be totally disordered.

We imagine then that we start with the ordered phase $\langle z \rangle \neq 0$ and construct the skyrmion living in this phase. In the field theory literature, an effective quantum operator creating a soliton can be constructed. In the limit in which the skyrmion size goes to zero, or equivalently if we are only interested in long distance physics, the skyrmion may be described by some effective complex scalar field φ. The field has to be complex because the skyrmion carries a conserved change. At long distances, the skyrmions are described by an effective Lagrangian with the terms

$$\mathcal{L}_{\text{eff}} = \ldots - M^2(\varphi^\dagger\varphi) + \lambda(\varphi^\dagger\varphi)^2 + \ldots \tag{6.2}$$

where the mass squared of the vortex M^2 is some function of g^2 that can be calculated perturbatively. We conjecture that when g^2 exceed g_c^2, the parameter M^2 in (6.2) becomes negative and the vacuum develops an instability towards skyrmion condensation.

What about the ordered phase? If we are correct in describing the disordered phase as one in which a complex scalar field φ condenses and acquires a non-zero vacuum expectation value, then in this phase we can construct a vortex out of the field φ. According to the duality relation (6.1), the statistics of this vortex is equal to precisely that of the z-field. We believe that this vortex constructed out of the skyrmion field φ represents precisely the original z field.

In the ordered phase, z has a non-zero vacuum expectation value by definition, and the gauge potential a_μ acquires a mass according to the usual Higgs mechanism. On the other hand, in the disordered phase, z has zero vacuum expectation value and a_μ remains massless. In contrast, the effective skyrmion field φ has a non-zero vacuum expectation value, and it is A_μ that becomes massive. We summarize the situation in a table.

	order	disorder
a_μ	massive	massless
A_μ	massless	massive

Using this dual language, we have a very simple picture for the T and P broken states discussed earlier. The solitons or skyrmions may condense into the many possible Laughlin states described by the wave function of the solitons

$$\Phi(w_1, w_2, \ldots) \propto \prod_{i,j}(w_i - w_j)^{2n} e^{-\sum_k |w_k|^2/\xi^2} \tag{6.3}$$

where $w_j = x_j + iy_j$ is the coordinates of the jth skyrmion. When we exchange w_1 and w_2, say, we obtain a phase $(-1)^{2n}$. Here n is an integer when the skyrmion is boson as in the model. This is also consistent with the analysis of Ref. 41 where the fractional statistics of the skyrmion was determined to be characterized by the phase $e^{i8\pi^2(n/4\pi)} = e^{i2\pi n} = (-1)^{2n}$. The Laughlin state (6.3) obviously break T and P and may correspond to the quantum disordered states discussed here.

But unfortunately the duality picture described here is not complete and the above argument is only heuristic. It would be very interesting to find a complete duality picture for the CP^1 (or CP^{N-1}) model. Perhaps by going to a lattice formulation we can formulate this duality picture more rigorously. We will come back to the duality picture later.

VII. THE ANYON SUPERFLUID

Now we shift gears and forget about physics on the lattice, the spinon (which is a high energy excitation), and all that stuff. We proceed to study a fluid of holons obeying semion statistics. Our goal is to show that such a fluid is a superfluid. We will treat the semions as electrically neutral, and turn on its coupling to the electromagnetic potential later.

The semion superfluid is a beautifully clean problem in theoretical physics. There is no coupling parameter. We are given a collection of particles such that when two of them are interchanged the wave function gets a phase of i. It may well be that there is an exact solution.

Begin with a naive dimensional analysis. The problem is characterized by the mass m and the density n of the semions. Thus, the transition temperature, which has dimension of energy, must be equal to

$$T_c = (\text{pure number})\hbar^2 n/m \tag{7.1}$$

If put in a value for n corresponding to 10/oxide, we obtain something like $T_c \sim$ (pure number) $(me/m) \times 200^\circ\ K$, where m is the unknown mass of the holon. The fact that this numer is in the right neighborhood may be taken to be encouraging.

This analysis is naive, of course. The fact that semions are of finite size and other solid state effects would presumably change the phase boundary to the dotted line shown in Fig. (8). Compare with Fig. (2).

We will restrict ourselves here to the zero temperature problem. What makes the problem difficult is that unlike the case for fermion or for boson, the N-body wave function cannot be written in terms of the 1-body wave function. We have a strongly correlated quantum many-body problem.

While the exact solution to this problem is unknown, the basic physics is rather easy to understand. Indeed, I believe that it is quite well established by now that the semion fluid is a superfluid.

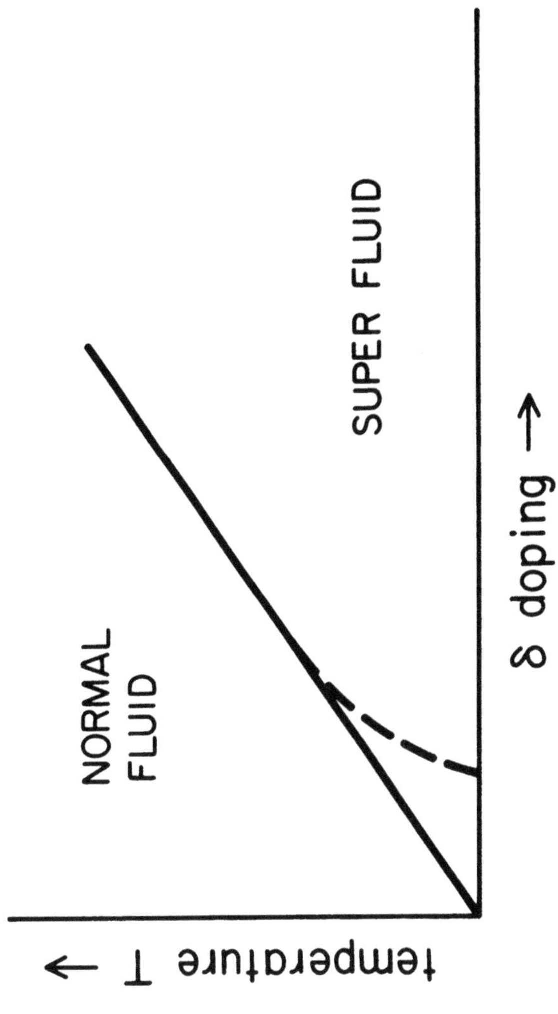

FIGURE 8

Progress has been made starting with a mean field approximation suggested by Arovas, Schrieffer, Wilczek, and Zee[50]. Consider moving an anyon through a big loop enclosing an area A. Since in the mean, there are nA anyons in the loop, the wave function picks up the phase $e^{i(\theta/\pi)(2\pi)nA}$. But we know that a charged particle moving in a magnetic field b picks up a phase e^{ibA}. Thus, we can consider an approximation: the particles act as if they are moving in a mean magnetic field given by

$$b = 2n\theta \tag{7.2}$$

and generated by all the other particles.

In the quantum Hall effect we have electrons moving in a plane in the presence of an external (real) magnetic field. Thus, in the mean field approximation the semion fluid is closely related to the Hall fluid.

Let us then give a lightning review of the quantum Hall effect. The single body energy states of a charged particle moving in a magnetic field B are grouped into Landau levels at energies

$$E_n = \left(n + \frac{1}{2}\right) B/m, \qquad n = 0, 1, 2, \ldots. \tag{7.3}$$

Each Landau level has degeneracy $BA/2\pi$ where A is the area of the system. (The physical origin of this vast degeneracy is clear. Classically, the particles move in Larmor orbits with radius $\sim 1/\sqrt{B}$. Because of the translation invariance these orbits may be placed anywhere and thus we would expect a degeneracy proportional to A.)

The filling factor is defined as the ratio of the number of particles to the number of states in a Landau level

$$\nu = \frac{N}{(BA/2\pi)} = \frac{2\pi n}{B} \tag{7.4}$$

where n is the density of particles. Note that

$$\nu^{-1} = \frac{(BA/N)}{2\pi} \tag{7.5}$$

is equal to the flux per particle measured in fundamental units of flux. (In our choice of units with $\hbar, c$, and e set equal to 1, the fundamental unit of flux is $hc/e = 2\pi$.)

The integer Hall effect states that for ν an integer we have an incompressible quantum fluid. This is relatively easy to understand since the number of electrons is first so as to fill ν Landau levels exactly. There is a big gap in the excitation spectrum. (The actual theory is more complicated;[18] one needs to argue that the presence of impurities does not vitiate this simple argument, *etc.*) We see that in the mean field approximation a fractional statistics liquid with statistics paramater θ is related to a Hall fluid with filling factor

$$\nu = \frac{\pi}{\theta}. \tag{7.6}$$

Here we have simply substituted for B in (7.4) the mean field b given in (7.2). We are now ready to turn to a fluid of semions. The first remark is notice that two semions do not make a fermion. You may think that semions are half way between bosons and fermions and so if I put two semions together they might make a fermion. But in fact, suppose we move a bound state of two semions half way around another bound state of two semions, thus interchanging the two bound states. Then we get a phase. This semion gets a phase of i going around that semion but he also gets a phase from going around the other semion. Now we must do a mathematical calculation: $2 \times 2 = 4$. We get the phase four times. Each time a semion goes around another semion we get a phase of i. Since we get this phase 4 times we get i^4, which is equal to 1. Thus, two semions actually make a boson.

We may now give some handwaving arguments. We all know that fermions like to stay apart and bosons like to stick together. Semions, being half way between are more likely than fermions to pair. When they do pair, they form bosons, whose condensation can then lead to superfluidity and superconductivity. Just think, if you were a fermion and you want to condense into the ground state, what could you do? You have basically two strategies: (1) Find another fermion and pair with him or her, or (2) turn yourself into a boson. Strategy (2) is an attractive possibility, but in the systems under discussion you can apparently make it only half way and turn yourself into a semion.

Much of the physics of anyons is contained in the following deep mathematical identity $\theta = 0 + \theta = \pi - (\pi - \theta)$. What is the mathematics trying to tell us? The first equality says that an anyon with statistics θ is effectively a boson (statistics 0) with a gauge interaction of "strength" θ; the second equality says that it is also a fermion (statistics π) with a gauge interaction of "strength" $-(\pi - \theta)$. The physics is made completely clear by looking at the 2-anyon problem. In the center of mass, the Schrödinger equation has a centrifugal potential like $(\ell + \theta/\pi)^2/r^2$ with ℓ an integer. The smallest possible value of this potential is thus $0/r^2$ for bosons, $1/r^2$ for fermions, and $1/4r^2$ for semions. Anyons are like bosons with "centrifugal repulsion" between them or fermions with a "centripetal attraction". Thus, theorists have a choice. They can begin either with a fermi gas[51,52,53] — certainly not a superfluid — and show that the attraction gives pairing and superfluidity, or with a boson gas,[54,55,56] — which with a short ranged repulsion is a superfluid as was shown by Bogoliubov[1] ages ago — and show that the repulsion does not destroy superfluidity.

The literature on the subject bifurcates according to the choice made. Laughlin[57] used the mean field approximation to be described below to make a first attempt in understanding the statistical mechanics of the anyon fluid, treating the particles as fermions. Later, Fetter, Hanna and Laughlin[51] did a random phase calculation in which they went beyond the mean field approximation. They concluded that the excitation is gapless and that the system behaves as a superfluid.

The calculation of Fetter, *et al.* is technically involved and the physical origin of the gapless excitation is far from clear. Wen and I were motivated[54] by a desire to understand the appearance of this gapless excitation in more physical terms. We chose to treat the particles as bosons as we felt that it would be valuable to look at the underlying physics from a different starting point than that of Fetter, *etal.*

The work of Fetter, *et al.* was later clarified and extended by Chen, Wilczek, Witten and Halperin.[53] They were able to prove a number of rigorous statements for ideal anyons and to provide some additional insights on the origin of superfluidity. Related work was also carried out by Hosotani and collaborators.[52]

Of course, if our treatments are exact then it wouldn't matter whether we start from the bose end or the fermi end. But after various approximations, it would certainly make a difference. Consider for instance the energy of the ground state. Treating the anyons as bosons with a gauge repulsion between them, we can, in the mean field approximation, put all the particles into the lowest Landau level. Using (7.2) and (7.3), we find that the energy density is given by

$$\epsilon = n(b/2m) = \theta n^2/m. \tag{7.7}$$

This represents an energy due to quantum statistics. The phase acquired by the wave function upon the exchange of two anyons implies that the anyon system is frustrated and the anyons have non-zero kinetic energies even in the ground state. For comparison, the exchange energy of a free fermion gas in two dimensional space is $\epsilon = \frac{\hbar^2 \pi n^2}{m}$ which coincidentally happens to be reproduced by (7.6) for $\theta = \pi$.

In contrast, if we had treated the semions as fermions, the energy would be exactly twice as high. We can see this easily by noting that for $\theta = \pi/2$, $\nu = 2$ according to (7.6). Since the energies of the Landau levels go as $\frac{1}{2}\hbar\omega_c, \frac{3}{2}\hbar\omega_c, \frac{5}{2}\hbar\omega_c$, and so on, we have $\left(\frac{3}{2} + \frac{1}{2}\right) = 2\left(\frac{1}{2}\right)$. The true energy density of a gas of semions is presumably $\xi\left(\frac{\hbar^2 \pi n^2}{2m}\right)$ with ξ between 1 and 2.

The hallmark of the Hall fluid is its incompressibility. The basic physics of this striking phenomenon may be understood heuristically. The electrons are forced by the external fixed magnetic field to move in circles. Even though the electrons are point particles, they have to occupy a minimal and finite amount of room. Because of fermi statistics, the circles can't "overlap." The combination of fermi statistics and the magnetic field makes the system incompressible. Superficially, the semion fluid and the Hall fluid sound similar. We have particles and magnetic fields (real in one case, fake in the other). What's the difference?

The crucial difference is that in one case the magnetic field is external and nailed down by the experimentalists, while in the other, particles carry their own private (statistical) magnetic field around with them in bags. This makes all the difference in the world. (It is sort of like the difference between a fixed political system imposed on the people and a dynamical political system generated by the people.) There is at least no *a priori* reason for the semion fluid to be incompressible and in fact it isn't.

The naive mean field approximation is too drastic. We have to go to an extended or dynamical mean field approximation in which the mean field is allowed to do its thing. The mean field should respond to the particles and not be nailed down. We can summarize this in a nutshell by saying that the same mathematical equation can express different physics according to which quantity is being fixed. In the Hall fluid, the equation $B = 2n(\pi/\nu)$ is to be read with B held fixed. The same equation (with the change of notation $B \to b$ and with π/ν equated to θ by (7.4)), $b = 2n\theta$, describes the anyon fluid but only if we insist on θ, but not b, as being fixed.

The essential physics behind the compressibility of the semion fluid can be captured, I believe, by a hydrodynamic approximation[54] treating the semions as bosons with gauge repulsion between them. To say a fluid is compressible is to say that it has gapless excitations, that is, excitations that cost infinitesimally little energy. Consider a long wavelength density fluctuation. Where the density is higher, the effective (fake) magnetic field is also higher according to $b = 2n\theta$, and so the (Landau or Larmor) circles the particles move in are correspondingly smaller, just so that they don't bump into each other any more than in the low density regions. As the wavelength becomes longer and longer, the energy of the density wave becomes smaller and smaller. In the gauge theory language, the massless mode corresponds to the gauge boson.

Another way of approaching the physics is to calculate the effective energy density. Since in a magnetic field the Landau or Larmor energy is $\frac{1}{2}\hbar\omega_c \sim b$, the energy density is $\sim nb \propto n^2$. The quadratic dependence of the energy density on the number density is as if there is a short ranged repulsion between the bosons. We may then invoke Bogoliubov's work[1] to say that we have a superfluid.

Let us thus consider a hydrodynamic treatment. Including the kinetic energy we write down the energy functional

$$E = \int d^2x \left\{ \frac{1}{2}mnv^2 + \gamma(n-n_0)^2 - \gamma n_0^2 \right\} \tag{7.8}$$

with $\gamma = \hbar^2\theta/m$. We have added a chemical potential term to fix the total number of particles in the system.

The pressure is easily calculated

$$P = -\frac{\partial}{\partial A}\gamma A n^2 = \gamma n^2 \tag{7.9}$$

The hydrodynamic equation is thus

$$\frac{\partial \vec{v}}{\partial t} + (\vec{v}\cdot\vec{\partial})\vec{v} = -\frac{1}{mn}\vec{\partial}P = -\frac{2\gamma}{m}\vec{\partial}n \tag{7.10}$$

Linearizing (7.10) around the ground state with uniform density n_0 and the equation of continuity

$$\frac{\partial n}{\partial t} + \vec{\partial}(n\vec{v}) = 0 \tag{7.11}$$

we find the wave equation

$$\frac{\partial^2 n}{\partial t^2} + \frac{2\gamma n_0}{m}\nabla^2 n = 0 \tag{7.12}$$

so that the speed of sound is given by

$$c_s^2 = 2\gamma n_0/m = 2\hbar^2\theta\, n_0/m^2 \tag{7.13}$$

Thus, the low lying elementary excitation of the system is expected to be a gapless phonon. The form of this result could have been determined by dimensional analysis. The θ dependence is consistent with the fact that $c_s = 0$ for an ideal Bose gas. It was crucial here that the potential energy is proportional to n^2 rather than to n, (in which case c_s would clearly vanish). This is a direct consequence of the "magnetic" interaction between the bosons, or in other words, the exchange pressure of the fractional statistics as discussed above.

Subsequently, Mori[55] using a random phase approximation, and Trugenberger[56] using a formalism in which the Hamiltonian is written in terms of density and current operators, have also calculated c_s. While these calculations are substantially more involved than the simple hydrodynamics calculation, their results for c_s agree with the hydrodynamical result (7.13), as they must, since the hydrodynamic approach incorporates general physical priniples such as Galilean invariance. In the literature, the hydrodynamics calculation presented above is sometimes regarded to as "too simple" and hence "inadequate". I would like to suggest the contrary: a simple calculation based on general physical principles is perfectly adequate in its domain of validity.

The speed of sound c_s calculated by treating the semions as bosons with a gauge repulsion is smaller by a factor of $\sqrt{2}$ from the speed of sound calculated by treating the semions as fermions with a gauge attraction. I believe that this merely reflects the factor of 2 in the ground state energy, as explained above. (In other words, the constant γ in (7.8) would be scaled up by a factor of 2.)

VIII. EFFECTIVE LAGRANGIANS FOR THE SEMION SUPERFLUID

We can also formulate our discussion using the language of effective field theory. One particular simple approach involves starting with the equation of motion derived from (4.6):

$$4\alpha\epsilon^{\mu\nu\lambda}\partial_\nu a_\lambda = j^\mu \tag{8.1}$$

The time component of this equation relates the "magnetic" field to the density

$$2\alpha\epsilon_{ij}f_{ij} = n \tag{8.2}$$

and the space components realtes the "electric" field to the current

$$4\alpha\epsilon_{ij}f_{oj} = nv_i \tag{8.3}$$

Incidentally, we recognize that (8.2), when applies to a situation in which the density n is a constraint, forst reproduces the mean field realtion (7.2). Plugging (8.2) and (8.3) for n and $\vec{v}$ into the Lagrangian corresponding to (7.8), we find almost immediately (that is, after droping some irrelevant higher dimensional term and after choosing length and time units so that the speed of the phonon is equal to 1) we obtain

$$\mathcal{L} = -\frac{1}{16\pi g^2}f_{\mu\nu}^2 \tag{8.4}$$

with $g^2 = \theta m/2\pi\hbar^2$.

The long distance effective theory is just the Maxwell Lagrangian. The phonon is represented by the "photon" of this Lagrangian! In a propagating electromagnetic wave of wave number $\vec{k}$ the electric field is oscillating perpendicular to the direction of propagation. According to (8.3) the particles are oscillating back and forth along $\vec{k}$. Notice that what we are describing here is the "guiding-center motion" of the particles. The much faster (at the cyclotron frequency ω_c) cyclotron motion around the guiding center has been averaged out, so to speak.

As Wen and I emphasized in Ref. 54, it is crucial that the Chern-Simons term $\epsilon_{\mu\nu\lambda}a^\mu f^{\nu\lambda}$ has disappeared.[58] If not, a gauge invariant mass for a_μ would be generated and we would obtain an incompressible fluid rather than a superfluid.

In effect, what we have done here is to integrate out the "matter fields", that is the particles, and represent the low frequency long wavelength dynamics solely in terms of the gauge fields. This may be described more formally as follows. The Lagrangian for a system of non-interacting anyons by

$$\mathcal{L} = \mathcal{L}_m - \alpha\epsilon_{\mu\nu\lambda}a^{\mu}f^{\nu\lambda} - \frac{1}{4g^2}f_{\mu\nu}^2 \tag{8.5}$$

with the "matter Lagrangian"

$$\mathcal{L}_m = \psi^{\dagger}i(\partial_0 + ia_0)\psi + \frac{1}{2m}\psi^{\dagger}(\partial_i + ia_i)^2\psi. \tag{8.6}$$

Integrating out the matter fields ψ we obtain the effective action

$$S = S_m(a) - \int d^3x\left[\alpha\epsilon_{\mu\nu\lambda}a^{\mu}f^{\nu\lambda} + \frac{1}{4g^2}f_{\mu\nu}^2\right]. \tag{8.7}$$

In mean field approximation, we expand $a_{\mu} = a_{\mu}^0 + a_{\mu}'$ where a_{μ}^0 is the solution of the mean field equation

$$\left.\frac{\delta S_m}{\delta a_{\mu}}\right|_{a^0} = 2\alpha\epsilon_{\mu\nu\lambda}f^{0\nu\lambda} - \frac{1}{g^2}\partial^{\nu}f_{\mu\nu}^0. \tag{8.8}$$

This is, of course, essentially (8.1) again. To second order in a_{μ}', we have the action

$$\begin{aligned} S = S\Big|_{a^0} &+ \int\int \pi_{\mu\nu}(a^0)a'^{\mu}a'^{\nu} \\ &- \int d^3x\left[\alpha\epsilon_{\mu\nu\lambda}a'^{\mu}f'^{\nu\lambda} + \frac{1}{4g^2}f_{\mu\nu}'^2\right] \end{aligned} \tag{8.9}$$

where $\pi_{\mu\nu}(a^0) = (\delta^2 S_m/\delta a_{\mu}\delta a_{\nu})\Big|_{a^0}$. Expanding $\pi_{\mu\nu}(a^0)$ in powers of derivatives, we finally have the effective action

$$\begin{aligned} S = S|_{a^0} &+ \int d^3x\left[\beta\epsilon_{\mu\nu\lambda}a'^{\mu}f'^{\nu\lambda} + \frac{1}{4g'^2}f_{\mu\nu}'^2 + \ldots\right] \\ &- \int d^3x\left[\alpha\epsilon_{\mu\nu\lambda}a'^{\mu}f'^{\nu\lambda} + \frac{1}{4g^2}f_{\mu\nu}'^2\right]. \end{aligned} \tag{8.10}$$

But the mean field a^0 describes a constant "magnetic" field. Thus, the determination of $\pi_{\mu\nu}(a^0)$ is precisely the quantum Hall problem. The physics of the quantum Hall effect and the physics of anyon superfluidity are intimately related.

The coefficient β may be determined from the theory of the Hall effect or, more heuristically, simply by counting. We note that the β term first says that in a constant magnetic field there is a current

$$j_i = 4\beta\epsilon_{ij}f'_{0j} \tag{8.11}$$

in response to an external electric field $f'_{0j} \equiv e_j$. We can now determine β as follows. Balancing the Lorentz force against the electric force gives $v_j b = \epsilon_{ij}e_j$ and so the current $j_i = nv_i = (n/b)\epsilon_{ij}e_j$. Thus $\beta = n/4b = n/(8n\theta) = 1/8\theta = \alpha$ where we have used (4.7) and (7.2). The induced and bare Chern-Simons terms in (8.10) cancel each other exactly and we have

$$S = S|_{a^0} + \text{Maxwell action} + \ldots. \tag{8.12}$$

The low lying excitation is described by a Maxwell Lagrangian as in (8.4). Thus, the incompressible Hall fluid "begets" to the anyon superfluid.

This discussion is rigorous except for the implicit assumption that we can express S_m in a gradient expansion. This is justified if the quantum Hall state has a finite gap, which corresponds precisely to our assumption earlier that when a small repulsion is introduced between the bosons in a magnetic field a unique N-body ground state separates off with a gap between it and the next state. In other words, we have to prove the incompressibility of the boson Hall fluid. In contrast this point is easily proved were we to treat semions as fermions. In that case, the gap in question is just the Landau gap.

Finally, let us turn on electromagnetism and verify that we have superconductivity. Recall from (8.1) that the current of the semion is given by $j^\mu = \frac{1}{2\pi}\epsilon^{\mu\nu\lambda}\partial_\nu a_\lambda$ and so the effective Lagrangian is

$$\mathcal{L} = -\frac{1}{2g^2}f_{\mu\nu}^2 + \frac{1}{2\pi}A_\mu\epsilon^{\mu\nu\lambda}\partial_\nu a_\lambda. \tag{8.13}$$

We will now derive superconductivity the "particle physicist's way", namely by demonstrating that we obtain the Meissner effect. Let us integrate out a in (8.13). Doing it schematically, we have

$$\begin{aligned}\mathcal{L} &\sim a\partial^2 a + a\epsilon\partial A \\ &\sim \epsilon\partial A \, \frac{1}{\partial^2} \, \epsilon\partial A \\ &\sim A \, \frac{\epsilon\partial\epsilon\partial}{\partial^2} \, A \\ &\sim A^2\end{aligned} \tag{8.14}$$

In the first step we integrated out a noting that a couples to $\epsilon\partial A$. In the second step, we integrated by parts. This is of course just the Anderson-Higgs mechanism: the photon eats the statistical gauge field. You may be confused that a gauge boson has eaten another gauge boson, rather than a massless scalar. The point is that in (2+1)-dimensional spacetime, a massless scalar is equivalent to a gauge boson, precisely because of the duality transform. (See (9.2) below.)

We will see later that the effective Lagrangian for the quantum Hall fluid is given

$$\mathcal{L} = \frac{k}{4\pi}\epsilon^{\mu\nu\lambda}a_\mu\partial_\nu a_\lambda + \frac{1}{2\pi}A_\mu\epsilon^{\mu\nu\lambda}\partial_\nu a_\lambda \tag{8.15}$$

Indeed, this is almost obvious if we recognize that in (8.5) we simply cross out the Chern-Simons term which endows the anyons with fractional statistics. In contrast to (8.13) we end up with a Chern-Simons Lagrangian rather than the Maxwell Lagrangian. If we now integrate out a, we obtain schematically

$$\begin{aligned}\mathcal{L} &\sim a\epsilon\partial a + a\epsilon\partial A \\ &\sim \epsilon\partial A\frac{1}{\epsilon\partial}\epsilon\partial A \\ &\sim A\epsilon\partial A\end{aligned} \tag{8.16}$$

Compare this to (8.13). Instead of the Meissner effect, we obtain the Hall effect. (To see that we obtain the Hall effect, we simply vary $\mathcal{L}$ with respect to A_μ to obtain the electromagnetic current $J_\mu \sim \epsilon_{\mu\nu\lambda}\partial_\nu A_\lambda$. With an external electric field, we have $J_i \sim \epsilon_{ij}E_j$, *i.e.*, a current perpendicular to the electric field.)

For completeness and for later use, we note that if for some reason the gauge potential acquires a gauge symmetry breaking mass, we would have

$$\begin{aligned} \mathcal{L} &\sim a^2 + a\epsilon\partial A \\ &\sim \epsilon\partial A\epsilon\partial A \\ &\sim F^2 \end{aligned} \tag{8.17}$$

We obtain the Maxwell term.

All of this can be summarized as follows:

$a\partial^2 a$	A^2	Meissner
$a\epsilon\partial a$	$A\epsilon\partial A$	Hall
a^2	$A\partial^2 A$	Maxwell

We see duality at work.

IX. VORTICES

The astute reader may notice that our low energy effective "Maxwell" Lagrangian (8.1) does not violate T and P. This makes sense: due to semion pairing we should see T and P violation only at short distances or high frequencies and we have not included any higher energy excitations. In a superfluid, the higher energy excitations are vortices. In our "fake Maxwell" description, the vortices correspond to "electric charges". This is easy to see. Around an electric charge there is an electric field $f_{0i} \propto x_i/r^2$. According to (8.3), this describes a vortex flow $v_i \propto \epsilon_{ij}x_j/r^2$. Thus, we can incorporate vortices by introducing a complex scalar field ϕ coupled to a_μ via the usual minimal coupling $(\partial_\mu - iqa_\mu)\phi$. The "charge" of ϕ, q , may be determined by appealing to the quantization of fluid flow around the vortex. As we will now see, this is at the heart of the duality picture we discussed earlier.

At the same time we show how to determine q we will also go through the instructive exercise of showing how our effective description of phonons and vortices may be derived directly from a Ginzburg-Landau Lagrangian. In this section we will find it convenient to absorb q into a_μ so that the vortex has unit charge. The coupling g^2 is suitably renormalized. This changes the normalization of the electromagnetic current of course. See below.

In this so-called duality picture[59] we represent the phonon or the superfluid component by a gauge potential a_μ and the vortex or the normal fluid component by a complex scalar field ϕ carrying "electric" charge with respect to a_μ. Let us briefly review how this comes about. In the standard Ginzburg-Landau theory, the potential for the complex order parameter field Φ, has the form $V(\Phi) = a|\Phi|^2 + b|\Phi|^4$. In the superconducting phase, a is negative, while in the insulating phase a is positive.

In the superconducting phase, we have a Goldstone mode η corresponding to the phase of $\Phi : \Phi = |\Phi|e^{i\eta}$. The dynamics of η is described by the Lagrangian

$$\mathcal{L} = \frac{1}{2h^2}(\partial_\mu \eta)^2 \tag{9.1}$$

where we have chosen spacetime units so that the phonon velocity is unity. Also, in spite of my earlier disparaging remarks about relativistic physicists using relativistic formalism to describe condensed matter systems, here I have shamelessly ignored my own injunction and written down a relativistic Ginsburg-Landau theory. I have done this for pedagogical reasons: the formulas are so much simpler and the results are essentially the same. At the end of this section, I will give a non-relativistic treatment. (Throughout this section we will treat the electromagnetic gauge field A_μ as a background classical field that we can switch on at the end of our calculation.)

In (2+1)-dimensional spacetime, we can go to a dual representation by introducing a gauge potential a_μ and a gauge field $f_{\mu\nu} = \partial_\mu a_\nu - \partial_\nu a_\mu$ related to η by

$$\partial_\mu \eta = \beta h^2 \epsilon_{\mu\nu\lambda} f^{\nu\lambda} \tag{9.2}$$

We will now explain a choice of the arbitrary overall constant β, which sets the scale of $f_{\mu\nu}$, so that the electromagnetic current J_μ takes on a convenient form. Had we switched on electromagnetism through the covariant derivative $(\partial_\mu - iA_\mu)\Phi$ we would have gotten, instead of (9.1), $\mathcal{L} = \frac{1}{2h^2}(\partial_\mu \eta - A_\mu)^2$. Thus, we identify the electromagnetic current (or equivalently the velocity current in the superfluid) $J_\mu = \partial_\mu \eta / h^2$ (this is standard, of course) and hence $J_\mu = \beta \epsilon_{\mu\nu\lambda} f^{\nu\lambda}$ according to (9.2). (Note that we have suppressed the charge ($2e$) carried by the order parameter Φ; this factor will be restored at the end. We also set $\hbar = 1$.)

In addition to the gapless excitation η, the superfluid also contains a gapful excitation, namely the vortex, one at each place where $|\Phi|$ vanishes. In the dual picture, the vortex is described by an "electric" charge coupled to a_μ. As was just mentioned we will now choose β so that the vortex carries unit "electric" charge. First, note that with (9.2), the Lagrangian (9.1) in the dual picture is just good old Maxwell

$$\mathcal{L} = -\beta^2 h^2 f_{\mu\nu} f^{\mu\nu} \tag{9.1$'$}$$

as in (8.4). Thus, if we couple a_μ to a current j_μ by adding $a_\mu j^\mu$ to (9.1$'$) we obtain the Maxwell equation

$$4\beta^2 h^2 \partial^\mu f_{\mu\nu} = j_\nu. \tag{9.3}$$

Integrating the time component of this equation over space, normalizing the vortex charge $\int d^2x j_0 = 1$, and keeping in mind the quantization of vortex circulation, namely $\oint dx_i \partial_i \eta = 2\pi$, we determine that $\beta = 1/(4\pi)$ and thus

$$J_\mu = \frac{1}{4\pi} \epsilon_{\mu\nu\lambda} f^{\nu\lambda}. \tag{9.4}$$

In summary then, the following effective Lagrangian

$$\mathcal{L} = \frac{1}{4g^2}(2f_{0i}^2 - c^2 f_{ij}^2) + |(\partial_0 - ia_0)\phi|^2 + v^2|(\partial_i - ia_i)\phi|^2 - m^2\phi^\dagger\phi \tag{9.5}$$

describes the long distance dynamics of the superfluid. This is to be supplemented by the equation for the electromagnetic current

$$J_i = \frac{1}{2\pi}\epsilon_{ij} f_{0j}. \tag{9.6}$$

We have written a non-relativistic Lagrangian in (9.5) with the phonon velocity c and the vortex velocity v. We have taken the low momentum dispersion of the vortex to have the form $\omega = \sqrt{m^2 + v^2k^2} \simeq m^2 + \frac{v^2}{2m^2}k^2$. Note that m^2 represents the gap for creating a vortex-antivortex pair and m/v the effective "mass" of the vortex.

The Lagrangian does not yet violate T and P. To see that it does not violate P, for instance, we note that according to (9.4), the gauge potential a_μ should transform as an axial vector under parity. This leaves the Lagrangian (9.5) invariant provided we also interchange ϕ and $\phi^\dagger$, as is physically reasonable. Thus, this Lagrangian may be useful even outside the context of semion superconductivity. Wen and I have invoked this duality picture to calculate the universal conductance[60,61] observed in thin films at the superconductor-insulator phase transition.

In the field theory literature it is known that a gauge invariant T and P violating coupling[62] may be introduced as follows:

$$\mathcal{L} = \frac{1}{4g^2}(2f_{0i}^2 - c^2 f_{ij}^2) + |(\partial_0 - ia_0 - i\eta\epsilon_{ij} f_{ij})\phi|^2 + v^2|(\partial_i - ia_i - i\xi\epsilon_{ij} f_{0j})\phi|^2 \tag{9.7}$$

where η and ξ are phenomenological parameters. The coupling of the vortex current $\sim \phi^\dagger(\partial_i - ia_i)\phi$ to the superfluid current $\sim \epsilon_{ij} f_{0j}$ implies that the superfluid density would be different near the core of a vortex and the core of an anti-vortex. Thus, a vortex and an anti-vortex have different energies.

We can integrate out ϕ to obtain an amended low energy effective Lagrangian. Clearly, it will have the form "Maxwell Lagrangian" plus a "higher Chern-Simons term" where the higher Chern-Simons term denotes the non-relativistic form of $\epsilon^{\mu\nu\lambda} f_{\mu\sigma}\partial_\nu f_{\lambda\sigma}$. (We know from our construction that this term must be gauge invariant.) That it is a higher derivative term implies that T and P violation vanish at low frequency. Clearly, this conclusion will propagate itself into the effective Lagrangian for A_μ when we turn on electromagnetism.

Here we would like to mention a rather subtle point. Since the gauge potential a_ν originated in a phase on the lattice, we should really be talking about compact quantum electrodynamics in (8.4). As is well known, compact QED contains an instanton — the monopole in the Euclidean version of the theory. In a celebrated paper, Polyakov[63] has shown that the non-perturbative effects of instantons can change the character of the theory. Would our conclusions here be completely modified by the monopoles? Wen and Zee[54] answered this question as follows. The theory is more than non-compact QED: it also contains (8.1), telling us that $j^\mu = 2\alpha\epsilon^{\mu\nu\lambda} f_{\nu\lambda}$. If we insist that the current is conserved, $\partial j = 0$, then this implies that $\epsilon^{\mu\nu\lambda}\partial_\mu f_{\nu\lambda} = 0$. Monopoles are not allowed. Alternatively, we can allow monopoles and give up current conservation. Now we have $\partial j = 2\alpha\epsilon^{\mu\nu\lambda}\partial_\mu f_{\nu\lambda} =$ monopole density $\neq 0$ which we can interpret by saying that a monopole creates normal fluid. the current j represents only the superfluid and only the total current is conserved $\partial(j + j_{\rm normal}) = 0$. In Euclidean spacetime, we can have virtual quantum processes in which a monopole creates some normal fluid which propagates for some distance before it is annihilated, or absorbed, by an antimonopole. The characteristic distance is governed by the inverse of the difference in energy density between the normal fluid and the superlfuid. Thus, monopoles and antimonopoles are effectively bound together.

At some finite temperature monopoles and antimonopoles are liberated. The normal fluid becomes an increasingly dominant component. We recall from (8.17) that when the gauge potential a_μ becomes massive, superconductivity disappears. (To be precise, we note that Polyakov's mechanism generates a non-local term of the form $f\frac{1}{\partial^2}f$. In a suitable gauge, this becomes a^2.)

As promised, I will now give a non-relativistic treatment of duality. The treatment given here follows essentially that given by Fisher and Lee.[49] Let us start, as in section I, with bosons interacting via some short range repulsion

$$\mathcal{L} = i\phi^+\partial_0\phi - \frac{1}{2m}\partial_i\phi^+\partial_i\phi - \lambda(\phi^+\phi - \bar{\rho})^2 \tag{9.8}$$

We have written the usual quartic potential in a form emphasizing that these bosons are moving in a "neutralizing" background as is the case in actual applications to condensed matter systems.

We substitute

$$\phi = \rho^{1/2}e^{i\eta} \tag{9.9}$$

in (9.8) to obtain

$$\mathcal{L} = -\rho\partial_0\eta - \frac{\rho}{2m}(\partial_i\eta)^2 - \lambda(\rho - \bar{\rho})^2 + \ldots \tag{9.10}$$

which we rewrite as

$$\mathcal{L} = -\xi_\mu\partial^\mu\eta + \frac{m}{2\rho}\xi_i^2 - \lambda(\rho - \bar{\rho})^2 + \ldots \tag{9.11}$$

In (9.10) we have dropped a term $\sim (\partial_i\rho^{1/2})^2$. In (9.11) we have defined

$$\xi_0 \equiv \rho \tag{9.12}$$

Integrating out ξ_i in (9.11) we recover (9.10).

We introduce vortices by decomposing the angle variable into two parts:

$$\eta = \eta_{\text{smooth}} + \eta_v \tag{9.13}$$

The field η_{smooth} represents the phonon while η_v is such that when we go around a place where a vortex is located, η_v changes by 2π. The first term in (9.11) become $\xi_\mu \partial^\mu \eta_{\text{smooth}} + \xi_\mu \partial^\mu \eta_v$. Integrating out the phonon field η_{smooth}, we obtain the constraint

$$\partial^\mu \xi_\mu = 0 \tag{9.14}$$

which is solved by writing

$$\xi^\mu = \epsilon^{\mu\nu\lambda} \partial_\nu \hat{a}_\lambda \tag{9.15}$$

The hat ˆ on a_μ is for later convenience.

It is at this point that the gauge field comes in. Note that

$$\xi_0 \equiv \rho = \epsilon_{ij} \partial_i \hat{a}_j \equiv \hat{f} \tag{9.16}$$

is the "magnetic" field strength while

$$\xi_i = \epsilon_{ij} (\partial_0 \hat{a}_j - \partial_j \hat{a}_0) \equiv \epsilon_{ij} \hat{f}_{0j} \tag{9.17}$$

is the "electric" field strength.

The coupling to the vortex angle η_v is given by

$$\begin{aligned} \xi_\mu \partial^\mu \eta_v &= \epsilon^{\mu\nu\lambda} \partial_\nu \hat{a}_\lambda \partial_\mu \eta_v \\ &= \hat{a}_\lambda \left(\epsilon^{\mu\nu\lambda} \partial_\nu \partial_\mu \eta_v \right) \end{aligned} \tag{9.18}$$

According to Leibniz and Newton, differentiation is a commuting operation and so the operator $\epsilon^{\mu\nu\lambda} \partial_\nu \partial_\mu$ should be zero. Here it is operating, however, on a topologically singular function. The quantity in parenthesis is non-zero and in fact equal to the vortex current $j^\lambda_{\text{vortex}}$ up to a factor of 2π. To see this we integrate j^0_{vortex} over a circle containing a vortex. Using Stoke's theorem, we find

$$\begin{aligned} \int d^2x \; j^0_{\text{vortex}} = \int d^2x \; \epsilon_{ij} \partial_i \partial_j \eta_v &= \oint dl_j \; \partial_j \eta_v \\ &= 2\pi \end{aligned} \tag{9.19}$$

Henceforth, we will ignore factors of 2π *etc.* in the interest of keeping the equations simple.

Putting all of this into (9.11) we have

$$\mathcal{L} = \hat{a}_\mu j^\mu_{\text{vortex}} + \frac{m}{\rho}\hat{f}^2_{0i} - \lambda(\hat{f} - \bar{\rho})^2 + \ldots \tag{9.20}$$

To "subtract out" the background $\bar{\rho}$, an obviously sensible move is to write

$$\hat{a}_\mu = \bar{a}_\mu + a_\mu \tag{9.21}$$

where we define the background gauge potential by $\bar{a}_0 = 0$ and

$$\epsilon_{ij}\partial_i \bar{a}_j = \bar{\rho} \tag{9.22}$$

The Lagrangian (9.20) then has the cleaner form

$$\mathcal{L} = \left(\frac{m}{\bar{\rho}} f^2_{0i} - \lambda f^2\right) + a_\mu j_\mu^{\text{vortex}} + \bar{a}_i j_i^{\text{vortex}} \tag{9.23}$$

We have expanded $\rho \approx \bar{\rho}$ in the first term. The theory contains a propagating mode with speed

$$c = (\lambda\bar{\rho}/m)^{1/2} \tag{9.24}$$

The reader with a good memory would recognize this as just (1.12) revisited.

In suitable units in which $c = 1$, we now have the Maxwell Lagrangian

$$\mathcal{L} = -\frac{1}{2g^2} f^2_{\mu\nu} + a_\mu j_\mu^{\text{vortex}} + \bar{a}_i j_i^{\text{vortex}} \tag{9.25}$$

Compare this with (9.5). Interestingly, the spatial part f^2_{ij} of the Maxwell Lagrangian comes from the short ranged repulsion between the original bosons.

The one thing we missed with our (pseudo-)relativistic treatment is the last term in (9.25), for the simple reason that we didn't put in a background. Recall that a term like $A_i J_i$ in ordinary electromagnetism means that a moving particle associated with the current J_i sees a magnetic field $\vec{\nabla} \times \vec{A}$. Thus, a moving vortex will see a "magnetic field" equal to the sum of the density of the original bosons and a fluctuating field

$$\epsilon_{ij}\partial_i(\bar{a}+a)_j = \bar{\rho} + \epsilon_{ij}\partial_i a_j \tag{9.26}$$

Condensed matter physicists customarily take the Coulomb gauge wherein

$$\partial_i a_i = 0 \tag{9.27}$$

Thus,

$$(f_{0i})^2 = (\partial_0 a_i)^2 + (\partial_i a_0)^2 \tag{9.28}$$

where the crossterm $(\partial_0 a_i)\ (\partial_i a_0)$ vanishes upon integration by parts because of (9.27). Integrating out the Coulomb field a_0 we have the Lagrangian as written in the condensed matter literature

$$\begin{aligned}\mathcal{L} = &\frac{1}{2g^2}\left[(\partial_0 a_i)^2 - (\partial_i a_j - \partial_j a_i)^2\right] \\ &+ g^2 \int\int d^2x d^2y\ j_0^{\text{vortex}}(x) \log|x-y| j_0^{\text{vortex}}(y) \\ &+ (a_i + \bar{a}_i) j_i^{\text{vortex}}\end{aligned} \tag{9.29}$$

Note that this discussion is closely related to our discussion from (7.8) through (8.4).

If we had taken the bosons to interact by an arbitrary potential $V(x)$ we would have, instead of the last term in (9.11),

$$\int\int d^2x d^2y\, (\rho(x) - \bar{\rho})\, V(x-y)(\rho(y) - \bar{\rho}) \tag{9.30}$$

It is easy to see that all the steps go through essentially as before, but now the second term in (9.23) becomes

$$\int\int d^2x d^2y f(x) V(x-y) f(y) \tag{9.31}$$

Thus, the gauge field propagates according to the dispersion relation

$$\omega^2 = (\bar{\rho}/m) V(k) \vec{k}^2 \tag{9.32}$$

where $V(k)$ is the Fourier transformation of $V(x)$. In the special case $V(x) = \lambda\delta^{(2)}(x)$ we recover the linear dispersion given in (9.24). Indeed, we have a linear dispersion $\omega \propto |\vec{k}|$ as long as $V(x)$ is sufficiently short ranged for $V(\vec{k}=0)$ to be finite.

An interesting case is when $V(x)$ is logarithmic. Then $V(k)$ goes like $1/k^2$ and so $\omega \sim$ constant: the gauge field a_i becomes massive. Referring to (9.29) we see that a theory of bosons with a logarithmic repulsion between them is in the long distance approximation self dual.

This discussion of duality is not quite rigorous. From standard topological discussions, we know that at the location of the vortex the magnitude field ρ must vanish. In other words, η_v and ρ should be coupled. This was ignored in the derivation.

Let us also sketch what happens when the electromagnetic potential is turned on. We replace ordinary derivative ∂_μ by covariant derivative $\partial_\mu - iA_\mu$. In (9.11) instead of $\xi^\mu\partial_\mu\eta$ we have $\xi^\mu(\partial_\mu\eta - A_\mu)$. Going through the same steps as before we find that we just have to drag along an extra term $A_\mu\epsilon_{\mu\nu\lambda}\partial_\nu\hat{a}_\lambda = A_0\bar{\rho} + A_\mu\epsilon_{\mu\nu\lambda}\partial_\nu a_\lambda$. The $A_0\bar{\rho}$ term is cancelled by the coupling of A_0 to the background. We see that the only modification in the Lagrangian in (9.29) is that the vortex current is replaced by $(j_\mu^{\text{vortex}} - \epsilon_{\mu\nu\lambda}\partial^\nu A^\lambda)$. When we integrate out a_μ, we obtain the Meissner effect just as in (8.14).

Incidentally, here we have also cleared up a potential source of confusion in the discussion following (9.2). For the Lagrangian $\mathcal{L} = \frac{1}{2h^2}(\partial_\mu\eta - A_\mu)^2$, we were rather sloppy in identifying the current to be $J_\mu = \partial_\mu\eta/h^2 = \beta\epsilon_{\mu\nu\lambda}f^{\nu\lambda}$. Some people may be confused by the quadratic A_μ^2 term in the Lagrangian and wonder if the current J_μ should depend on A_μ. (The subtlety here hinges on the gauge invariance of J_μ.) The procedure given here provides a resolution of this point. Let us then follow the steps outlined above and which, at the risk of repeating ourselves, we will sketch here. Starting with $\mathcal{L} \sim (\partial_\mu\eta - A_\mu)^2 \sim \xi_\mu(\partial_\mu\eta - A_\mu) + \xi_\mu^2$ we integrate out η to obtain the constraint $\partial^\mu\xi_\mu = 0$

which we solve by writing $\xi_\mu = \epsilon_{\mu\nu\lambda}\partial_\nu a_\lambda$. Plugging in, we find $\mathcal{L} \sim f^2 + A_\mu \epsilon_{\mu\nu\lambda}\partial_\nu a_\lambda$. The current is indeed as given above. (All of this is to done at the level of the path integral of course and may be made completely rigorous by latticizing spacetime.)

X. EXPERIMENTAL TESTS

We turn finally to our fourth "big question", on whether there are experimental tests. Certainly there are! In contrast to a number of theories of superconductivity which make more or less generic predictions, this theory goes out on a limb and says there should be T and P violation – it's certainly a falsifiable theory.

To look for T and P violation, we can study the polarization of the light reflected from a high-T_c material when the incident light is linearly polarized[64,65]. Thus far, three experiments[66] have been done with two showing a rotation of the polarization axis and one showing no effect. Of course there are numerous solid state effects to worry about. Also, it is important to see both T and P violation with TP conserved. A rotation of the polarization axis shows only P violation, but not T violation. Another worry is that the statistics parameter θ may alternate in sign as we go from layer to layer, in which case the effect would be washed out.[67] An alternative experiment is to look at the precession of muons injected into the material[68]. In contrast to the polarized light experiment, the muon is a local probe. The experiment has been done and fails to show a signal at the expected level. Thus, we have now two negative experiments and two quasi-positive experiments.

In this context I would also like to emphasize a generic consequence of these gauge theories of high temperature superconductivity. In these theories, there are two gauge symmetries, one associated with the gauge potential a_μ responsible for the fractional statistics, which may be denoted by $U(1)_a$, and another associated with the electromagnetic gauge potential A_μ, denoted by $U(1)_A$. As far as the gauge symmetries are concerned there are four possible states: (1) both gauge symmetries are unbroken, (2) $U(1)_a$, but not $U(1)_A$, is broken, (3) $U(1)_A$, but not $U(1)_a$, is broken, and (4) both gauge symmetries are broken.

Superconductivity is obtained only when no unbroken gauge symmetry is left and thus only in state (4). Thus, generically, there may be three normal states. This point was emphasized and studied by Wen and I in the context of a generalized Landau-Ginzburg theory.[64] Note that the very general remarks given here do not depend on whether or not T and P are violated and on whether or not fractional statistics plays a role. It only depends on the appearance of another gauge field. The violation of T and P is logically independent. Recently, Sachdev[69] has studied the possible occurence of three normal states in more detail. I mention this point because experimentally the normal state properties of the high temperature superconductors are more extraordinary than the superconducting state properties appear to be.

To summarize let me list the current answers to the "four questions". (1) Does the ground state violate T and P? Possible (but unlikely?). (2) What are the quantum numbers of the excitations? If the answer to (1) is yes, then we have semions. (3) Is a semion fluid a superfluid? Almost certainly yes. (4) Are there experimental tests? Definitely a falsifiable theory.

I am fond of reminding people of an elementary point of logic: Even if the presently known high temperature superconductors are not semion superconductors, it does not follow that semion superconductors do not exist. In particle physics, there is a rule known as Gell-Mann's rule stating that whatever is not forbidden is required. I was told by condensed physicists that they have a priority claim: they have long known this rule as Herring's rule. (And of course the humanists know it as T.H. White's rule.) Unlike particle physicists who have only one darn universe to look at, condensed matter physicists have a wealth of systems to study, particularly in this era of great progress in materials science when you can almost have designer materials made to order. Let us then end with the credo of an extreme optimist. It would certainly be lots of fun if the presently known superconductors turn out to be semionic superconductors, but won't it be even more exciting if the presently known superconductors turn out not to be semionic superconductors? That would mean that there may be a new class of high temperature superconducting materials out there waiting to be discovered!

XI. TOPOLOGICAL FLUIDS

Quite independent of whether semionics describe high temperature superconductivity, we have learned a great deal in developing this subject. Indeed, as was mentioned earlier, an interesting new field of study, that of strongly correlated electron systems in two dimensions, has emerged. One fascinating feature of such systems is that they may have "global" and "topological" properties not reachable, or at least not easily reachable, with conventional condensed matter physics methods. The long distance physics of the ground state and its low lying excitations is described by a new class of field theories known as topological field theories. Wen[70] has emphasized that these systems enjoy a new type of order called "topological order." Topological field theories were first investigated for their possible relevance to string theory.[71] Thus, they may well be field theories in search of physical realization. It appears that at least some watered-down version may be realized in condensed matter systems. There are now three examples of such "topological fluids", namely the Hall fluid,[7] the chiral spin fluid,[8] and the anyon superfluid.

The Hall fluid is realized experimentally, while the other two remain, at this point, figments of the theorists' imagination. As we will see, the structure of these three fluids are intimately related both physically and mathematically. One may be tempted to think that it would be "unnatural" for only one of them to be realized experimentally.

As I will explain, the simplest example of these field theories consists of a single gauge potential α whose dynamics is described by a Chern-Simons term:

$$\mathcal{L} = \frac{k}{4\pi}\epsilon^{\mu\nu\lambda}\alpha_\mu\partial_\nu\alpha_\lambda. \tag{11.1}$$

For $k = 1, 3, \ldots$, an odd integer, this Lagrangian describes the long distance physics of a Hall fluid with filling factor $\nu = 1/k = 1, 1/3, \ldots$ For $k = 2, 4, \ldots$, an even integer, it describes a chiral spin fluid with statistics parameter $\theta/\pi = 1/2, 1/4, \ldots$. We find this an elegantly unified picture of the Hall fluid and the chiral spin fluid. Indeed, were the chiral spin fluid unknown it may have to be invented in order to fill in the "gaps" at the even integers. (The integer k is sometimes referred to as the level.[71])

While the mathematical structure of (1) may be similar for the Hall fluid and the chiral spin fluid the detailed physical contents of these two fluids are quite different of course. In particular, the excitation in the chiral spin fluid, the spinon, is electrically neutral.

Field theorists have been greatly enthused over the Chern-Simons term[72,73] in recent years. Well, why is it so special? We will mention several properties of the Chern-Simons term that are particularly relevant for our purposes.

(1) For the Chern-Simons term, gauge invariance implies Lorentz invariance. (Once again, be reminded that condensed matter physics is not Lorentz invariant.) *A priori*, one may think that there are two separate Chern-Simons terms, $\epsilon_{ij}a_i\partial_0 a_j$ and $\epsilon_{ij}a_0\partial_i a_j$. Gauge invariance under $\alpha_\mu \to \alpha_\mu - \partial_\mu\Lambda$, however, implies that the two terms must combine into the Chern-Simons term $\epsilon^{\mu\nu\lambda}\alpha_\mu\partial_\nu\alpha_\lambda$. In contrast, the Maxwell term would in general be non-relativistic, consisting of two terms, $(f_{0i})^2$ and $(f_{ij})^2$, with an arbitrary relative coefficient between them. This is one reason that relativistic field theorists find this subject of topological Hall and anyon fluids so appealing. As long as we are interested only in long distance physics we can ignore the Maxwell term (see the next remark) and play with a relativistic theory.

(2) The Chern-Simons term dominates the Maxwell term f^2 at long distances: it contains only one derivative while the Maxwell term contains two. Note that this picks out (2+1)-dimensional spacetime as special. In (3+1)-dimensional spacetime the generalization of the Chern-Simons term is $\epsilon\partial a\partial a$, the $f\tilde{f}$ term much studied in connection with quantum chromodynamics, it has the same scale dimension as the Maxwell term f^2. In higher dimensions, the Chern-Simons term, for example $\epsilon a\partial a\partial a$ in (4+1)-dimensional space, is less important at long distances than the Maxwell term f^2. (However, there is an alternative generalization of the Chern-Simons term to higher dimensions.[74])

(3) Most importantly, the Chern-Simons term depends on only the antisymmetric symbol $\epsilon^{\mu\nu\lambda}$ and not on the metric $g^{\mu\nu}$; in other words, it is topological. We have entered a realm without rulers and without clocks. Suppose we study the Hall fluid or the chiral spin liquid on a two-dimensional surface of genus G. (The genus simply measures the number of holes or handles. Thus, the sphere has genus 0, the torus has genus 1, and so on.) This is not as far fetched as it may sound: if we impose periodic boundary conditions on a system with length L_1 and width L_2, we are effectively dealing with a system defined on a torus. In contrast to the Chern-Simons term, the Maxwell term cannot be written without the metric $g_{\mu\nu}$. Thus, in curved space the action takes the form

$$S = \int d^3x \left[\frac{k}{4\pi} \epsilon^{\mu\nu\lambda} a_\mu \partial_\nu a_\lambda - \sqrt{g} \frac{g^{\mu\lambda} g^{\nu\sigma}}{2g^2} f_{\mu\nu} f_{\lambda\sigma} + \ldots \right] \tag{11.2}$$

(I trust the reader not to confuse the gauge coupling g with the determinant of the metric.) The pure Chern-Simons theory depends only on the topology of the underlying surface, not on the metric: it is a topological field theory of the type first studied by Witten.[71]

This is a powerful statement: it suggests those properties of the microscopic theory that depend on the metric cannot affect the Chern-Simons term. In particular, disorder and impurities cannot affect the long distance physics of the Hall fluid or the chiral spin fluid. As discussed earlier, the coefficient of the Chern-Simons term describing the chiral spin fluid simply counts the number of Dirac zeroes in the energy spectrum of a single particle hopping on the lattice in the presence of flux. For a flux per plaquet of $2\pi p/q$, the number of Dirac zeroes is q. This number is guaranteed by an index theorem[21,22] and is thus robust against small perturbations. For instance, we can change the lattice hopping amplitudes in the x and y directions t_x and t_y and let them depend on location without changing the number of Dirac zeroes. These changes my be regarded as changes in the metric.

(4) Since the theory does not depend on the metric $g^{\mu\nu}$, the stress energy tensor, which is defined as the variation of the action with respect to $g^{\mu\nu}$, is identically zero. After all, the concepts of energy and momentum depend on rulers and clocks. The Hamiltonian is exceedingly simple: it is just zero. All states have zero energy.

Thus, as was pointed out by Witten, all we can ask about a topological field theory is how many such zero energy states there are. This ground state degeneracy depends on the topology of the compactified two-dimensional surface. (More precisely, the scale of energy (or mass) is set by the Maxwell coupling g^2. All local excitations have mass of order g^2.) In the context of condensed matter physics, Wen has emphasized that this implies a new kind of order. We have long been familiar with the fact that a spontaneously broken symmetry implies ground state degeneracy. The ground state degeneracy in a topological theory, however, cannot be attributed to the breaking of any symmetry. As we will see, the degeneracy depends on topology. In contrast, the finite degeneracy due to the breaking of a discrete symmetry does not care about the topology. (Consider the two-fold degeneracy of the ground state of the Ising model.) This notion of topological order represents a potentially important new concept in condensed matter physics.

Let us now compute the ground state degeneracy on a torus, say. We will follow the discussion given by Wen.[70] The basic idea goes back to an observation by Wu and Zee[75] that a topological term in field theory may be regarded as due to the presence of a gauge field in field configuration space. This is a real mouthful but is easy to understand. Suppose we are given a Lagrangian describing the motion of a particle with coordinates q_i. If in the Lagrangian we find a term linear in the time derivative $\dot{q}_i$, then we would know that the particle is moving under the influence of an electromagnetic potential $\mathcal{A}_i$, where $\mathcal{A}_i$ is just the coefficient of $\dot{q}_i$.

We have precisely this situation in a Chern-Simons theory. In the $a_0 = 0$ gauge the Lagrangian is

$$\mathcal{L} = \frac{k}{4\pi}\epsilon_{ij}a_i\dot{a}_j \tag{11.3}$$

Thus, we see that in the configuration space of the gauge potentials, a magnetic field is present. The analog of Gauss' law, obtained by varying with respect to a_0, tells us that

$$\epsilon_{ij}\partial_i a_j = 0 \tag{11.4}$$

This tells us that $a_i = a_i(t)$ in the gauge $\partial_i a_i = 0$. In the $a_0 = 0$ gauge, we can still make time independent gauge transformations. On a torus, the allowed gauge transformation must have the form

$$u = e^{i2\pi\left(\frac{n_1 x^1}{L_1}+\frac{n_2 x^2}{L_2}\right)} \tag{11.5}$$

where n_1 and n_2 are integers, since u cannot depend on whether we coordinatize the same point as $(0, x^2)$ or (L_1, x^2), for example. The gauge potentials $a_i' = a_i - iu^{-1}\partial_i u = a_i + 2\pi n_i/L_i$ is equivalent to a_i. Purely for the sake of convenience, let $a_i(t) = (2\pi/L_i)q_i(t)$. We have just learned that q_i and $q_i + 1$ are equivalent. In other words, the dynamics may now be described in terms of a point particle living on a torus of length 1 and width 1.

Plugging in, we find the action becomes

$$\begin{aligned} S =& \int d^3x \frac{k}{4\pi}\epsilon^{\mu\nu\lambda}a_\mu\partial_\nu a_\lambda \\ =& k\frac{L_1 L_2}{4\pi}\int dt \frac{(2\pi)^2}{L_1 L_2}\epsilon_{ij}q_i\dot{q}_j \end{aligned} \tag{11.6}$$

It is convenient to regulate the theory by adding a kinetic energy term:

$$S = \int dt(k\pi\epsilon_{ij}q_i\dot{q}_j + \frac{1}{2}m\dot{q}_i^2) \tag{11.7}$$

We can let the mass m go to zero at the end. Indeed, if we had included the Maxwell term in the field theory, it would generate this kinetic energy term with $m \sim 1/g^2$ ($f_{\mu\nu}^2 \sim \dot{a}_i^2 \sim \dot{q}_i^2$). As $g^2 \to \infty$, the Maxwell term disappears and $m \to 0$.

We have a point particle moving on a torus in a magnetic field $B = \epsilon_{ij}\partial_i\epsilon_{lj}q_l(k\pi) = 2\pi k$. The total flux going through the surface of the torus is $\Phi = BA = 2\pi k$ since the area is normalized to 1. This quantum mechanical problem can be solved explicitly[76] and the result is that the ground state is k-fold degenerate. The degeneracy does not depend on m.

The degeneracy can also be seen heuristically by arguing that the torus is locally flat, but then we know that in the planar problem the the degeneracy of each Landau level is precisely $BA/2\pi = k$.

It can be shown that on a surface of genus G the ground state degeneracy is equal to k^G. Heuristically, this may be argued by topologically deforming the surface into G "quasi-torus" connected by long tubes. In effect, the dynamics of the gauge potential $a_i(t)$ on such a surface is the dynamics of G independent particles, one on each "quasi-torus". The degeneracy is thus k^G.

For a chiral spin liquid, the "actual" degeneracy is $2k^G$. There is a non-topological 2-fold degeneracy due to the spontaneous breaking of T and P. The two states, in which the order parameter $\left\langle \vec{S}_i(\vec{S}_j \times \vec{S}_k) \right\rangle$ takes on $+E$ and $-E$ respectively, are connected by time reversal and parity. (There is of course a similar "degeneracy" in the Hall fluid upon reversing the magnetic field or the charge of the electron, but we do not count that as a degeneracy.)

To an "old-fashioned" field theorist, the parameters in a field theory, such as the electric charge or the mass of the electron in quantum electrodynamics, could be varied continuously, without affecting the qualitative behavior of the theory, at least within a certain range. Here we really have a new type of field theory. In particular, the degeneracy formula $D = k^G$ indicates that the Chern-Simons theory with G not equal to an integer cannot be defined on a compact surface. On the non-compact plane, the theory is apparently defined for any value of G.

It is well known that for the non-abelian Chern-Simons theory the coefficient k, known as the level of the theory, is quantized by the requirement of gauge invariance to be an integer. The level 1 theory is qualitatively different from the higher level theories.

We have already seen, in our derivation of the chiral spin state, that the theory is qualitatively different for different k, which is determined by the number of Dirac fields.

With this long preamble about the properties of the Chern-Simons theory, I am now ready to talk about the effective theories[77] of the Hall fluid and the anyon superfluid. The subtitle here might well be "millions of gauge fields", for reasons that will become clear. The material represents work that I have done with X.G. Wen.[78] We begin with an effective theory of the fractional Hall effect.[79] Consider a fractional Hall system at some filling fraction ν described in second quantized formalism by

$$\mathcal{L} \;=\; \psi^{+}\, i(\partial_0 - i\, A_0)\, \psi + \frac{1}{2m}\, \psi^{+} (\partial_i - i\, A_i)^2 \psi \tag{11.8}$$

where A_μ denotes the external fixed electromagnetic potential. Jain[80,81] has proposed a construction that relates the fractional Hall effect to the integer Hall effect. Imagine moving some of the flux lines so that p units of flux are attached to each electron. For p even, the additional Dirac-Bohm-Aharonov phase associated with moving one electron around another is $e^{i\pi p} = 1$ and so the statistics of the electron is unchanged. The electrons are now moving in a reduced magnetic field $B_{\text{eff}} = B - 2\pi p n$ where n is the number density of electrons. (In our convention, $\hbar = 1 = c = e$ and the unit of flux $hc/e = 2\pi$.) The filling function has been increased to $\nu_{\rm eff}$, given by $\nu_{\rm eff}^{-1} = (B - 2\pi pn)/2\pi n = \nu^{-1} - p$. For $\nu_{\rm eff} = m$ an integer, we have $\nu^{-1} = p + m^{-1}$ and

$$\nu = \frac{m}{mp+1}. \tag{11.9}$$

Thus, fractional Hall systems with $\nu = m/(mp+1)$ (p even) may be adiabatically changed into an integer Hall system with filling factor m, as emphasized by Greiter and Wilczek.[82]

This construction may be implemented by considering the Lagrangian

$$\begin{aligned} \mathcal{L}' &= \psi^{+'} i(\partial_0 - i(\alpha_0 - A_0))\psi' + \frac{1}{2m}\psi^{+'}(\partial_i - i(\alpha_i - A_i))^2\psi' \\ &+ \frac{1}{4\pi p}\epsilon\,\alpha\,\partial\,\alpha\,. \end{aligned} \tag{11.10}$$

Here α_μ denotes a gauge potential. We have introduced and will use henceforth the compact notation $\epsilon\alpha\partial\beta \equiv \epsilon^{\mu\nu\lambda}\alpha_\mu\partial_\nu\beta_\lambda = \epsilon\beta\partial\alpha$ for two vector fields α_μ and β_μ. The gauge potential α_μ may be eliminated from $\mathcal{L}'$ by the equation of motion

$$\epsilon^{\mu\nu\lambda}\,\partial_\nu\,\alpha_\lambda = 2\pi\,p\,j_\mu \tag{11.11}$$

where j_μ is the (nonrelativistic) current of electrons. We expect the current j_i to be zero in the ground state. The magnetic field associated with α_μ is then $b = \varepsilon_{ij}\,\partial_i\,\alpha_j = 2\pi pn$ where n is the density of electrons. In the mean field approximation the electrons are then moving in an effective magnetic field $B_{\rm eff} = B - b = B - 2\pi pn$.

The integer Hall system may be described by the effective theory

$$\begin{aligned} \mathcal{L} = \sum_{I=1}^{m}\Big\{&\phi_I^+ i(\partial_0 - i(\alpha_0 + a_{I0} - A_0))\phi_I + \frac{1}{2M}\phi_I^+(\partial_i - i(\alpha_i + a_{Ii} - A_i))^2\phi_I \\ &- V_I(\phi_I^+\phi_I) + \frac{1}{4\pi}\varepsilon a_I\partial a_I\Big\} + \frac{1}{4\pi p}\varepsilon\alpha\partial\alpha\,. \end{aligned} \tag{11.12}$$

Here ϕ_I is a bose field corresponding to the $\mathrm{I^{th}}$ Landau level, and a_I is a gauge potential that converts the bose field ϕ_I into effectively a fermionic degree of freedom. The potential

$V_I(\phi_I^+\phi_I)$, whose detailed form we don't need to specify, represent a possible short distance interaction between electrons in the Ith Landau level. We have stacked m copies of the effective theory for the $\nu = 1$ Hall system on top of each other in order to describe the $\nu = m$ Hall system.[79]

We next write the theory in the dual representation as explained earlier.[83] The Landau-Ginsburg theory $\mathcal{L} = \phi^+ i\partial_0\phi + \frac{1}{2M}\phi^+\nabla^2\phi - V(\phi^+\phi)$ in the broken symmetric phase has a gapless Nambu-Goldstone excitation η and a gapful vortex excitation. The dual representation seeks to describe the physics directly in terms of these physical degrees of freedom rather than the original complex scalar field.

We can thus write the Lagrangian (11.12) in the dual representation as

$$\mathcal{L} = \sum_{I=1}^{m} \left\{ \frac{2}{4\pi}\varepsilon\,(\alpha + a_I - A)\,\partial\alpha_I + \frac{1}{4\pi}\varepsilon\, a_I\,\partial\, a_I + \alpha_{I\mu}\, j^{\mu}_{I\,\text{vortex}} \right\} + \frac{1}{4\pi p}\varepsilon\alpha\partial\alpha + \cdots . \tag{11.13}$$

We have introduced m gauge potentials $\alpha_{I\mu}$, one for each of the Nambu-Goldstone mode contained in the m complex scalar fields ϕ_I. The vortex degrees of freedom are not shown explicitly but are contained in $j_{I\mu}$, the Ith vortex current to which α_μ couples. Notice that we have also not explicitly displayed the Maxwell terms $f_{I\mu\nu}^2$ for the gauge potentials $\alpha_{I\mu}$. Compared to the various Chern-Simons terms we displayed, the Maxwell term has one power of mass dimension higher and so is unimportant at long distances. Our effective theory does not describe the short distance physics in any case. The long distance physics is now described entirely in terms of gauge fields.

We can now integrate out a_I. Note that the Chern-Simons term is "self-reproducing". Schematically, if we have $\mathcal{L} \sim a\epsilon\partial a + a\epsilon\partial b$, then integrating out a, we would get $\mathcal{L} \sim (\epsilon\partial b)(\epsilon\partial)^{-1}(\epsilon\partial b) \sim b\epsilon\partial b$. Basically, once again, we have the statement that the Chern-Simons term is the unique gauge and scale invariant term we can write down. We obtain thus

$$\mathcal{L} = \frac{1}{4\pi}\left[\sum_I \{2\varepsilon(\alpha - A)\partial\alpha_I + \varepsilon\alpha_I\partial\alpha_I\} + \frac{1}{p}\varepsilon\alpha\partial\alpha\right] + \sum_I \alpha_I j_I^{\text{vortex}}. \tag{11.14}$$

Next the gauge potential α may be integrated out to give

$$\mathcal{L} = \frac{1}{4\pi}\Big[\sum_I \varepsilon\alpha_I\partial\alpha_I + p\sum_{IJ}\varepsilon\alpha_I\partial\alpha_J\Big] + \frac{2}{4\pi}\sum_I\Big(-\varepsilon\alpha_I\partial A + 2\pi\alpha_I\, j_I^{\text{vortex}}\Big). \tag{11.15}$$

This may be written more compactly by introducing the m by m matrix $K \equiv I + p\,C$ where C is the matrix in which every entry is equal to 1, sometimes called the pseudo-identity. Explicitly,

$$K = \begin{pmatrix} p+1 & p & \cdots & p \\ p & p+1 & & . \\ \vdots & & \ddots & \vdots \\ p & \cdots & \cdots & p+1 \end{pmatrix} \tag{11.16}$$

Then,

$$\mathcal{L} = \frac{1}{4\pi}\,\varepsilon\alpha K\partial\alpha + \alpha\tilde{j} \tag{11.17}$$

where α now denotes the column or row vector $(\alpha_1, \alpha_2, \ldots, \alpha_m)$. The current $\tilde{j}$ is a reduced current defined as the vortex current corrected by the external electromagnetic field; we can read it off from (11.15):

$$\tilde{j}_{I\mu} = j_{I\mu}^{\text{vortex}} - \frac{1}{2\pi}\,\epsilon_{\mu\nu\lambda}\,\partial^\nu A^\lambda. \tag{11.18}$$

Integrating out the gauge fields entirely we obtain a matrix version of the Hopf interaction[15]

$$\mathcal{L} = \pi\tilde{j}_\mu K^{-1}(\epsilon^{\mu\nu\lambda}\partial_\nu/\partial^2)\tilde{j}_\lambda. \tag{11.19}$$

We now have "the world's most compact" representation of the fractional quantum Hall effect for the filling factor $\nu = m/(mp+1)$. Note that m and p come in quite differently: K is an m-by-m matrix in which the even integer p appears. It is perhaps good to keep in mind that for $m = 1$, *i.e.*, for the classic odd denominator filling factor $\nu = \frac{1}{(p+1)} = 1/(\text{odd integer})$ this discussion simplifies drastically: there is only one gauge potential α and the matrix K collapses into an odd integer $= \nu^{-1}$.

We can now easily read off the physics. The Lagrangian (density) contains three types of terms: AA, Aj_I, and $j_I j_I$. We have

$$\mathcal{L} = -\frac{1}{4\pi}\epsilon A\partial A\left(\sum_{IJ} K_{IJ}^{-1}\right). \tag{11.20}$$

Compare this with (8.16). Varying with respect to A to find the electromagnetic current

$$J_{em}^{\mu} = \frac{1}{2\pi}\epsilon^{\mu\nu\lambda}\partial_\nu A_\lambda\left(\sum_{IJ} K_{IJ}^{-1}\right). \tag{11.21}$$

An electric field would produce a current in the orthogonal direction. The Hall conductance is thus

$$\sigma_H = \sum_{IJ} K_{IJ}^{-1}. \tag{11.22}$$

The Aj_I term comes out to be

$$\mathcal{L} = A^\mu \sum_{IJ} K_{IJ}^{-1} j_{J_\mu}. \tag{11.23}$$

Thus, the electric charge of the Ith type of vortices is determined to be

$$q_I = \sum_J K_{JI}^{-1}. \tag{11.24}$$

Finally, the vortices interact with each other via

$$\mathcal{L} = \pi \sum_{IJ} j_I^\mu K_{IJ}^{-1}\epsilon_{\mu\nu\lambda}\partial_\nu/\partial^2 j_J. \tag{11.25}$$

Thus, a bound state of vortices consisting of l_I vortices of the Ith type would have statistics given by

$$\frac{\theta}{\pi} = \sum_{IJ} l_I K_{IJ}^{-1} l_J. \tag{11.26}$$

The three formulas (11.22), (11.24), and (11.26) for the Hall conductance, the vortex charge, and vortex statistics, hold for any K. Let us apply it to the specific K we have in

(11.16). Using the fact that C is proportional to a projection ($C^2 = mC$), we can readily invert $K = I + pC$ to be

$$K^{-1} = I - \left(\frac{p}{1+mp}\right) C. \tag{11.27}$$

We find immediately that

$$\sigma_H = \frac{m}{mp+1}, \tag{11.28}$$

$$q_I = \frac{1}{mp+1}, \tag{11.29}$$

and

$$\frac{\theta}{\pi} = \left(1 - \frac{p}{mp+1}\right)\left(\sum_I l_I\right)^2 \tag{11.30}$$

Note that q_I is independent of I.

In the simplest case of $m = 1$, we have the easily remembered and by now classic Laughlin result that

$$\sigma_H = \frac{1}{p+1}, \tag{11.28$'$}$$

$$q = \frac{1}{p+1}, \tag{11.29$'$}$$

and

$$\frac{\theta}{\pi} = \frac{1}{p+1}. \tag{11.30$'$}$$

Now onwards to the anyon superfluid. We have to take two steps. (1) Turn off the electromagnetic potential A_μ. (2) Couple in a gauge potential a_μ and add a Chern-Simons term $(1/4\theta)\varepsilon a\partial a$ to change the electrons in the Hall system into anyons.

We thus obtain immediately the effective theory for the anyon superfluid

$$\mathcal{L} = \frac{1}{4\pi}\Big[\sum_I (2\varepsilon\, a\,\partial\alpha_I + \varepsilon\alpha_I\partial\alpha_I) + p\sum_{IJ}\varepsilon\alpha_I\partial\alpha_J + \left(\frac{m}{mp+1}\right)\varepsilon a\partial a\Big] + \ldots . \tag{11.31}$$

This is obtained from (11.15) by replacing A_μ by a gauge potential a_μ and adding a Chern-Simons term for a_μ so that it changes the statistics of the electron by (recall (4.7))

$$\frac{\Delta\theta}{\pi} = (mp+1)/m = p + 1/m . \tag{11.32}$$

We write $\Delta\theta$ to remind us that it measures the change in statistics from fermi statistics. (Recall also that p is even. Thus, we are discussing anyons with statistics $\theta = \pi(1+\frac{1}{m})$ or equivalently $\pi(1-\frac{1}{m})$. Semions correspond to $m = 2$.)

The long distance physics of the semion fluid is thus described by $m+1$ gauge potentials, $a, \alpha_1, \alpha_2, \ldots \alpha_m$ interacting via the Chern-Simons term. Defining $\alpha_0 \equiv a$, we may write (11.31) in a more compact form

$$\mathcal{L} = \frac{1}{4\pi}\sum_{AB} \Lambda_{AB}\, \varepsilon\, \alpha_A\, \partial\, \alpha_B + \ldots \tag{11.33}$$

where $A, B = 0, 1, \ldots, m$ and Λ is an interesting looking $(m+1)$ by $(m+1)$ matrix

$$\Lambda = \begin{pmatrix} \frac{m}{mp+1} & 1 & \cdots & \cdots & 1 \\ 1 & p+1 & p & \cdots & p \\ \cdot & p & p+1 & & \cdot \\ \vdots & \vdots & & \ddots & \vdots \\ 1 & p & \cdots & \cdots & p+1 \end{pmatrix} \tag{11.34}$$

As is well-known by now, the Chern-Simons term together with the Maxwell term gives a gauge invariant mass[73] to the gauge field. Thus generically, we would expect to have $(m+1)$ massive gauge fields. The signal for superfluidity, however, is the appearance of a gapless linearly dispersing excitation.

Lo and behold! The matrix Λ indeed has a zero eigenvalue. By inspection, we can see that the corresponding eigenvector is $(-(mp+1), 1, 1, \ldots, 1)$. Thus, one of the gauge fields is massless. A linearly dispersing gapless mode is present and we have superfluidity (and superconductivity for charged anyons).

We can now put out a sign announcing "Have effective Lagrangian, will study long distance physics". I will describe several applications[79] here.

(A) In the literature, the fractional statistics between particles is defined in vacuo. In a semion superfluid, however, the semions are hardly moving in a vacuum, but in a complicated environment consisting of all the other semions. Thus, screening effects must be taken into account, and the effective statstics of a semion may well be quite different from its intrinsic statistics.

Fractional statistics is the manifestation of a long ranged gauge interaction between particles, and this gauge interaction could be screened just as the Coulomb interaction is screened. We find that semions are screened by vortices in the superfluid. Heuristically speaking, this makes sense, since the gauge interaction responsible for fractional statistics operates in the azimuthal direction rather than in the radial direction, as is the case with the more familiar Coulomb interaction.

We will sketch how the calculation of statistics screening goes. (1) We first find the massive modes c_i, $I = 1, \ldots, m$, that is, those linear combinations of α_A that correspond to eigenvectors of Λ with non-zero eigenvalues. (2) In the new basis, we re-write (11.33) as

$$\mathcal{L} = \frac{1}{4\pi} c \Omega \epsilon \partial c + (\ldots) \tag{11.35}$$

where Ω is a matrix derived from Λ. (3) We have in (11.35) the coupling of the gauge potentials to the vortex currents

$$\begin{aligned}(\ldots) &= \alpha j + \sum \alpha_I j_I \\ &= c_0[(mp+1)j + \sum j_I] + \sum c_I[-j + (mp+1)j_I]\end{aligned} \tag{11.36}$$

The gauge potential c_0 is by definition the massless mode responsible for superfluidity. We thus have logarithmic confinement of the particles c_0 couples to. Mathematically, this means that the current to which c_0 is coupled to is effectively zero

$$(mp+1)j_\mu + \sum_I j_{I\mu} \text{ “}=\text{” } 0 \tag{11.37}$$

(4) Finally, to obtain the effective statistics of the currents that are not confined, we have to sum up the effects of the massive gauge potentials (more precisely, the gauge potentials with Chern-Simons terms, namely c_I). Thus, we have to invert Ω.

After some computation, we find that for $\theta = \pi(1 - 1/m)$, $m = 2, 3 \cdots$ the anyon is surrounded by vortices in precisely such a way so that the anyon-vortices bound state comes out to be a boson. This result is rather satisfying; we may suppose that Nature abhors "weird" statistics and acts to screen it!

(B) We can use our effective Lagrangian to calculate the quantum of flux in the anyon superconductor. When we turned on the electromagnetism, the "external" electromagnetic potential eats the massless gauge mode c_0 according to the Anderson-Higgs mechanism. Thus, there is no more logarithmic confinement by c_0. The previously logarithmically confined excitations are liberated and carry a vorticity given by

$$\Phi = \frac{2\pi}{m} \sum l_I \tag{11.38}$$

In a semion superconductor, $m = 2$ and we have the standard quantum of flux

$$\Phi = \frac{2\pi}{2} \tag{11.39}$$

which in some archaic system of units is equal to $\left(\frac{2\pi}{2}\right)\left(\frac{\hbar c}{e}\right) = \left(\frac{hc}{2e}\right)$

(C) In addition to the construction described above, Jain[81] has given a second construction whose validity is somewhat controversial. It involves "cutting up the electron." We can readily construct the corresponding effective Lagrangians, with more complicated matrices K, Λ, and Ω. It turns out that with this second construction the screened statistics does not always come out to be bose. We can obtain fermions, semions, and more complicated objects.

(D) The statistics parameter θ specifies only the long distance physics. Thus, for a given θ there may well be different universality classes depending on the short distance physics.

For the same θ, we would have in general different effective Lagrangians, distinguished by, for example, different screened statistics.

(E) We can also write down more general effective theories for anyons in a magnetic field. We simply do not turn off the electromagnetic gauge potential A_μ.

I would like to end this section by remarking that the effective Lagrangians written down by different authors often look quite different. The reason is simply that different authors choose to integrate out different combinations of the gauge potentials. For illustration, let us derive the Lagrangian first written down by Zhang *et al.*[84] for filling factor $\nu = 1/(p+1)$ with p even.

Go to (11.12) and specialize to $m = 1$. We have

$$\mathcal{L} = \phi^{+} i\left(\partial_0 - i(\alpha + a - A)_0\right)\phi + (\ldots) + \frac{1}{4\pi}\left(\epsilon a\partial a + \frac{1}{p}\epsilon\alpha\partial\alpha\right) \tag{11.40}$$

(To avoid clutter we have represented the term with the spatial covariant derivative acting on ϕ by $(\ldots)$.) It turns out that to obtain the Lagrangian of Zhang *et al.* we should integrate out the combination $(a - \alpha)$. Let us write $a = (\gamma+\beta)/2$ and $\alpha = (\gamma-\beta)/2$. The last two terms in (11.40) became

$$\frac{1}{4}\cdot\frac{1}{4\pi}\left[\frac{(p+1)}{p}(\epsilon\gamma\partial\gamma + \epsilon\beta\partial\beta) + \frac{2(1-p)}{p}\epsilon\alpha\partial\beta\right] \tag{11.41}$$

Integrating out β we obtain

$$\mathcal{L} = \phi^{+} i\left(\partial_0 - i(\gamma - A)_0\right)\phi + (\ldots) + \frac{1}{4\pi}\frac{1}{(p+1)}\epsilon\gamma\partial\gamma \tag{11.42}$$

This is the original Lagrangian of Zhang *et al.* The reader would realize that this Lagrangian may be written down immediately: the Chern-Simons term for p even simply changes the statistics of ϕ to fermionic. The point is that the more elaborate Lagrangian in (11.12) allows us to reach more complicated filling fractions.

XII. ON THE EDGE

The matrix K (and its cousin Λ) control the long distance physics of topological fluids. Thus it behooves us to study the physical and mathematical meaning of K in more detail. Fröhlich and I[85] have taken as general an approach as possible to address these questions.

Any system of electrons has a conserved electromagnetic current, of course, but in two spatial dimensions the conservation law $\partial^\mu J_\mu = 0$ implies that the current can be written as the curl of a vector potential, *i.e.*,

$$J^\mu = \frac{1}{2\pi}\epsilon^{\mu\nu\lambda}\partial_\nu \alpha_\lambda \tag{12.1}$$

The factor of $1/(2\pi)$ defines the normalization of α_μ. Clearly, α_μ is defined only up to the transformation $\alpha_\mu \to \alpha_\mu - \partial_\mu \Lambda$, *i.e.*, α_μ is a gauge potential. Thus, the existence of a gauge potential follows from completely general considerations.

A field theoretic description of a physical system may be regarded as a means of organizing various aspects of the relevant physics in a systematic way according to their relative importance at long distances and according to symmetries. We classify terms in a field theoretic Lagrangian according to powers of derivatives, powers of the field variables, and so forth. A general scheme for classifying terms is according to their (naive) scaling dimensions. In (2+1) dimensional spacetime, the current has mass dimension 2, and thus the gauge potential α_μ has dimension 1, as is the case for any gauge potential coupled to matter fields according to the gauge principle. Since gauge invariance forbids the dimension 2 term $\alpha_\mu \alpha^\mu$ in the Lagrangian, the simplest possible term is in fact the dimension 3 Chern-Simons term $\epsilon^{\mu\nu\lambda}\alpha_\mu \partial_\nu \alpha_\lambda$. Thus, the Lagrangian (density) is simply

$$\mathcal{L} = \frac{\eta}{4\pi}\epsilon^{\mu\nu\lambda}\alpha_\mu \partial_\nu \alpha_\lambda + \ldots \tag{12.2}$$

where η is a dimensionless coefficient that we have yet to determine. The terms indicated by $(\ldots)$ include the dimension 4 Maxwell term $\frac{1}{g^2}f_{\mu\nu}^2$ (with $f_{\mu\nu} = \partial_\mu \alpha_\nu - \partial_\nu \alpha_\mu$) and other terms with higher dimensions. An important observation is that these higher dimensional terms are less important at long distances. The long distance physics is determined purely by the Chern-Simons term, (provided $\eta \neq 0$).

Note that the current J_μ has dimension 2, and thus the simplest term constructed out of the currents, $J_\mu J^\mu$, is already of dimension 4; indeed, this is just the Maxwell term. There is no way of constructing a dimension 3 local interaction out of the currents directly. Of course, we can fix a gauge (for instance, $\partial \alpha = 0$) and solve (12.1) for α_μ in terms of J_μ. Then the Chern-Simons term simply becomes the "Hopf interaction" between currents[15]

$$\left(\frac{\pi}{\eta}\right) J_\mu \epsilon^{\mu\nu\lambda} \partial_\nu \frac{1}{\partial^2} J_\lambda \tag{12.3}$$

The question "why gauge field?" often asked by condensed matter physicists may be answered by saying that the introduction of gauge fields allows us to avoid dealing with non-local interactions.

Note that in general the coefficient η may well be zero, in which case the physics is determined by the short distance terms represented by the $(\ldots)$ in (12.2). Put differently, a Hall fluid may be defined as a two dimensional electron system for which the coefficient of the Chern-Simons term does not vanish, and consequently is such that its long distance physics is largely independent of the microscopic details that define the system. Indeed, we may classify two dimensional electron systems according to whether η is zero or not.

Of course, the presence of a Chern-Simons term ($\eta \neq 0$) implies that parity P and time reversal T are violated. Violation of P and T may be spontaneous, as may conceivably happen in high temperature superconductors, or may be introduced by an experimenter turning on a magnetic field, as in the Hall effect. Here we will discuss the Hall fluid.

To determine η we couple the system to an "external" or "additional" electromagnetic gauge potential A_μ. From (12.1) and (12.2) we now have

$$\mathcal{L} = \frac{\eta}{4\pi} \epsilon^{\mu\nu\lambda} \alpha_\mu \partial_\nu \alpha_\lambda - \frac{1}{2\pi} \epsilon^{\mu\nu\lambda} A_\mu \partial_\nu \alpha_\lambda = \frac{\eta}{4\pi} \epsilon^{\mu\nu\lambda} \alpha_\mu \partial_\nu \alpha_\lambda - \frac{1}{2\pi} \epsilon^{\mu\nu\lambda} \alpha_\mu \partial_\nu A_\lambda \tag{12.4}$$

(In writing the second equality sign, we have dropped a total derivative.) Note that the gauge potential of the magnetic field responsible for the Hall effect should not be included in A_μ; it is implicitly contained already in the coefficient η. We can now integrate out α to obtain an effective Lagrangian for A. From (12.4), it is clear that the result may be formally read off from (12.3) by replacing J^μ by $\frac{1}{2\pi}\epsilon^{\mu\nu\lambda}\partial_\nu A_\lambda$. We obtain (just as in (8.16))

$$\mathcal{L}_{\text{eff}} = -\frac{1}{4\pi\eta}\epsilon^{\mu\nu\lambda}A_\mu\partial_\nu A_\lambda. \tag{12.5}$$

The electromagnetic current that flows in response to A_μ is thus

$$J^\mu \equiv -\frac{\delta\mathcal{L}_{\text{eff}}}{\delta A_\mu} = \frac{1}{2\pi\eta}\epsilon^{\mu\nu\lambda}\partial_\nu A_\lambda. \tag{12.6}$$

The time component of this equation tells us that an excess density δn of electrons is related to a local fluctuation of the magnetic field by

$$\delta n = \frac{1}{2\pi\eta}\delta B. \tag{12.7}$$

Since we are describing a system with filling factor $\nu = 1$, we should have η equal to 1.

The reader may be puzzled, initially, over what it is that tells us our Lagrangian describes a Hall system, and not some other electron system in two dimensions. The answer is that the coefficient of the Chern-Simons term may well be zero, in which case the physics is determined by the short distance terms represented by the $(\ldots)$ in (12.2). Put differently, a Hall fluid is a two dimensional electron system for which the coefficient of the Chern-Simons term does not vanish, and consequently is such that its long distance physics is largely independent of the microscopic details that define the system.

We thus have here a powerful statement of universality. Any two dimensional system whose effective theory is such that the coefficient of the Chern-Simons term does not vanish has universal long distance properties, properties that are unaffected by short distance details.

Thus far, we have discussed only non-interacting electrons. Let us now introduce interactions between the electrons. We have just argued that the Chern-Simons term is the only possible term at long distances. Thus, the only effect of the interaction can be to modify the coefficient of the Chern-Simons term.

We find it convenient to immediately study a more general situation. Consider non-interacting electrons at a density such that they fill m Landau levels: $\nu = 2\pi n/B = m$. We argue that the current of the electrons belonging to each Landau level is separately conserved. Physically, the large gap between the Landau levels implies that it is a good approximation to take the separate levels as dynamically independent. Thus, we introduce m gauge potentials as in (12.1)

$$J_I = \frac{1}{2\pi}\epsilon\partial\alpha_I \tag{12.8}$$

for $I = 1, 2, \ldots, m$. We have suppressed indices. The Lagrangian (12.2) is then generalized to

$$4\pi\mathcal{L} = \sum_I \alpha_I\epsilon\partial\alpha_I \tag{12.9}$$

We now take into account the interaction between the electrons. The interaction couples the different Landau levels, or equivalently the α_I's. We insist on the physical argument that the electrons ultimately can only interact electromagnetically. The interaction should only involve the total electromagnetic current $J = \sum J_I = \frac{1}{2\pi}\epsilon\partial\left(\sum \alpha_I\right)$, not the individual currents J_I. Thus, interaction can only change the Lagrangian to

$$4\pi\mathcal{L} = \sum_I \alpha_I\epsilon\partial\alpha_I + p(\sum_I \alpha_I)\epsilon\partial(\sum_J \epsilon\partial\alpha_J), \tag{12.10}$$

where p is some real constant to be determined. Or more compactly, we can write the Lagrangian as in (11.17) with the matrix K precisely as before except that we do not know what p is.

To determine the real constant p, we would now like to argue that the electron or hole should appear somewhere in the excitation spectrum.[86] After all, the theory is supposed to describe a system of electrons and thus far our rather general Lagrangian does not contain any reference to the electron!

Let us now look for the electron (or hole). Here we will illustrate the procedure by considering the simplest case $m = 1$. We note from (11.29') that a bound state of $(p+1)$ vortices would have charge equal to 1. This is perhaps the hole. The statistics of this bound state is from (11.30')

$$\frac{\theta}{\pi} = \frac{1}{(p+1)}(p+1)^2 = (p+1). \tag{12.11}$$

The hole is a fermion and thus we require $\frac{\theta}{\pi}$ to be an odd integer. This fixes p to be even, implying from (11.28') that the Hall conductance must be equal to 1/(odd integer). By finding the electron we have also recovered the classic "1 over odd denominator" fractional quantum Hall fluid! We will leave it to the reader to work out the more complicated cases. (A general treatment will be given later.)

An instructive way of understanding the Hall fluid is to confine it to a finite size container, say a disk or an annulus and to study the currents that flow on the edges of the container. The physical origin of these edge currents was elucidated some years ago by Halperin.[87] Basically, the potential energy that represents the presence of the edge pushes the Landau levels up, causing them to cross the Fermi surface at the edge. Since a given Landau level would in general cross the Fermi surface linearly, a physicist "living on the edge" would see a linearly dispersing, that is, a massless excitation. Also, since there is only one branch in the dispersion, the excitation is chiral and travels in only one direction. We have reproduced using more formal methods what we had just argued on physical grounds.

Is the Chern-Simons term smart enough to know about edge currents? Indeed, it is. The Chern-Simons term is not, strictly speaking, gauge invariant. Under a gauge transformation it changes by a boundary term:

$$\delta \int_{\Omega} ada = \int_{\Omega} d\Lambda da = \int_{\partial\Omega} \Lambda da. \tag{12.12}$$

Here we use the more compact notation of differential forms and $\partial\Omega$ denotes the boundary of the region Ω over which the theory is defined. This apparent "defect" of the theory actually represents one of its virtues. To have a gauge invariant theory we must impose some suitable condition on the behavior of a_μ on the boundary and this in turn leads to edge currents.[88,89] Indeed, the dynamics in $\partial\Omega$ may be described by the (1+1)-dimensional action

$$\int dtdu\frac{1}{2}\left[(\partial_t\varphi)^2 - v^2(\partial_u\varphi)^2\right] \tag{12.13}$$

supplemented by the constraint

$$\partial_t\varphi = v\partial_u\varphi \tag{12.14}$$

Here u is a length coordinate defined along the edge. The velocity v is not determined by the long distance topological Chern-Simons theory. The theory doesn't, and shouldn't, know about v since it is a property of the edge potential rather than the fluid in the bulk. The constraint (12.14) means that the edge excitations propagate in only one direction. We have thus reproduced using more formal methods what we had just argued on physical grounds.

The action (12.13), describing a massless chiral scalar field, is the simplest example of a conformal field theory. The theory has a conserved current $J_\mu = \epsilon_{\mu\nu}\partial^\nu\varphi$ (in spacetime units such that v is equal to one). Here $\mu, \nu = 0, 1$ denote the (1+1)-dimensional spacetime indices. The Fourier transforms J_n of J_0 (assuming that the edge $\partial\Omega$ is bounded) generate a Kac-Moody algebra defined by

$$[J_m, J_n] = m\delta_{m,-n}. \tag{12.15}$$

In the more general case in which we have the matrix Chern-Simons theory (11.17) the Kac-Moody algebra in (12.15) is obviously generalized to

$$[J_m^I, J_n^J] = m\delta_{m,-n}K^{IJ}. \tag{12.16}$$

We cannot simply diagonalize the matrix K because the gauge potentials α_I couple to the vortex current j_I^{vortex} and the vorticity defined by $\int d^2x\, j_I^{\text{vortex}}$ is quantized by topological considerations to be integers. Thus, we are allowed to transform K only to $X^T K X$ where X is an element of $SL(m, Z)$, that is, a matrix whose entries are all integers.

Remarkably, with

$$X = \begin{pmatrix} 1 & 0 & \dots & \dots & 0 \\ -1 & 1 & & & \vdots \\ 0 & -1 & 1 & & \vdots \\ \vdots & & & \ddots & \vdots \\ 0 & \dots & \dots & -1 & 1 \end{pmatrix} \tag{12.17}$$

we find

$$X^T K X = \begin{pmatrix} 2 & -1 & 0 & \dots & \dots & 0 \\ -1 & 2 & -1 & & & \vdots \\ 0 & -1 & 2 & & & \vdots \\ \vdots & & & \ddots & & 0 \\ \vdots & & & & 2 & -1 \\ 0 & \dots & \dots & 0 & -1 & p+1 \end{pmatrix} \tag{12.18}$$

Lo and behold (once again), if we knock off the last row and the last column we obtain the $(m-1)$ by $(m-1)$ matrix

$$\begin{pmatrix} 2 & -1 & 0 & \dots & 0 \\ -1 & 2 & -1 & & \vdots \\ 0 & -1 & 2 & & 0 \\ \vdots & & & \ddots & -1 \\ 0 & \dots & 0 & -1 & 2 \end{pmatrix} \tag{12.19}$$

which we recognize as the Cartan matrix for the Lie algebra $SU(n)$! Recall that the Cartan matrix is defined by the scalar products of the basic roots of $SU(m)$

$$\tilde{K}_{IJ} = \alpha^I \cdot \alpha^J / 2 \tag{12.20}$$

for $I, J = 1, \ldots, m-1$. This remarkable fact has also been noticed independently by Blok and Wen,[90] and by Read.[91]

Somehow the apparently simple problem of electrons moving in a magnetic field contains within it a non-abelian symmetry.

The transformation (12.17) corresponds to going to a new basis defined by $\tilde{\alpha}_1 = \alpha_1 - \alpha_2$, $\tilde{\alpha}_2 = \alpha_2 - \alpha_3$, $\tilde{\alpha}_{m-1} = \alpha_{m-1} - \alpha_m$, and $\tilde{\alpha}_m = \alpha_m$. Physically, the gauge potentials $\tilde{\alpha}_I$, $I = 1, \ldots, m-1$, are electromagnetically neutral in contrast to $\tilde{\alpha}_m$. The $SU(m)$ symmetry transforms these $m-1$ gauge potentials into each other.

We can go on and generalize the construction given here. The physical idea is simple. Given a bunch of Hall fluids, we can put them together and let them interact with Chern-Simons dynamics.

Let us generalize (12.9) and take each of the incompressible fluids to be described by

$$4\pi \mathcal{L}_I = k_I \alpha_I \epsilon \partial \alpha_I. \tag{12.21}$$

Then, when we put them together we obtain

$$\begin{aligned} 4\pi \mathcal{L} &= \sum_I k_I \alpha_I \epsilon \partial \alpha_I + p(\sum_I \alpha_I)\epsilon\partial(\sum_J \alpha_J) \\ &= \alpha K \epsilon \partial \alpha \end{aligned} \tag{12.22}$$

with the matrix

$$K = \begin{pmatrix} p+k_1 & p & \ldots & p \\ p & p+k_2 & & \vdots \\ \vdots & & \ddots & \vdots \\ p & \ldots & \ldots & p+k_m \end{pmatrix} \tag{12.23}$$

Applying the formulas (11.22), (11.24), and (11.26), we determine the filling factor and conductance to be

$$\nu = \sigma_H = \frac{1}{p + \frac{1}{\sum_I \frac{1}{k_I}}}, \tag{12.24}$$

the charge to be

$$q = \frac{1}{1+tp} \sum_I l_I / k_I, \tag{12.25}$$

and the statistics to be

$$\frac{\theta}{\pi} = \sum_I l_I^2/k_I - p(1+tp)q^2. \tag{12.26}$$

Here we have defined $t = \sum_I k_I^{-1}$. The filling factors reachable with (12.24) include 2/3, 2/7, 8/15, 8/31, and so on.

We can keep on going and take for each of the component incompressible fluids we want to put together to be the incompressible fluid we have just obtained. We then arrive at

$$4\pi\mathcal{L} = \sum_S \alpha^S K_S \epsilon\partial\alpha^S + +2a\epsilon\partial\sum_S \alpha^S - \frac{1}{p}a\epsilon\partial a \tag{12.27}$$

where K_S denotes an m_S by m_S matrix (as in (12.23))

$$K_S = \begin{pmatrix} p_S + k_1^S & p_S & \cdots & p_S \\ p_S & p_S + k_2^S & & \vdots \\ \vdots & & \ddots & \vdots \\ p_S & \cdots & \cdots & p_S + k_m^S \end{pmatrix} \tag{12.28}$$

A short computation using (11.22) shows that the filling factor is

$$\nu = \sigma_H = \frac{1}{p + \frac{1}{\sum_S p_S + \frac{1}{\sum_{I_S} \frac{1}{k_{I_S}^S}}}} \tag{12.29}$$

It is intriguing that a continued fraction emerges naturally. Clearly, we can keep on trucking and construct ever more elaborate incompressible quantum fluids in this way.

It may be verified that by this construction we obtain states outside the hierarchy construction. Thus, this construction is more general.

Here I would like to describe another "natural" generalization.[92] We can consider the possibility[93] that each of the incompressible fluids carries a charge t_I, not necessarily equal to 1. Thus, in the effective Lagrangian we will have the couplings of the electromagnetic gauge potential A_μ to the gauge potentials α_I described by[93]

$$4\pi\mathcal{L} = 2A\sum_I t_I \epsilon\partial\alpha_I + \ldots \tag{12.30}$$

One motivation is simply "why not?" Another motivation is that the short distance physics we have consistently ignored, and which we are unable to treat, may produce bound states. For instance, two electrons may bind into a charge 2 boson. The filling factor $\nu = \frac{2\pi n}{eB}$ then scales to $\nu = \nu_{\text{old}}(1/2)(1/2) = \nu_{\text{old}}/4$. Since the density is half of what it was and the charge (which we have indicated explicitly) is doubled. Thus, for instance, the $\nu = 1/2$ state is transformed into a $\nu = 1/8$ state for which a Laughlin-type wave function may be written down readily

$$\psi \sim \prod_{i>j}(z_i - z_j)^8 e^{-\sum|z|^2} \tag{12.31}$$

This was discussed by Halperin long ago and by Wen and me[93] in the language of effective Lagrangians more recently. The effective Lagrangian is then

$$4\pi\mathcal{L} = 8\alpha\epsilon\partial\alpha + 2A\ (2\epsilon\partial\alpha) \tag{12.32}$$

Note that we cannot simply scale the gauge potential α because of its coupling to the vortices.

With (12.30) we find, as generalizations of (11.22), (11.24), and (11.26), the filling factor and Hall conductance to be

$$\nu = \sigma_H = \sum t_I K_{IJ}^{-1} t_J, \tag{12.33}$$

the charge to be

$$q = \sum t_I K_{IJ}^{-1} l_J, \tag{12.34}$$

and the statistics to be

$$\frac{\theta}{\pi} = \sum l_I K_{IJ}^{-1} l_J. \tag{12.35}$$

I find these formulas more symmetric and attractive than those in (11.22), (11.24), and (11.26) where we see by hindsight that we had set the t_I's to 1. Here the t_I's and the l_I's appear on equal footing. The conductance has to do only with the t_I's, the statistics only with the l_I's, the charge with both.

Let us take l_I to be the columns in K, in other words, take

$$l_J^{(L)} = K_{JL} \tag{12.36}$$

Then, we find from (12.32) and (12.35)

$$q^{(L)} = \sum_I t_I \delta_{IL} = t_L \tag{12.37}$$

and

$$\frac{\theta^{(LL')}}{\pi} = \sum_{IJ} K_{IL'} K_{IJ}^{-1} K_{JL} = K_{LL'} \tag{12.38}$$

These formulas give the physical meaning of t and K. We also now see clearly and immediately that there is a hole in the excitation spectrum if some t_L is equal to 1 and if K_{LL} (no sum on L) is an odd integer. We also see that the statistical phase acquired by an "electron" of the Lth type moving around a vortex of the L'th type is given simply by

$$\frac{\theta^{(LL')}}{\pi} = \delta_{LL'} \tag{12.39}$$

We merely computed θ using (12.35) and taking $l_J^{(L)} = K_{JL}$ and $l_J^{(L')} = \delta_{JL}$. We put quotation marks around the word electron because this excitation *would* be an electron (or a hole) only if $t_L = 1$ and K_{LL} is odd.

Now that we have learned that an "electron" of the Lth type consists of a bound state of K_{JL} vortices of the Jth type, we can go to a new basis. Write

$$j^J_{\text{vortex}} = K_{JL}\, j^L_{\text{“electron”}} \tag{12.40}$$

Then the coupling of the gauge potentials to the vortices may be written as

$$\alpha_J\, j^J_{\text{vortex}} = \alpha_J\, K_{JL}\, j^L_{\text{“electron”}} \tag{12.41}$$

In other words, we should use the gauge potentials

$$\beta_{\mu L} = \sum_J K_{JL}\alpha_{\mu J} \tag{12.42}$$

In this basis, the Chern-Simons Lagrangian governing the gauge interaction

$$4\pi\mathcal{L} = \alpha K\epsilon\partial\alpha \tag{12.43}$$

becomes

$$\begin{aligned} 4\pi\mathcal{L} &= (\beta K^{-1})K\epsilon\partial(K^{-1}\beta) \\ &= \beta K^{-1}\epsilon\partial\beta \end{aligned} \tag{12.44}$$

The matrix K has exchanged places with its own inverse K^{-1}! We have discovered the duality between the vortex basis and the "electron" basis.

This emerging subject of topological fluids is just beginning to be explored. One interesting issue is how to characterize and label the orders in these systems. Traditionally, one uses broken symmetries and their associated order parametes to classify order and universality classes. Here, it appears that these topological orders have to be characterized by the matrix K.

We end by saying that this area of research, starting with semionics and ending with topological fluids, has proved to be extraordianrily rich. We believe that, quite independent of whether the present generation of superconductors turns out to be semionic or not, this subject will continue to be of some interest.

ACKNOWLEDGEMENT

I would like to thank my various collaborators, including D. Arovas, J. Fröhlich, J.R. Schrieffer, X.G. Wen, F. Wilczek, and Y.S. Wu. I would also like to thank those condensed matter physicists who have taken the effort over the years to educate me on various topics.

These include I. Affleck, P. Anderson, G. Baskaran, H. Fukuyama, K. Ishikawa, S. Kivelson, R. Laughlin, M. Rice, D. Scalapino, R. Scalettar, J.R. Schrieffer, P. Wiegmann, and others.

REFERENCES

1. N.N. Bogoliubov, *J. Phys.* (USSR) **11** (1947) 23.

2. S. Coleman and E. Weinberg, *Phys. Rev.* **D7**, 1883 (1973).

3. I. Dzyaloshinski, A. Polyakov, and P. Wiegmann, *Phys. Rev. Lett.* **127A**, 112 (1988).

4. J.R. Schrieffer, X.G. Wen, and S.C. Zhang, *Phys. Rev. Lett.* **60**, 944 (1988); *Phys. Rev. B* **39**, 11663 (1989); A. Kampf and J.R. Schrieffer, *Phys. Rev. B* **41**, 6399 (1990).

5. P.W. Anderson, *Science* **235**, 1196 (1987).

6. G. Baskaran and P.W. Anderson, *Phys. Rev.* **B37**, 580 (1988); G. Baskaran, Z. Zou and P.W. Anderson, *Solid State Comm.* **63**, 973 (1987); I. Affleck, Z. Zou, T. Hsu and P.W. Anderson, *Phys. Rev.* **B38**, 745 (1988); E. Dagotto, E. Fradkin and A. Moreo, *Phys. Rev.* **B38**, 2926 (1988).

7. *The Quantum Hall Effect*, edited by R.E. Prange and S.M. Girvin (Springer-Verlag, New York 1987).

8. X.G. Wen, F. Wilczek, and A. Zee, *Phys. Rev.* **B39**, 11413 (1990).

9. I. Affleck and J.B. Marston, *Phys. Rev. B* **37**, 3774 (1988).

10. G. Kotliar, *Phys. Rev. B* **37**, 3664 (1988).

11. L.B. Ioffe and A.I. Larkin, *Phys. Rev. B* **39**, 8988 (1989).

12. R. Laughlin and Z. Zou, *Phys. Rev. B* **41**, 664 (1990).

13. V. Kalmeyer and R. Laughlin, *Phys. Rev. Lett.* **59**, 2095 (1987).

14. J. M. Leinaas and J. Myrheim, *Nuovo Cimento* **37B**, 1, (1977); F. Wilczek, *Phys. Rev. Lett.* **48**, 1144 (1982); *Phys. Rev. Lett.* **49**, 957 (1982).

15. F. Wilczek and A. Zee, *Phys. Rev. Lett.* **51**, 2250 (1983).

16. X.G. Wen and A. Zee, *J. de Physique* **50**, 1623 (1989).

17. K. Ishikawa and T. Matsuyama, *Z. Phys.* **C33**, 41 (1986); *Nucl. Phys.* **B280**, 523 (1987).

18. N. Imai, K. Ishikawa, T. Matsuyama, and I. Tanaka, *Phys. Rev. B* **42**, 10610 (1990), and references therein.

19. S.A. Kivelson, D.S. Rokshar, and J.P. Sethna, *Phys. Rev. Lett.* **60**, 821 (1988); *Phys. Rev. B* **35**, 8865 (1987).

20. H. Fukuyama, O. Narikiyo and K. Kuboki, *J. Mag. & Mag. Mat.*, **90-91**, 607, (1990); O. Narikiyo, K. Kuboki, and H. Fukuyama, *J. Phys. Soc. Jpn.* **59**, 7 (1990); *J. Phys. Soc. Jpn.* **59**, 807 (1990).

21. M. Kohmoto, *Phys. Rev.* **B39**, 1989 (1194)3; Y. Hatsugai and M. Kohmoto, *Phys. Rev.* **B42**, 8282 (1990); and references therein.

22. X.G. Wen and A. Zee, *Nucl. Phy.* **B326**, 619 (1989).

23. X.G. Wen, *Phys. Rev. B* (1991), to appear.

24. A. Zee, *Mod. Phys. Lett. B.* **5**, 1339 (1991).

25. P. Wiegmann, IAS preprint (1991).

26. A. Zee, *Mod. Phys. Lett.*, **B5**, 1339, (1991).

27. R. Laughlin, Z. Zou and S. Libby, *Nucl. Phys.* **B348**, 693 (1991).

28. A. Zee, to appear in Memorial Volume to M.A.B. Bég, edited by A. Ali (1991); A. Zee, *Int. J. Mod. Phys.* **B5**, 529 (1991).

29. H. Hasegawa, *J. Phys. Soc. Japan*, **59**, 4384, (1990).

30. Z. Kunszt and A. Zee, *Phys. Rev. B* **44**, 6842 (1991).

31. C. Pryor, and A.Zee, UCSB preprint 1991.

32. Y. Hasegawa, P. Lederer, T.M. Rice and P.B. Wiegmann, *Phys. Rev. Lett.* **63**, 908 (1989).

33. D.R. Hofstadter, *Phys. Rev.* **B14**, 2239 (1976).

34. Y. Hasegawa, O. Narikiyo, K. Kuboki, and H. Fukuyama, *J. Phys. Soc. Jpn.* **59**, 822 (1990).

35. Z. Zou, H. Levy, and R. Laughlin, Stanford preprint (1991).

36. B. Douçot and X.G. Wen, *Phys. Rev. B* **40**, 2719 (1989).

37. D. Haldane, UCSD-report; *Phys. Rev. Lett.* **57**, 1488 (1986).

38. A.A. Belavin and A.M. Polyakov, *JETP Lett.* **22**, 245 (1975).

39. S. Chakravarty, B.I. Halperin and D.R. Nelson, *Phys. Rev. Lett.* **60**, 1057 (1988); *Phys. Rev. B* **39**, 2344 (1989).

40. Y.S. Wu and A. Zee *Phys. Rev. Lett.* **147B**, 325 (1984).

41. X.G. Wen and A. Zee, *Phys. Rev. Lett.* **63**, 461 (1989).

42. A. D'Adda, P. DiVecchia, and M. Lüscher, *Nucl. Phys.* **B146**, 63 (1978); E. Witten, *Nucl. Phys.* **B149**, 285 (1979).

43. X.G. Wen, *Phys. Rev. B* **39**, 7227 (1989).

44. X.G. Wen and A. Zee, *Nucl. Phys.* **B326**, 619 (1989).

45. X.G. Wen and A. Zee, *Phys. Rev. Lett.* **62**, 1937 (1989).

46. G. 't Hooft, Lectures on Gauge Symmetry in Physics, 1983.

47. J. Cardy and E. Rabinovici, *Nucl. Phys.* **B205**, 1 (1982); J. Cardy, *Nucl. Phys.* **B205**, 16 (1982).

48. S.J. Rey and A. Zee, *Nucl. Phys.* **B352**, 897 (1991).

49. D.H. Lee and M.P.H. Fisher, IBM report 1990.

50. D. Arovas, R. Schrieffer, F. Wilczek and A. Zee, *Nucl. Phys.* **B251**, 117 (1985).

51. A. Fetter, C. Hanna, and R. Laughlin, *Phys. Rev. B* **39**, 9679 (1989).

52. Y. Hosotani and S. Chakravarty, *Phys. Rev. B* **42**, 342 (1990); J.E. Hettrick, Y. Hosotani, and B.-H. Lee, *Ann. Phys. (N.Y.)*, **209**, 151 (1991); Y. Hosotani, *World Scientific*, 383, (1991); J. Hetrick and Y. Hosotani, UMN-TH-936/91 *Phys. Rev. B*, to appear.

53. Y.H. Chen, F. Wilczek, E. Witten, and B.I. Halperin, *Int. J. Mod. Phys.* **B3**, 1001 (1989).

54. X.G. Wen and A. Zee, *Phys. Rev. B* **41**, 240 (1990).

55. H. Mori, Hiroshima preprint (1990).

56. C.A. Trugenberger, Los Alamos preprint 1991.

57. R. Laughlin, *Phys. Rev. Lett.* **60**, 2677 (1988).

58. X.G. Wen and A. Zee, Ref. 54; P. Wiegmann, private communication; E. Fradkin, private communication, T. Banks and J.D. Lykken, SLAC preprint.

59. X.G. Wen and A. Zee, *Phys. Rev. Lett.* **62**, 1937 (1989); *Phys. Rev.* **B41**, 240 (1990); *Int. J. Mod. Phys.* **B4**, 437 (1990); *Phys. Rev.* **B39**, 2756 (1989).

60. X.G. Wen and A. Zee, *Int. J. Mod. Phys.* **B4**, 437 (1990); M.P.H. Fisher and D.H. Lee, *Phys. Rev. Lett.* **63**, 903 (1989). M.C. Cha, M.P.A. Fisher, S.M. Girvin, M. Wallin, and A.P. Young, IBM report 1991.

61. X. Wen and A. Zee, *Int. J. Mod. Phys.*, **B4** (1990) 437.

62. S.K. Paul and A. Khare, *Phys. Lett.* **B174**, 420 (1986); *Phys. Lett.* **193**, 253 (1987). In Ref. 54, for the sake of simplicity we carried out this computation only in a relativistic form.

63. A.M. Polyakov. *Nucl. Phys.* **B120**, 429 (1977).

64. X.G. Wen and A. Zee, *Phys. Rev. Lett.* **62**, 2873 (1989).

65. X.G. Wen and A. Zee, *Phys. Rev. B* **43**, 5595 (1991).

66. For a review, see the talks by H.J. Weber, A. Kapitulnik, and K.B. Lyons in *Physics and Mathematics of Anyons*, edited by S.S. Chern, *et al.*, (World Scientific Publishing, 1991).

67. A.G. Rojo and A. Leggett, University of Illinois preprint.

68. B. Halperin, J. March-Russell and F. Wilczek, *Phys. Rev. B* **40**, 8726 (1989).

69. S. Sachdev, Yale preprint (1991).

70. X.G. Wen, *Phys. Rev.* **B40**, 7387 (1989); *Int. J. Mod. Phys.* **B2**, 239 (1990).

71. E. Witten, *Comm. Math. Phys.* **117**, 353 (1988); **121**, 351 (1989).

72. A. Schwarz, *Lett. Math. Phys.* **2**, 201 (1978); *Comm. Math. Phys.* **67**, 1 (1979); W. Siegel, *Nucl. Phys.* **B156**, 135 (1979).

73. J.F. Schönfeld, *Nucl. Phys.* **B185**, 157 (1981); S. Deser, R. Jackiw, and S. Templeton, *Phys. Rev. Lett.* **48**, 975 (1982).

74. Y.S. Wu and A. Zee, *Phys. Lett.* **B207**, 39 (1988).

75. Y.S. Wu and A. Zee, *Nucl. Phys.* **B258**, 157 (1985); *Nucl. Phys.* **B272**, 322 (1986).

76. D. Haldane and E.H. Reyazi, *Phys. Rev. Lett.* **54**, 327 (1985); *Phys. Rev. B* **31**, 2529 (1985).

77. S.C. Zhang, T.H. Hansson and S. Kivelson, *Phys. Rev. Lett.* **62**, 82 (1989). See also S. Girvin and A.H. McDonald, *Phys. Rev. Lett.* **58**, 1252 (1987); and S. Girvin, in Ref. 7. For a more recent formulation, see J. Frölich and T. Kerler, ETH-TH/90-26. X.G. Wen and Q. Niu, *Phys. Rev. B* **41**, 9377 (1990). B. Blok and X.G. Wen, *Phys. Rev. B* **42**, 8133 (1990).

78. X.G. Wen and A. Zee, *Nucl. Phys.* **B15**, 135 (1990); A. Zee, in *High Temperature Superconductivity*, edited by K. Bedell, D. Coffey, D. Pines and J.R. Schrieffer (Addison-Wesley 1990).

79. X.G. Wen and A. Zee, *Phys. Rev. B* **44**, 274 (1991).

80. J.K. Jain, *Phys. Rev. Lett.* 199, (1989).

81. J.K. Jain, *Phys. Rev.* **B40**, 8079 (1989).

82. M. Greiter and F. Wilczek, *Mod. Phys. Lett. B* **4**, 1063 (1990).

83. M.P.H. Fisher and D.H. Lee, *Phys. Rev. Lett.* **63**, 903 (1989); X.G. Wen and A. Zee, *Phys. Rev. Lett.* **62**, 1937 (1989); *Phys. Rev.* **B41**, 240 (1990); *Int. J. Mod. Phys.* **B4**, 437 (1990); *Phys. Rev.* **B39**, 2756 (1989).

84. S.C. Zhang, T.H. Hansson, and S. Kivelson, *Phys. Rev. Lett.* **62**, 82 (1989).

85. J. Fröhlich and A. Zee, *Nucl. Phys.* **B364**, 517 (1991).

86. R. Tao and Y.S. Wu, *Phys. Rev.* **B39**, 1087 (1984).

87. B. Halperin, *Phys. Rev.* **B25**, 2185 (1982).

88. X.G. Wen, *Phys. Rev.* **B40**, 7387 (1989); *Int. J. Mod. Phys.* **B2**, 239 (1990).

89. J. Fröhlich and T. Kerler, ETH-TH/90-26.

90. B. Blok and X.G. Wen, *Phys. Rev. B* **42**, 8133 (1990).

91. N. Read, *Phys. Rev. Lett.* **65**, 1502 (1990).

92. X.G. Wen and A. Zee, to be published.

93. This possibility was raised but not investigated in Ref. 78 and Ref. 79.

ORGANIZING COMMITTEE

LIST OF PARTICIPANTS

Daniel Altschuler	: Arecibo Observatory and University of Puerto Rico, Department of Physics Río Piedras, PR
Manuel Aquino	: Centro de Investigación y Estudios Avanzados, IPN Mexico D.F., Mexico
Don Backer	: University of California Department of Astronomy, Berkeley, CA
John Barrow	: University of Sussex School Math. Phys. Sci., Sussex, England
Bari Bhuiyan	: University of Puerto Rico Department of Physics, Rio Piedras, PR
William Bruckman	: University of Puerto Rico Department of Physics and Electronics, Humacao, PR
Max Chávez	: Universidad de Costa Rica Escuela de Física, San José, Costa Rica
J. Chengalur	: Arecibo Observatory Arecibo, PR
Pedro Colin	: University of Chicago Department of Physics, Chicago, IL
Dennis Collins	: University of Puerto Rico Department of Mathematics, Mayagüez, PR
Luiz da Costa	: Harvard-Smithsonian Center for Astrophysics Cambridge, MA

Michael Davis	: Arecibo Observatory Arecibo, PR
Ernesto Esteban	: University of Puerto Rico Department of Physics and Electronics, Humacao, PR
Anamaria Font	: Universidad Central de Venezuela Caracas 1051, Venezuela
Wolfram Freudling	: Arecibo Observatory Arecibo, PR
Carlo Giovanardi	: Arcetri Astrophysical Observatory Florence, Italy
Riccardo Giovanelli	: Arecibo Observatory Arecibo, PR
Alejandro Gonzalez	: University of Sussex School Math. Phys. Sci., Sussex, England
Krishnaswamy Gounder	: University of South Carolina Department of Physics and Astronomy, Columbia, SC
Rodrigo Huerta	: Centro de Investigación y Estudios Avanzados, IPN Mexico D.F., Mexico
Hoshik Kang	: Ohio State University Department of Physics, Columbus, OH
Ernest Ma	: University of California Department of Physics, Riverside CA
Loris Magnani	: Arecibo Observatory Arecibo, PR

William Marciano	: Brookhaven National Laboratory Upton, NY
José F. Nieves	: University of Puerto Rico Department of Physics, Rio Piedras, PR
Burt Ovrut	: University of Pennsylvania Department of Physics, Philadelphia, PA
Carmen Pantoja	: University of Oklahoma Department of Physics and Astronomy, Norman,OK
Philip Pennance	: University of Puerto Rico Department of Mathematics, Rio Piedras, PR
Jaime Ponce de León	: University of Puerto Rico Department of Physics, Rio Piedras, PR
K. Rozga	: University of Puerto Rico Department of Mathematics, Mayagüez, PR
Zbigniew Ryzak	: Harvard University Department of Physics, Cambridge, MA
Eve Mariel Santangelo	: Universidad Nacional de La Plata Departamento de Física, 1900 La Plata, Argentina
David Schramm	: University of Chicago Department of Physics, Chicago, IL
Daniel Sudarsky	: University of Chicago Department of Physics, Chicago, IL
Jesus Tharrats	: University of Puerto Rico Department of Physics, Rio Piedras, PR

Alfredo Torruella	: University of Puerto Rico Department of Physics, Rio Piedras, PR
Neil Turok	: Princeton University Department of Physics, Princeton, NJ
Marcelo R. Ubriaco	: University of Puerto Rico Department of Physics, Rio Piedras, PR
Esov Velazquez	: University of Puerto Rico Department of Physics, Mayagüez, PR
Anthony Zee	: University of California Institute of Theoretical Physics Santa Barbara, CA

www.ingramcontent.com/pod-product-compliance
Lightning Source LLC
LaVergne TN
LVHW051925120926
847902LV00009B/76